ARTIFICIAL INTELLIGEN(
REAL-TIME CONTROL 1998

A Proceedings volume from the 7th IFAC Symposium,
Grand Canyon National Park, Arizona, USA, 5 - 8 October 1998

Edited by

YOH-HAN PAO
Case Western Reserve University, Cleveland, Ohio, USA

and

S.R. LeCLAIR
Materials Process Design Branch, Air Force Research Laboratory,
Wright-Patterson AFB, Ohio, USA

Published for the

INTERNATIONAL FEDERATION OF AUTOMATIC CONTROL

by

PERGAMON
An Imprint of Elsevier Science

UK	Elsevier Science Ltd, The Boulevard, Langford Lane, Kidlington, Oxford, OX5 1GB, UK
USA	Elsevier Science Inc., 660 White Plains Road, Tarrytown, New York 10591-5153, USA
JAPAN	Elsevier Science Japan, Tsunashima Building Annex, 3-20-12 Yushima, Bunkyo-ku, Tokyo 113, Japan

First edition 1999

Library of Congress Cataloging in Publication Data
A catalogue record for this book is available from the Library of Congress

British Library Cataloguing in Publication Data
A catalogue record for this book is available from the British Library

ISBN: 9780080432274

Printed and bound in Great Britain by
CPI Antony Rowe, Chippenham and Eastbourne

IFAC SYMPOSIUM ON ARTIFICIAL INTELLIGENCE IN REAL-TIME CONTROL 1998

Sponsored by:
IFAC - International Federation of Automatic Control
US Air Force, European Office of Aerospace Research & Development, London, England

Organized by:
US Air Force, Air Force Research Laboratory, USA
Case Western Reserve University, USA

International Program Committee

Yoh-Han Pao, Co-Chairman (USA)
David W. Russell, Co-Chairman (USA)
K.J. Astrom (Sweden)
J. Bayne (USA)
P. Borne (France)
T. Chandra (Australia)
C.L.P. Chen (USA)
H.J. Chizeck (USA)
J. Efstathiou (United Kingdom)
I. Grabec (Slovenia)
V. Haase (Austria)
C.J. Harris (United Kingdom)
S. Kahne (United States)
R. Karba (Slovenia)
Mateja Kavcic (Slovenia)
J. Kocijan (Slovenia)
V. Krebs (Germany)
S.R. LeClair (USA)
T.H. Lee (Singapore)
R.R. Leitch (United Kingdom)
I. MacLeod (South Africa)
J. Maguire (United States)
Jianbo H. Meng (PR China)
L. Motus (Estonia)
Y.H. Pao (USA)
J.F. Peters (Canada)
S. Ramanna (Canada)
M.G. Rodd (United Kingdom)
D. Russell (USA)
T. Samad (USA)
G. Schmidt (Germany)
Chang-Yun Shen (PR China)
You-Xian Sun Shen (PR China)
H.B. Verbruggen (The Netherlands)
R.A. Vingerhoeds (The Netherlands)
You-Shou Wu (PR China)
Xiao-Ming Xu (PR China)

FOREWORD

This Proceedings volume contains the full papers presented and discussed at the 1998 IFAC International Symposium on Artificial Intelligence in Real-Time Control, held at the Grand Canyon National Park, Arizona, USA, October 5-8, 1998.

This Symposium is the seventh of a series of very successful symposia and workshops in this field bringing together control technology specialists, artificial intelligence experts and process control practitioners. Since the commencement of this Symposia and Workshop series, there have been a number of very positive developments in the topical area of 'Intelligent Control'. Visions have broadened and new methodologies such as Neural-Net computing, Genetic Algorithms and Evolutionary Programming, and Fuzzy Logic have all been investigated in some depth and have been found to be helpful in various appropriate circumstances.

Over the past decade yet another area, referred to as 'situated control', has stimulated the formation of new perspectives towards real-time intelligent systems. The performances of such artificial species as walking cockroaches, maze-negotiating mice, coke-can collecting robots and the like have encouraged the exploration of yet more adaptive control perspectives. Such explorations are certain to generate intrigue and widespread discussions regarding the maturation and application of situated control.

Perhaps it is helpful to think of intelligent control in terms of the combined use of judicious amounts of reasoning and pattern-processing. Reflex control based on hard won knowledge of appropriate reflex response might serve very well if monitored and coordinated with suitably juxtaposed logic. In this Symposium, there is a strong wind of change bringing more consideration of the roles of learning, evolution, hybrid systems and so on under many diverse labels and for many different systems and circumstances.

But one consolidates gains and builds upon strengths. Accordingly there are five interesting tutorials which are balanced between that which is well accepted and that which will be viewed as new insights and advances. In addition, a set of eleven sessions address a broad context of developments in control methodology ranging from generic to specific and theoretical to real world. Noteworthy is a special session on advanced industrial practice which has also been assembled for presentation and discussion.

CONTENTS

ARCHITECTURES FOR REAL-TIME EXPERT SYSTEMS

HYBRID SYSTEMS

EVOLUTIONARY I

REAL-WORLD TASKS I

SPECIAL SESSION ON DISTRIBUTED CONTROL

VISUALIZATION AND IMAGING

EVOLUTIONARY II

THEORETICAL ISSUES AND TOPICS

FUZZY OR ROUGH SETS

GENERIC PRINCIPLES AND STRATEGIES

REAL-WORLD TASKS II

HANDLING TIMING IN A TIME-CRITICAL REASONING SYSTEM – A CASE STUDY

T. Naks [1,2], L. Motus [1]

[1] *Tallinn Technical University, Chair of Real-time Systems, Ehitajate tee 5, Tallinn EE0026, Estonia. E-mail: Tonu.Naks@ttu.ee, leo.motus@dcc.ttu.ee.*
[2] *MiTime Ltd, The Innovation Centre, Singleton Park, Swansea SA2 8PP.*

Abstract: An attempt has been made to integrate a case-based reasoning system with a software timing formal analysis system so as to develop a tool for building complex diagnostic systems for time-critical applications.
The focus of this paper is on the elicitation of timing requirements and constraints and on pre-run-time verification of system's timing properties together with run-time monitoring and detection of potential timing violation.
The basic ideas presented, have been separately tested in many occasions. The integration of the two methodologies is carried out in ESPRIT 22 154 project. *Copyright © 1998 IFAC*

Keywords: time critical reasoning, case-based reasoning, timing requirements and constraints, formal verification of timing, run-time monitoring of timing

1. INTRODUCTION

The information technology is rapidly penetrating into more and more sensitive spheres of engineering and other human activities. For instance, fly-by wire airplanes, drive-by-wire cars, automated highways, chemical process control, medical and biomedical engineering systems, and others. Potential risks caused by malfunctioning, or failure, of software, communication, sensors, actuators, and/or computer hardware is increasing rapidly. This evolution is accompanied by a constant increase in complexity of models and algorithms required and used.

The complexity of modelling, and related algorithmic problems, can often be reduced by applying artificial intelligence (AI) based methods. Quite often these methods lead to well-founded and fast approximate solutions to computationally hard but well-defined mathematical problems. In other cases the AI-based methods introduce empirical heuristic assumptions about unknown, or incompletely known causal relations. The introduction of empirical knowledge

and heuristics often leads to reasonably fast approximate solutions to otherwise computationally intractable problems. Successful application examples of AI ideas in engineering are neural networks, neuro-fuzzy control methods, expert systems, a variety of reasoning methods (case-based, progressive, abductive, probabilistic, etc reasoning). Examples of such systems can be found in Harris, et al (1966), Wirth et al (1966), Zhu (1994).

However, the reduction of complexity reduces, in the most cases, also determinism in the quality of obtained solution and in the time required to obtain the solution. These unwanted side-effects of AI-based methods may cause serious additional safety risks in real-time applications and should be specially addressed. Quality of solution has traditionally played a central role in computer and systems science, and has been carefully studied, with some success, by many researchers. Timeliness of the solution has only comparatively recently started to attract theoreticians. Therefore, in spite of the outstanding importance of timeliness, theoretical achievements in studying the methods of guaranteeing timely solutions are still rather modest. Real understanding of the role of time correctness in embedded (real-time) systems is only now emerging from the subconsciousness of computer science and control communities. Some authors (e.g. Kopetz (1997)) classify time-critical systems as a subclass of safety critical systems, hence timing violations should be considered as direct threat to safety of the system. First books on the essence of timing analysis have been published (see, for example, Motus and Rodd (1994)).

Violations of timing constraints and requirements may happen in all parts of a computer control and/or monitoring system. In addition to obvious user requirements and design inconsistencies, timing violations may be related, for instance:

- in the environment, to device faults and/or malfunctioning, accidents, incorrect mathematical models, and (indirectly) to technological condition violation

- in the communication between the environment and the computer system, to the presence of electromagnetic noise; too heavy traffic in the communication network; errors in, or not appropriate properties of, the communication protocol; physical damage of communication media

- in the computer system, to hardware performance problems; hardware faults; not quite correct priority, interrupt handling and/or scheduling policy; too large execution time indeterminacy of algorithms (e.g. iterations, convergence problems, traditional reasoning algorithms); inconsistencies in specified timing requirements for interprocess communication; inconsistent or erroneous timing requirements provided by the user.

The abovegiven list of causes for violation of timing requirements can clearly be partitioned into two classes-- pre-run-time causes (user requirements, specification, design, algorithms), and run-time causes (faults, unexpected situations, physical design and implementation, systems software). Intuitively one needs different methods for different classes of causes. For the pre-run-time class of causes one needs some insight into the system analysis, requirement specification, and system design process. In order to be able to detect run-time timing violations, one should be able to verify the proper timing constraints and requirements in the pre-run-time stage.

This paper does not consider the timing problems specific to applications and algorithms. Some ideas about methods of handling those problems, can be found in Netten and Vingerhoeds (1995) and Jones (1995). The main attention of this paper is focused on studying the correctness of overall system's timing, with special attention to reasoning subsystem. The problem is approached in two separate steps:

- verifying the timing correctness at specification and design stage of the system and its software, and

- on-going run-time checking of the verified timing behaviour.

To be able to verify timing correctness, one has to concentrate on the impact of intrinsic timing indeterminacy upon systems behaviour, leaving the other, not less important, issues of solution quality to be studied elsewhere by different methods. A system may comprise many algorithms, in this paper the focus is on a reasoning component of a system – any system that uses artificial intelligence based algorithm has a reasoning part that interacts with the rest of the system. The reasoning subsystem receives source data on which the reasoning is based. The source data must be valid in time and each data item should be consistent in time with the other used data items. The reasoning system provides decisions, control recommendations, forecasts, responses – they all must be properly timed in order to influence the environment in a wanted manner.

In diagnostic applications, for example, the output from the reasoning system often invokes tests in the environment, the results of those tests are then used to complete the reasoning process. In such cases the time-validity requirements on input data depend on the joint functioning of the environment, computer system and communication system. This indicates how sophisticated and intertwined can be the definition of time constraints that are to be imposed upon various components of a system.

Characteristic features of reasoning in real-time applications are:

- dependence on directly measured data with limited validity time

- active data elicitation during the reasoning process (by invoking tests and using the test results to complete the reasoning)

- deadlines on the overall reasoning time, and on various stages of reasoning.

Although the reasoning in real-time applications is a conceptually generic task, it is quite difficult to explain the process on an abstract application. Therefore this paper explains the use of a specific timing verification tool and the related ideas on an example of building an intelligent real-time diagnosis tool (BRIDGE). The paper gives an overview of the BRIDGE tool, then describes the timing issues of a diagnostic application. The basic focus of the paper is on handling the time constraints when designing, re-configuring and running the diagnostic system. Many details of the timing verifications, especially those regarding the use of timing verification tool LIMITS are left out of this paper. Those details have been published elsewhere, the references are given in the text.

2. THE BRIDGE TOOL

The BRIDGE tool is a generic fault diagnosis tool for large technical systems, based essentially on case based reasoning. The tool is being developed by a consortium of six partners from four countries as an ESPRIT project no. 22 154. For each application, the BRIDGE tool is tuned – the application case base is compiled into a specific network structure so as to guarantee deterministic worst case reasoning time,

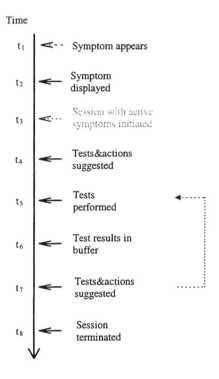

Fig.1. Simplified sequence of actions for the Bridge Run-time tool

links to the technical system are specially re-designed together with the specification of time-constraints imposed upon the behaviour (determined by the technical system and communication media).

The BRIDGE tool should be embedded into, or at least linked with a communication line to, a technical system with inbuilt monitoring capabilities of its own behaviour. This means that the technical system should, at least, be able to collect information about its current state. In the case of detecting the events or measurements referring to abnormal situation, the corresponding *symptom* code is generated, either by the monitoring subsystem or by the operator, and sent to the Bridge tool.

Using an internal inference engine, the Bridge tool will find all *cases* in the case-base that are possibly related to the newly discovered symptom(s). Each case may have a number of attached *tests* and/or *corrective actions.*. The tests are performed to strengthen or weaken the originally computed match with one or more of the existing cases by performing additional measurements or observations on the technical system. After the additional information is

gained from the tests, the inference engine generates a refined list of related cases. When the best matching case is finally identified, the corresponding corrective actions should be performed to fix the problem in the technical system.

This is a simplified description of the BRIDGE tool. The following section will describe some of the timing aspects of reasoning in BRIDGE. Please note that BRIDGE is a generic tool and may work in an automatic, as well as in automated mode, depending on the requirements and capabilities of the specific technical systems.

3. TIMING IN BRIDGE APPLICATION

BRIDGE application is an interacting complex comprising a technical system that is to be diagnosed and the BRIDGE tool with customised case-base.

Timing constraints are quantitatively determined by technical system's tolerance to a particular failure, validity period of the symptoms, and response of the operator. The timing constraints can roughly be partitioned into two groups. The first group, constraints on performance (*performance requirements*), is often used and is the most thoroughly studied group of timing requirements. This group comprises, for instance, deadlines for execution of separate tasks, response times of complex sequences of activities in the embedded system and in its environment. The second group, constraints on the validity of data and events (*data validity requirements*), includes the admissible usability times of data items, meaningful lifetimes of messages, and the consistency of validity constraints with the system's overall performance requirements.

A realistic behavioural scenario of a technical system with embedded (linked) BRIDGE tool is as follows. A failure occurs in the technical system (t_1 in Fig.1). This event may or may not be directly visible to the operator. For instance, the operator notices immediately when alarm lights on a control panel switch on, without receiving any measurements from the system. At the same time, when fuel is critically low, it can be noticed only when the fuel monitoring system gives an alarm signal. At time instant t_2 either the monitoring subsystem, or the operator detects the symptom.

After the symptoms have been discovered and placed into the symptom buffer for the Bridge tool, the operator may, if considered necessary, change the symptom's status (e.g. mark it as active, inactive, or primary) and at time instant t_3 initiate a diagnostic session. Regardless of, whether the session was initiated by the operator of the technical system or not, the real-time kernel of the BRIDGE tool checks the contents of the buffer periodically and passes all the detected symptoms to the inference engine. The latter creates a list of recommended tests and actions, required to achieve a better match with a case from the case-base (at time instant t_4 at the latest), and displays the list to the operator. The operator selects (at least) some of the recommended tests and performs them. The results from the performed tests will be received by the operator at time instant t_5. The operator, or the monitoring system, then has to make the test results known to the Bridge tool by placing them in the symptom's buffer. The inference engine considers the additional information and produces a new, updated list of recommended tests and corrective actions and terminates. Now the operator may choose to restart the cycle (see, Fig.1), perform tests, transfer the results and start inference engine again), or perform corrective actions and terminate the session (time instant t_8).

Some of the diagnosis related activities are performed automatically by the BRIDGE tool and/or technical system, and some others by the operator. It is clear that only the assessment of time characteristics for the automatically functioning part of the system provides reasonable estimates. The behaviour of a human operator is difficult to predict. In the ideal case one can count on physically feasible human response times. In the not so ideal case the operator may be sleeping, unable (e.g. because of injuries, mental stress), or just unwilling to act. Preferably the system should carry on, as long as possible, independently of the human operators activities, although this is not possible for many applications.

Four time intervals will be useful in the development of timing analysis for the abovedescribed high-level sequence of actions in Fig. 1.:

- $[t_1, t_2]$ -- time from the actual occurrence of a failure symptom till its detection by the operator or by the monitoring subsystem, i.e. how long it takes to notice and display the symptoms by the technical system, further in this paper referred to as T_{notice}

- $[t_3, t_4]$ -- how long it takes to receive the list of suggested tests and actions, after placing the symptoms into the buffer, selecting the primary symptom and active symptoms -- i.e. the response time of the BRIDGE when searching the customised case-base for actions corresponding to the symptoms; this interval is denoted by $T_{suggest1}$

- $[t_6, t_7]$ -- how long it takes to update the list of suggested actions and tests, after receiving the additionally performed test results -- i.e. response time of the BRIDGE when searching the case-base for actions corresponding to the symptoms and extended test results, denoted further by $T_{suggest2}$

- $[t_1, t_4]$ -- time from the actual occurrence of a symptom till the first suggestion of additional tests and/or corrective actions; this interval describes primary response time to a symptom and is denoted by T_{resp}.

These four intervals are usually constrained by user requirements and those imposed by the technical system. In principle, there should not be any difference between the $T_{suggest1}$ and $T_{suggest2}$. However, they are considered separately since creation of a list and updating the list may have different performance characteristics.

3.1. Timing requirements on performance

The main goal for the operator (and/or the technical system) is to solve a problem related to a particular potential failure. From the viewpoint of timing, the problem should be resolved before the failure leads to serious consequences. As seen from the discussion above, the time required to resolve the problem is only partially predictable. Only tasks performed by the hardware and software can be given definite worst case response time estimates. The worst case execution time estimates for tasks performed by human operators are, strictly speaking, unpredictable. The diagnostic system may set requirements on the operator's response time, but cannot guarantee their satisfaction.

Thus, realistically the user can impose the following timing requirements upon the system's performance:

- **Req1**: I want to be alarmed within a certain interval, after a symptom indicating a potential failure has emerged (T_{notice}).

- **Req2**: I want a list of suggested corrective actions and additional tests within a fixed time interval ($T_{suggest1}$) after the symptoms have been made available for the BRIDGE tool.

- **Req3**: I want the updated list of suggested actions and tests within a fixed time interval ($T_{suggest2}$) after the additional test results have been made available for the BRIDGE tool.

- **Req4**: I want the first response from the BRIDGE tool to a failure symptom in a fixed time interval (T_{resp})

The *Req1* applies only to the symptoms that are discovered automatically by the monitoring subsystem of the technical system. The satisfaction of this requirement depends on processing times within monitoring and alarm subsystems, and human-computer interface, and also on transport and non-transport delays of the messages transferring symptom in the technical system.

The intervals related to *Req2* and *Req3* characterize explicitly the whole diagnostic system, and depend on tolerance of the technical system (or on that of the operator, or both) to the potential failures; or more specifically, the upper bound for these intervals are defined by validity intervals of the involved data.

The integral performance requirement for the diagnostic system is represented by *Req4* and is the sum of *Req1* and *Req2*. Please note that *Req4* includes some of the timing characteristics of the technical system operator. It sets a time constraint on the first iteration of technical system diagnosis – i.e. on the time starting from actual occurrence of the failure symptom until getting the first advice on corrective actions, or additional tests from the BRIDGE tool. In the case when additional tests are not needed, *Req4* expresses the available time to resolve the problem.

3.2. Data validity requirements

Failures are different, some of them are urgent and demand fast reaction (e.g. milliseconds, seconds, minutes), the others may wait until the next routine maintenance of the technical system. In complex technical systems, it seems reasonable to assign explicit and individual time tolerance to each failure. This leads to the necessity of defining and monitoring the satisfaction of reasonable life-time of symptoms. Those values that do not satisfy the validity requirements of the technical system can not be used in the reasoning process.

It is suggested that the following three factors express the user's expectations on the reasonable lifetime of different data items in the Bridge system:

- *Validity interval* defines the period during which the value assigned to a variable, at the left endpoint of the interval, can be used as a characteristics of the state of the technical system, or that of the BRIDGE tool

- *Response deadline* specifies the right endpoint of an interval (upper bound) within which the BRIDGE tool is to provide a response to a symptom in the form of explicitly specified corrective actions and/or additional tests

- *Equivalence interval* fixes a period within which the results of two assignments (e.g. two occurrences of the same event, two measured or computed values of the same variable) are considered to be equivalent and the latter assignment is practically neglected.

Two types of symptoms (at least) should be distinguished when deriving symptoms from measurements and events, and attaching occurrence time-stamps to them -- *instantaneous symptoms* ("explosion", "power failure", etc.), and *continuous symptoms* ("door is open", "fuel is low", etc.).

Instantaneous symptoms refer to alarming events in a system. These events can be detected in a very short time interval although their influence may last significantly longer than the period when they could be detected. For instance, the sound of an explosion can be detected in a comparatively short time interval. However, if a symptom "high temperature" occurs soon after the sound of an explosion, one should remember about recent "explosion sound". This will help to select a correct case in a case-based reasoning system. On the other hand, if one discovers "high temperature" in the technical system *long* after the "explosion sound", it is quite reasonable to conclude that these two events are not related – the validity time of an event "explosion sound" may have expired already. The actual quantitative value of *long* is critical for correct reasoning.

The number of occurrences may be significant for instantaneous symptoms. It is necessary to distinguish whether the two detected occurrences of an instantaneous symptom refer to the same event, or to different events. For instance, did the two explosions occurred in reality, or was the same explosion detected twice? In order to distinguish between the multiple occurrences of symptoms, the *equivalence interval* is important. However, in order to reach the right conclusion one has quite often, in addition to using the equivalence interval, to use sensor fusion. Two measurements within the equivalence interval refer probably to the same event, unless the measurements are geographically apart from each other.

Continuous symptoms refer to a certain erroneous state of the system. They are usually significant only during their actual existence in a system. For instance, if "door open" is a symptom that indicates a problem with the doors, then it is necessary to keep this symptom active only while it can be measured in a system. After the door is closed, or the other characteristics of a system are changed in a way that it is normal to have open doors, the problem ceases to exist and the symptom "door open" should be removed from the buffer. The validity interval, in this case, lasts until the instant when the next measurement's value is assigned to the variable, after that the latest measured value is valid.

For *continuous symptoms* the number of occurrences is not significant -- the consecutive measurements only confirm or reject the results form the previous measurements. Therefore the equivalence interval is not essential in this case although, if specified it can be used, for instance, for comparing the quality of multiple sensors measuring a single variable.

The validity intervals are essential for recommended actions (suggested by the BRIDGE tool) and additional test results (obtained from the technical system). The response deadlines for the technical system are necessary for tests and actions recommended by the BRIDGE tool. The tests and corrective actions should be performed within a certain prefixed time limit, before the technical system's state changes radically.

The following parameters fully determine the validity of essential messages in most of the applications of the BRIDGE tool:

- *Symptom validity interval* describes how long the symptom, after its detection, still reflects the actual state of the technical system

- *Symptom response deadline* defines how long after the symptom detection the recommended (by BRIDGE) actions and tests can be considered as appropriate

- *Symptom equivalence interval* fixes the time period within which the results of the two consecutive measurements reflect the same occurence of the symptom; this interval usually applies to *instantaneous symptoms* and determines granularity of time for them

- *Action validity interval* defines the period during which the suggested corrective action remains a valid suggestion; this interval should be equal to, or less than, the shortest symptom response deadline for symptoms related to this action.

- *Action response deadline* provides the upper bound of a time for the action to be performed; this deadline reflects the integral tolerance of the technical system and its environment to this kind of failure.

- *Test validity interval* gives the time period for performing the test; the right endpoint of this interval should be equal to or less than the shortest response deadline for the symptoms related to this test; also, the right endpoint of this interval should be considerably shorter than the response deadlines for the possibly related actions -- after the test is performed, there must remain time to update the goals and to perform corrective actions, if necessary.

Test response deadline is the upper bound of a time when the response to the completed test becomes available to the BRIDGE tool. Again, test response deadline must be shorter than the response deadlines for the potentially suggested corrective actions.

4. RUN-TIME HANDLING OF TIME CONSTRAINTS

The timing constraints and requirements specified during specification and design phases should be non-contradicting among themeselves, and consistent with the technical system properties. The required formal correctness of those constraints and requirements can be verified by using the LIMITS tool (see, for example, Motus and Naks (1998)). The consistency of constraints with the technical system properties should be approved by an expert, based on animation results of the specification and design. During the run-time of the diagnostic system, a special monitoring system assesses periodically the actual timing caracteristics and detects potential violations of the specified/designed timing constraints and requirements in due time. Additional assessments and

readjustments can be done during the reconfiguration phases of the BRIDGE tool, or during the build-up of the diagnostic system for a particular application.

The BRIDGE tool is suitable for time-critical applications for two reasons. First, the inference engine of the BRIDGE tool has a fully determined worst case response time. Second, the tool offers a possibility of tailoring timing requirements for each application, considering actual characteristics of the technical system at hand, and monitoring whether the requirements are satisfied or not. If the result of timing analysis is negative, it is usually possible to change the system's configuration, or to modify the case-base in order to meet the requirements. Also, the modification of timing requirements is possible in some cases.

4.1. Time-determinism of the BRIDGE inference engine

A necessary precondition for using reasoning in a time-critical system is the time-determinism of the reasoning process. Two steps have been taken to force time-deterministic behaviour upon the Bridge inference engine -- searching the case-base is replaced with sequential execution of declarative clauses, and event-driven activation of the inference engine is substituted by time-driven activation.

The main problem with commercial "real-time" expert system shells is that no upper bound on the inference completion time is fixed, since the rule- (or case-) base has a general network structure. The Bridge tool has, in principle, a well-defined and guaranteed upper bound on the time spent on inferencing. The specific value of this upper bound depends on the given rule-base (or case-base), and on the computer platform. The upper bound is achieved in BRIDGE by applying the following procedures.

Instead of using a conventional case-base for reasoning, an index in the form of fault network (Netten and Vingerhoeds (1994)) is generated, which covers the conventional case-base. This makes the search for the appropriate cases faster and the search time more deterministic.

As a next step, the fault network is converted into procedural form, comprising a "flat" set of IF-THEN-ELSE clauses (Jones (1995)). In addition to having achieved deterministic execution time of the inference engine, one can now analyse the code used by the inference engine. The worst-case execution time can be determined by any static analysis method, or by any monitoring technique (for the survey of code analysis methods and techniques see, for example, Olenyi (1996)). Within the Bridge tool, the mixture of static analysis and monitoring technique is used. In particular, the worst-case path is determined by analysing the source code. Then the code, corresponding to the worst-case path, is executed in the same soft- and hardware environment as it will operate in the real situation, and the actual performance characteristics are measured by a monitoring technique.

However, even if the execution time of the worst-case path is assessed by a monitoring technique in a laboratory environment, it does not guarantee that the actual execution time can not be longer under real operational conditions. In the laboratory conditions the number of concurrent processes is fixed. The actual operational conditions may be more dynamic. For instance, the number of concurrently activated processes may increase dynamically, tasks may be assigned different priorities than in a laboratory set-up, in an event-driven diagnostic system uncontrollably many concurrent copies of the same process may be invoked.

To reduce the influence of dynamic factors, the Bridge tool has adopted the state-based approach (as explained in Jones, 1995). This leads to strict scheduling of all the time-critical processes, instead of allowing to trigger them by external events at random time instants. In order not to neglect some occurrences of external events, all the occurrences of events are stored in special input event buffers. Each time a process is activated, it scans the input event buffers and interprets the buffer contents. For each activation of a time-critical process, the scheduler is requested to allocate worst-case execution time for it, so as to avoid pre-emption. Because rather realistic worst-case execution time estimates exist for the processes, it is possible to prepare an effective schedule for time-critical processes. The rest of the processes run in empty time-slots between time-critical processes -- it is important to reserve sufficient number of empty time-slots.

The applied methods result in longer than average response times achieved in event-based systems, and in systems that rely on dynamic scheduling policies. However, the used method guarantees that the assessed worst-case response time is not exceeded. For many applications the guaranteed maximum response time is the most important feature of the diagnostic system. Additional advantage of the assessed, and guaranteed worst case execution times is the possibility of formal verification that the actual execution times satisfy timing constraints and requirements imposed by the user and the technical system. Formal verification is done by a stand-alone LIMITS CASE-tool, and/or by embedding the analysis part of LIMITS into the BRIDGE tool for run-time checks.

4.2. Satisfaction of timing requirements

The basic structural correctness of the BRIDGE tool (e.g absence of deadlocks, structure of communications) is verified during specification and design stages. For that reason an adequate timing model of the tool is composed and the consistency of

the specified time constraints is verified (see Section 4.3). The timing analysis must continue during implementation and configuring of the diagnostic system – as soon as improved estimates for timing parameter values become available. In truly time-critical applications the final check for timing correctness is carried out at run-time. The tasks performed during implementation, configuration, and run-time, are:

- Assessment of worst-case performance characteristics for a particular configuration of the inference engine

- Analysing the consistency of the performance-bound and data validity requirements *per se*, and with the assessed timing characteristics

- Deriving the scheduler parameters from the requirements and assessed parameter values

- Monitoring the actual behaviour during operation, and warning the operator about the danger of timing violation, or performing automatic corrective actions.

As the covential application case-base and the BRIDGE tool's inference engine (real-time kernel) are not interacting directly, a new "flat" set of IF-THEN-ELSE clauses for the inference engine must be compiled each time a new case has been added to the conventional case-base, or a case is otherwise modified. The interaction of conventional case-base and the inference engine functions as follows. The inference engine provides a coded recommendation for actions and/or tests, and the conventional case-base is used to get the comments and explanations to that recommendation.

After the real-time kernel of the BRIDGE tool has been re-generated, its worst-case response times can be assessed in off-line situation. The timing models composed during the specification and design of the BRIDGE system can be complemented with the new parameters, and/or new parameter values. It is necessary to check again whether the newly assessed performance characteristics still satisfy the specified timing requirements. Based on the newly verified timing parameter values, the new parameters for the scheduler should be calculated.

After the scheduler's new parameters are calculated, the working inference engine can be replaced with a new one and activated. During the work of the inference engine, all the specified data validity parameters are monitored.

4.3. Timing Models

Timing models are used in several stages of the BRIDGE life-cycle. During the specification and design, adequate models for all software components are created so as to describe and verify the structure and behaviour of the BRIDGE tool. For time-critical parts, Q-model (Motus and Rodd (1994)) descriptions representing their behaviour in time, are composed. The Q-models are verified using the LIMITS tool (Motus and Naks (1998)). At this stage, mostly the structural correctness can be checked, since timing requirements and timing constraints are vaguely, or incompletely known. Better estimates for timing requirements become available when the BRIDGE tool is linked with a particular technical system, and the case-base is defined. Please note that the schedule for the time-critical processes can be derived based on the data available from the verified Q-models.

In order to check the satisfaction of timing constraints and requirements during operation of the diagnostic system, the Q-models and part of the LIMITS analyser are integrated into the Bridge tool. The Q-models are stored in the so-called *template model* form, where the application-dependent timing parameters have pointers to corresponding variables from the Bridge tool and the linked technical system. Values of those variables will represent the timing requirements of the technical system and the actual performance characteristics of the BRIDGE tool. As soon as those values have been assigned to models' variables, the LIMITS analyser is invoked and verification of the models starts.

For adequate description of all necessary timing characteristics (requirements and constraints), three layers of Q-models are suggested (see Fig). Lower layer models are implementation specific and usually give a more detailed description as compared with a higher-level model. The highest layer -- *requirements layer* – provides the user specified performance-bound requirements and constraints (see section 3.1). This is the most abstract layer with few (fairly abstract) processes.

The second layer from the top -- *scheduler layer* – describes the time-critical processes which are to be scheduled. In the beginning, this layer comprises processes without specified values for the timing parameters, except those obtained from the requirements layer as requirements or constraints. The other timing parameter values for these processes will be estimated by aggregating the parameter values from the lowest (implementation) layer. At the same time the adjustment of already given values for requirements and constraints may take place. The timing parameter values from this layer are used to build the scheduler.

The lowest layer -- *implementation layer* -- comprises the detailed model of the BRIDGE inference engine, models of the other parts of the BRIDGE tool, and parts of the technical system. The parameter values of processes are assessed on the basis of actual execution times of respective processes as measured in the working environment.

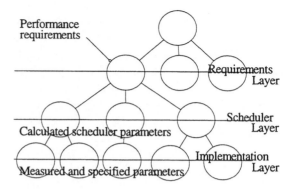

Fig. 2. Layers of Q-models

Such an architecture allows the combination of top-down and bottom-up approaches to parameter specification. The user requirements on timing are specified at the upper layer of hierarchy, the specific parameter values for each physical process appear at the bottom layer. The final verification process is organised in a bottom-up manner. First, performance assessment of the BRIDGE tool is made. The timing requirements of the technical system and required parameter values are to be specified before that. After the assessment is completed, the not-specified parameter values of the template Q-models can be filled in and the consistency of the timing characteristics is checked by formal and informal analysis according to the Q-methodology.

This schema of activities stems from the idea of checking consistency between different layers of the heredity tree (see Tekko and Tommingas (1992), Motus and Rodd (1994)), slightly adapted for diagnosis applications. Aggregate estimates of the actual performance are calculated from the assessed parameters of implementation layer model, and then compared to the corresponding user requirements. If the comparison gives positive results, the parameters of the scheduler layer processes will be calculated in a similar way. Finally, the schedule for time-critical processes will be generated automatically by the BRIDGE tool.

If the results of the comparison are negative – i.e. the system does not satisfy the requirements -- then the response time can be improved (usually) by modifying the case-base. Alternatively, the communication between the technical system and the BRIDGE tool can be modified so as to ensure suitable response times. The verification process should be invoked again after those modifications.

4.4. Modelling the user requirements on performance.

Specification of the Q-model for the requirements layer is described in this paper as an example of the modelling process. The resulting Q-model is shown in Fig. 3. The specification comprises three processes. The *Operator* of a technical system is modelled as an input-output selector process, having the ability to decide which combination of inputs and outputs to use when activated. Source for faults is shown explicitly (*ForceMajeure*) as a common Q-model process. This process gives the designer a possibility to model the delay between the actual occurrence of the fault and detection of its symptoms by the operator.

Neither *ForceMajeure*, nor *Operator* have specified activation periods and execution times in Figure 3. As mentioned earlier, guessing the frequency of fault occurrences, or the decision-making speed of the operator has a questionable value in the worst case. On the other hand, if those estimates would have been provided, the designer would have an opportunity to evaluate, for instance, maximum diagnosable frequency of faults, and/or the required performance capabilities of the operator. Technically the BRIDGE tool can readily handle these two problems (estimating the maximum diagnosable frequency of faults, or the worst case requirements to the oprator). It is the task of the researchers and system analysts to agree upon certain values for the worst case estimates for fault frequency and/or operator's decision-making capability.

The third process, *BridgeApplication*, comprises the BRIDGE tool and monitoring facilities of the technical system. *Bridge Application* is described as an *object process* (Naks (1996), Naks and Motus (1998)) in Figure 3. The behaviour of object processes must be specified by a separate Q-model that usually is a decomposition (or refinement) of the object process. If timing properties and/or constraints have been specified for the object process, they should comply with integral properties of the group of lower level processes that result from the refinement of the object process. The corresponding analysis is included in the BRIDGE tool.

In the *BridgeApplication* model, all the performance requirements are specified as response intervals – time intervals within which the response to an input message arrival at an input port appears at a specified output port. For instance, the right endpoint of interval T_{notice} describes the deadline for a response to a message received by the input port *BridgeApplication.fault*, the response is to appear at the output port *BridgeApplication.symptoms*. The right endpoint of the interval $T_{suggest1}$ describes the deadline for a response to message received by the input port *BridgeApplication.fault* that is to appear at the output port *BridgeApplication.actions&tests*. Please note that input/output ports are denoted as <process_name>.<port_name>.

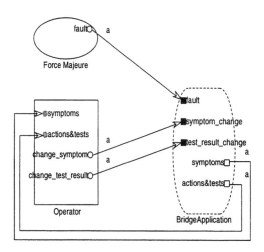

Fig. 3. Q-model description of the Requirements Layer

4.5. Checking data validity requirements

Performance requirements and capabilities of the diagnostic system are analysed, as a rule, using standard Q-methodology procedures. Almost the same applies to analysis of data validity requirements and constraints at the design and reconfiguration stages. A couple of application-specific checks, to be performed in the pre-run-time stages, have been added in the particular case of BRIDGE. These tests are actually focused on detecting timing problems at integration of BRIDGE with a technical system. For instance:

- searching the symptom base for symptoms whose validity is specified to expire before they can, in principle, be discovered by the diagnostic system; modifications are necessary if a symptom has validity interval, or response deadline shorter than the activation period of the inference engine

- searching the case base for cases whose validity will expire before the system is able to suggest them; in the other words, searching for cases which suggest tests and corrective actions with specified validity intervals and response deadlines that are inconsistent with the other timing characteristics of the system.

The run-time checks are designed to provide an operator with early warning of potential violation of deadlines, and suggest counter-measures (if possible). The first group of run-time checks monitors deadlines, suggested for actions and tests, and compares them with symptoms validity intervals. The operator can be warned of the coming problems, and possibly advised about the urgency and a reasonable order of counter-measures. Alternatively, in many situations the scheduler can resolve the potential deadline violations by automatically applying a standard in-built algorithm. The decision between the alternatives depends heavily on the nature of the application.

The second group of run-time checks monitors validity intervals and response deadlines of symptoms, and takes care that only valid symptoms are used for reasoning and actions. The examples of run-time checks are:

- Monitor that the symptoms in the symptom buffer still reflect the actual state of the technical system.

- Monitor that the suggested actions and tests are still applicable, i.e. whether their results can still be used in a specified time interval.

- Sort suggested actions according to their criticality level -- the action with the closest deadline is often, but not necessarily always, the most critical one.

- Advise the operator whether there is sufficient time to perform suggested tests before the corrective action. The operator often faces the dilemma – to elicit more additional information before the final decision can be made, or to make a decision based on incomplete information.

5. CONCLUSIONS

An attempt has been made to integrate two usually independently evolving research domains – case based reasoning and timing analysis of systems behaviour. The goal is to develop a tool that can be used for developing and run-time support of time-critical reasoning in an intelligent computer control, and/or diagnostic system.

Integration of various domain specific innovative results becomes important in the process of achieving the stated goal. For instance, increasing the intrinsic time determinism of reasoning methods, elicitation of time requirements and constraints from the users and experts in the application, verification the consistency of time requirements, measuring and/or assessing timing characteristics of software, run-time monitoring of system's behaviour in time, developing appropriate schedulers, etc. It is suggested that even if some of the listed issues have, not yet, a good theoretical solution, the use of empirical experience and expert guesses may give better results than just neglecting the timing issues.

The ideas described in this paper have been separately thoroughly tested on many occasions. Independent tools have been built to test the innovative case-base reasoning (see Netten (1997)), and timing analysis of software design (Motus and Naks (1998)). The results presented in this paper stem from a project which integrates the ideas used in separate tools.

6.ACKNOWLEDGEMENTS

Fruitful discussions with the BRIDGE and LIMITS consortium members have enormously influenced the development of the ideas presented in this paper. The financial support of EU grant CP-94 no.1577, ESPRIT 22 154, and ETF grant no.2849 is appreciated.

7. REFERENCES

Harris C.J., M.Brown, K.M.Bossley, D.J.Mills, F.Ming (1996) "Advances in neurofuzzy algorithms for real-time modelling and control", *Engineering Applications of Artificial Intelligence*, **vol.9**, no.1, 1-16

Netten, B.D., Vingerhoeds, R.A. (1995). Large-scale Fault Diagnosis for On-board Train Systems. *First International Conference on Case-Based Reasoning*, 23-26 October 1995, Sesimbra, Portugal. In: Case-Based Reasoning Research and Development, (eds) M. Veloso, A. Aamodt, Lecture Notes in Artificial Intelligence 1010, Springler

Jones A.V. (1995). *An Approach to the Design of Expert Systems for Hard Real-time Applications*. PhD-thesis, Dept. Electrical and Electronic Engineering, University of Wales Swansea.

Kopetz H. (1997) *Real-time systems. Design principles for distributed embedded applications*, Kluwer Academic Publishers, 338 pp.

Motus L., Naks T. (1998). Formal timing analysis of OMT designs using LIMITS. *Computer Systems – Science and Engineering*, **vol.13**, no.3, 161-170

Motus L., Rodd M.G. (1994). *Timing analysis of real-time software*. Elsevier Science/Pergamon, 212 pp

Naks T., (1996). *Towards Timed Object Modelling*. Diploma thesis in engineering, Tallinn Technical University, 57 pp.

Netten B.D. (1997) *Knowledge based conceptual design. An application to Fibre reinforced composite sandwich panels*. Ph.D thesis, Delft University of Technology, 178 pp.

Netten B.D., Vingerhoeds R.A., (1994) Automatic Fault-Tree Generation, A Generic Approach for Fault Diagnosis Systems. *IFAC Workshop Safety, reliability and Applications of Emerging Intelligent Control Techniques*, Hong Kong, 12-14 december, pp 182-187.

Olenyi A. (1996) *Measuring real-time execution time on PC using a Hardware Monitoring System*. MScthesis, Dept. Electrical and Electronic Engineering, University of Wales Swansea.

Tekko J., Tommingas T. (1992) Some ideas for advancement of the Q-model based embedded software engineering environment CONRAD. Workshop on tools and environments for developing control systems. Kääriku, Estonia.

Zhu Q. (1994) "Probabilistic reasoning in an augmented fact-proposition space and its applications", *Engineering Applications of Artificial Intelligence*, **vol.7**, no.6, 627-638

Wirth R., B.Berthold, A.Krämer, G.Peter (1996) "Knowledge-based support of system analysis for the analysis of failure modes and effects, *Engineering Applications of Artificial Intelligence*, **vol.9**, no.3, 219-230

DECOUPLING OF MULTIVARIABLE RULE-BASED FUZZY SYSTEMS

R. Babuška, A. Gegov and H.B. Verbruggen

*Delft University of Technology, Faculty of Information Technology and Systems
Control Engineering Laboratory, Mekelweg 4, P.O. Box 5031, 2600 GA Delft
the Netherlands, fax: +31 15 2786679 e-mail: r.babuska@its.tudelft.nl*

Abstract: Decoupling of interactions among variables in multiple-input, multiple-output (MIMO) rule-based systems is addressed. It is well known that by proper decoupling the design of a controller for a MIMO process can be substantially simplified. Also the computational costs are reduced which facilitates the real-time implementation of the control system. However, decoupling methods are well developed for linear systems only. No general approach seems to be available for nonlinear MIMO systems. It is shown in this article that if a nonlinear MIMO system is represented by a set of linguistic if-then rules, the decoupling can be accomplished in a systematic way. An algorithm for the design of a decoupler (pre-compensator) is presented and illustrated by an example. *Copyright ©1998 IFAC.*

Keywords: Multivariable (MIMO) systems, decoupling, linguistic models, fuzzy models.

1. INTRODUCTION

A major limitation of multivariable fuzzy systems based on if-then rules is that their complexity grows exponentially with the number of antecedent variables. Methods have been investigated to reduce this complexity either *directly* by reducing the number of antecedent variables and linguistic terms (Setnes, et al., 1998; Lacrose, 1997), or *indirectly* by changing the structure of rules (Jamshidi, 1997; Babuška, 1998). A common feature of these methods is that they reduce the complexity by simplifying the original system. This is achieved at the expense of neglecting some aspects of the system's behavior. The methods often use rather ad hoc principles such as the fusion of variables (Lacrose, 1997).

This article presents a different approach. Instead of simplifying the original system, a pre-compensator (decoupler) is designed in order to eliminate (reduce) the coupling among the variables. Then, the (partially) decoupled system can be represented as a set of simpler subsystems. The design procedure for the de-

coupler is completely systematic. It is, however, based on a pure linguistic fuzzy system and it completely abstracts from the numerical behavior. The paper is organized as follows: in Section 2, a set-theoretic representation of a linguistic rule base is given. Some preliminaries needed for the presentation of the decoupling algorithm are discussed in Section 3. The design of the decoupler is addressed in Section 4. An illustrative example is presented in Section 5, and Section 6 concludes the paper.

2. RULE BASE REPRESENTATION

In this section, a set-theoretic notation for a general linguistic MIMO rule base is introduced. This notation will be used in the presentation of the decoupling algorithm. Consider a MIMO linguistic fuzzy system in the following form:

$$\mathcal{R}_i: \textbf{If } u_1 \textbf{ is } A_{i1} \textbf{ and } \ldots \textbf{ and } u_p \textbf{ is } A_{ip}$$
$$\textbf{then } x_1 \textbf{ is } B_{i1} \textbf{ and } \ldots \textbf{ and } x_q \textbf{ is } B_{iq}, \quad (1)$$
$$i = 1, 2, \ldots, K.$$

Here $\mathbf{u} = [u_1, \ldots, u_p]^T$ with $u_j \in U_j \subset \mathbb{R}$ are the antecedent (input) variables, and $\mathbf{x} = [x_1, \ldots, x_q]^T$ with $x_j \in X_j \subset \mathbb{R}$ are the consequent (output) variables. $A_{ij} \in \{A_{jl}|l = 1, 2, \ldots, g_j\}$ and $B_{ij} \in \{B_{jl}|l = 1, 2, \ldots, h_j\}$ are linguistic terms (fuzzy sets). g_j and h_j are the number of terms defined for the jth antecedent and consequent variable, respectively. The number of rules in the rule base is denoted by K. For a complete rule base (2), $K = \prod_{j=1}^{p} g_j$. It is convenient to represent each of the if-then rules as an ordered pair of the antecedent and consequent ordered tuples of linguistic terms:

$$r_i = [a_i, b_i], \quad \text{with} \quad a_i = [A_{i1}, \ldots, A_{ip}]$$
$$\text{and} \quad b_i = [B_{i1}, \ldots, B_{iq}].$$

The entire rule base is denoted by $R = \{r_i|i = 1, 2, \ldots, K\}$. Further, the sets of all antecedents and consequents in R are denoted by $A = \{a_i|i = 1, 2, \ldots, K\}$ and $B = \{b_i|i = 1, 2, \ldots, K\}$, respectively. For a rule based system R with input $\mathbf{u}$ and output $\mathbf{x}$ the shorthand notation $\mathbf{x} = R(\mathbf{u})$ will be used. This structure is sufficiently general to represent both static and dynamic MIMO systems.

3. LINGUISTIC DECOUPLING

Linguistic decoupling is based on the notion of active decomposition of MIMO fuzzy systems (Gegov, 1996). The term *active decomposition* means that an intentional intervention is applied in the original system in order to simplify the design of a controller for this system and also to reduce the computational effort for the real-time implementation of the control algorithm. For this purpose, *decoupling (compensation)* of the interactions among input and output variables is introduced. The term *interaction* denotes that a particular input influences more than one output of the system. The decoupling methods use the following types of linguistic if-then rule bases (2):

3.1. Plant Rule Base

The plant rule base describes the input–output (input–state) mapping of a given nonlinear MIMO plant. This plant is assumed to be coupled, i.e., each output depends on more than one input. The plant rule base is denoted by:

$$\mathbf{x} = R^P(\mathbf{u}).$$

The individual rules and the antecedent and consequent sets are denoted by the same superscripts, such as r_i^P is the ith rule of the plant rule base, a_i^P (b_i^P) is the antecedent (consequent) of this rule, A^P is the set of all antecedents in R^P, K^P is the number of rules in R^P, etc.

3.2. Target (Desired) Rule Base

The aim of decoupling is to design a pre-compensator such that the series connection of the plant and this pre-compensator results in a decoupled system with a desired behavior. This target decoupled system is specified by the user, and it is denoted by

$$\mathbf{x} = R^T(\mathbf{z}).$$

The input (antecedent) variables $\mathbf{z}$ can be regarded as a reference for the plant output variables $\mathbf{x}$. The target rule base can be given in an *implicit form* which contains all combinations of the liguistic terms for the variables z_j (the above equation). It can also be presented in an *explicit form* which consists of q independent rule bases, one for each variable:

$$x_j = R_j^T(z_j), \quad j = 1, 2, \ldots, q.$$

For a fully decoupled rule base, these two forms are equivalent, but the explicit form is a more efficient representation. Depending on the scope of compensation, the specified decoupling may be *full* or *partial* as described in the subsequent paragraphs.

Full decoupling of interactions. This decoupling applies to the complete original MIMO system, and its purpose is to compensate the interactions among all inputs and outputs. As a result, selected pairs of input and output variables may be considered independently of each other as SISO subsystems (Figure 1). These pairs must cover the entire set of system variables.

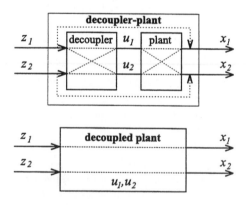

Fig. 1. Full decoupling of interactions.

Partial decoupling of interactions. The purpose of this decoupling is to compensate the interactions among selected inputs and outputs. As a result, some input-output pairs of variables may be considered independently from each other as SISO subsystems, while other parts remain coupled as smaller MIMO, MISO or SIMO subsystems, see Figure 2 for an example of a system with three inputs and three outputs.

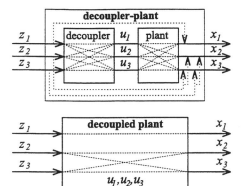

Fig. 2. Partial decoupling of interactions.

3.3. Pre-compensator (Decoupler) Rule Base

The decoupler can be regarded as a feedforward controller which is prepended to the original plant, in order to obtain the target (decoupled) system as a result of this series connection, see Figures 1 and 2. The decoupler is denoted by

$$\mathbf{u} = R^C(\mathbf{z}) \,.$$

Given the plant and the target system, the decoupler can be designed by the procedure presented in Section 4..

3.4. Decoupled Rule Base

The decoupled rule base is obtained as a series connection of the decoupler and the plant rule bases. It is denoted by

$$\mathbf{x} = R^D(\mathbf{z}) = R^P\big(R^C(\mathbf{z})\big) \,.$$

Ideally, the decoupled rule base is equall to the target rule base, i.e., $R^D = R^T$, but generally, it will be its subset: $R^D \subset R^T$.

4.   DESIGN OF THE DECOUPLER

This section presents a systematic algorithm to design the decoupler (pre-compensator) rule base from the rule bases of the plant and of the target (partially or fully decoupled) system. Given are the plant rule base

$$R^P = \{r_i^P | i = 1, 2, \ldots, K^P\}$$

and the target (desired) rule base

$$R^T = \{r_i^T | i = 1, 2, \ldots, K^T\} \,.$$

The decoupler (pre-compensator)

$$R^C = \{r_i^C | i = 1, 2, \ldots, K^C\}$$

is derived in the following two steps:

1. *Determine the decoupled rule base, R^D, by selecting those rule of R^T whose consequents are present in the plant rule base R^P:*

$$R^D = \{r_i^T | b_i^T \in B^P, i = 1, 2, \ldots, K^T\} \,. \quad (2)$$

Note that generally $R^D \subseteq R^T$, as the system rule base does not necessarily contain all the desired output combinations.

2. *Determine the pre-compensator (decoupler) rule base R^C:*

$$R^C = \Big\{ [a_i^D, a_j^P] \Big| i = 1, 2, \ldots, K^D,$$
$$\text{and } j \in \{1, 2, \ldots, K^P\} \quad (3)$$
$$\text{such that } b_j^P = b_i^D \Big\} \,.$$

The meaning of the above equation as as follows. The antecedents of R^C are equal to the antecedents of R^D. The consequents of R^C are the antecedents of those rules of R^P whose consequents coincide with the consequents of R^D. The pre-compensator is not unique. Several j's in $\{1, 2, \ldots, K^P\}$ can be found such that $b_j^P = b_i^D$ for some $i \in \{1, 2, \ldots, K^D\}$.

5.   EXAMPLE

The presented method has been applied in simulation to a laboratory system which consists of four cascaded tanks as shown in Figure 3. The control inputs are the two flow rates $\mathbf{u} = [Q_1, Q_2]^T$, and the outputs are the liquid levels $\mathbf{x} = [h_1, h_2]^T$. The decoupling has been applied to a steady state model of this system. The liquid levels in the upper tanks are not considered as they are a priori decoupled with respect to the two liquid flows.

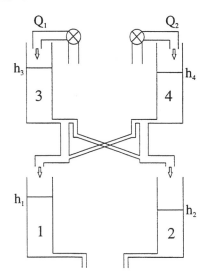

Fig. 3. Cascaded tanks.

The plant rule base is given in Table 1. All the variables are represented by three linguistic terms: S – small, M – medium, and B – big.

Table 1   Original (coupled) rule base R^P.

Rule	If	u_1	u_2	then	x_1	x_2
1		S	S		S	S
2		M	S		S	S
3		B	S		M	S
4		S	M		S	S
5		M	M		S	S
6		B	M		M	M
7		S	B		S	M
8		M	B		M	M
9		B	B		B	B

An example of a target (desired) rule base in the implicit form is given in Table 2. Note that this rule base is decoupled (x_1 only depends on z_1 and x_2 only depends on z_2).

Table 2   Desired (decoupled) plant rule base R^T.

Rule	If	z_1	z_2	then	x_1	x_2
1		S	S		S	S
2		M	S		M	S
3		B	S		B	S
4		S	M		S	M
5		M	M		M	M
6		B	M		B	M
7		S	B		S	B
8		M	B		M	B
9		B	B		B	B

One can already see that the decoupled plant rule base cannot be obtained in its complete form because not all combinations of the consequent terms are present in the plant rule base (such as $b_8^T = [M, B]$). Therefore, the decoupled rule base obtained in step 1 of the algorithm is a subset of the desired one (Table 3). The pre-compensator rule base, obtained in step 2 of the algorithm is presented in Table 4. One can easily verify that the series connection of R^C and R^P yields R^D, i.e., a decoupled system.

Table 3   Obtained (decoupled) rule base R^D.

Rule	If	z_1	z_2	then	x_1	x_2
1		S	S		S	S
2		M	S		M	S
3		S	M		S	M
4		M	M		M	M
5		B	B		B	B

6. CONCLUSIONS

The presented method of linguistic decoupling is a useful tool for the compensation of interactions in MIMO

Table 4   Pre-compensator (decoupler) rule base R^C.

Rule	If	z_1	z_2	then	u_1	u_2
1		S	S		S	S
2		M	S		B	S
3		S	M		S	B
4		M	M		B	M
5		B	B		B	B

fuzzy systems. It can be used to reduce the complexity of the original MIMO system by decomposing it into subsystems of smaller dimensions. This simplifies the design of controllers for MIMO systems. Also the real-time implementation of the control system is facilitated, as the number of rules can be reduced significantly.

One of the topics for future research is the analysis of the relationships between the decoupling in the linguistic domain (which is guaranteed by the presented algorithm) and in the numerical domain of the plant inputs and outputs (which is generally not guaranteed). The proposed method will also be applied to the real-time control of a laboratory scale model of the system presented in Section 5..

ACKNOWLEDGEMENT

This research has been supported by the ESPRIT long term research project FAMIMO (Fuzzy Algorithms for Control of Multi-Input-Multi-Output Processes), no.21911.

REFERENCES

Babuška, R. (1998). *Fuzzy Modeling for Control.* Kluwer Academic Publishers, Boston.

Gegov, A. (1996). *Distributed Fuzzy Control of Multivariable Systems.* Dordrecht, The Netherlands: Kluwer.

Jamshidi, M. (1997). *Large Scale Systems: Modelling, Control and Fuzzy Logic.* Englewood Cliffs, USA: Prentice-Hall.

Lacrose, V. (1997). *Complexity Reduction of Fuzzy Controllers: Application to Multivariable Control.* PhD dissertation, Laboratory of Systems Analysis and Architecture, Toulouse, France.

Setnes, M., R. Babuška, U. Kaymak and H.R. van Nauta Lemke (1998). Similarity measures in fuzzy rule base simplification. *IEEE Transactions on Systems, Man & Cybernetics* 28(3), 376–386.

NATURAL HEURISTIC DYNAMIC PROGRAMMING FOR DYNAMIC SYSTEMS CONTROL

K. Wendy Tang* and Jahangir Rastegar**

*Department of Electrical and Computer Engineering
**Department of Mechanical Engineering
State University of New York at Stony Brook
Stony Brook, NY 11794
wtang@ee.sunysb.edu and rastegar@motion.eng.sunysb.edu

Abstract: Heuristic Dynamic Programming (HDP) is the simplest kind of Adaptive Critic (Werbos, 1992).
It can be used to maximize or minimize any utility function, such as total energy or trajectory error, of a
system over time in a noisy environment. This article proposes a new version of HDP, called NHDP
(Natural Heuristic Dynamic Programming). This new version incorporates basic HDP algorithm with the
following features:(i) use of Trajectory Pattern Method to guarantee smoothness of trajectory and control
signals; (ii) use multiple critic networks to localize effect of each parameter mimicking the natural
biological model of human brain; and (iii) allow the controller to learn from slow to fast motion, analogous
to the natural learning behavior of humans. A simple dynamic system is used to illustrate NHDP.
Copyright©1998 IFAC

Keywords: Neural control, Dynamic systems, Learning control, Neural Networks, Backpropagation.

1. INTRODUCTION

Recently Adaptive Critic Design (ACDs) has
received increasing attention as a powerful form
of neuro-control. A good overview for a range
of ACDs can be found in (Werbos, 1992;
Prokhorov and Wunch, 1997). Basically, the
main advantage of ACD is its ability to
maximize or minimize a utility function such as
total energy or trajectory error over time.
Among the various ACDs, Heuristic Dynamic
Programming (HDP) is the simplest kind.

In this paper, the objective is to illustrate how
HDP can be effectively used to control dynamic
systems when combined with trajectory pattern
method. The trajectory pattern method was
proposed by Rastegar and Fardanesh (Rastegar
and Fardanesh, 1991). It allows motion of the
dynamic system be synthesized to contain
sinusoidals with a fundamental frequency and its
harmonics. The task of the controller is
essentially simplified to identifying the

coefficients of these harmonics that constitute
the inverse dynamic equations. Furthermore, by
learning the motion from slow to fast through
gradual increment of the fundamental frequency,
the controller learns *naturally* the characteristics
of the motion. The end result is a HDP
controller that learns naturally to control a
motion from slow to fast, hence the name
Natural HDP.

This article is organized as follows: Section 2 is
a description of *Natural Heuristic Dynamic
Programming (NHDP)*. Section 3 illustrates
how the algorithm is used for the control of a
simple dynamic system. Finally, conclusions and
a summary are included in Section 4.

2. NATURAL HEURISTIC DYNAMIC PROGRAMMING

In Natural Heuristic Dynamic Programming, the
desired motion to be controlled is first
synthesized as superpositions of sinusoidals

through the Trajectory Pattern Method. Once the motion is synthesized, it can be shown that the desired control action and the motion are signals with only a fundamental frequency and its higher harmonics. The task of the controller is simplified to identifying these coefficients. Basic HDP algorithm is then deployed for this task. Furthermore, the controller first learns control of the system in slow motion corresponding to a small fundamental frequency. We then gradually tune the fundamental frequency to a higher value for faster motion, mimicking the natural learning strategy of human beings. In this section, the Trajectory Pattern Method is first used to synthesis motion. Then a description of NHDP for controlling the synthesized motion is described.

2.1 Trajectory Pattern Method

The Trajectory Pattern Method was first proposed by Rastegar and Fardanesh (Rastegar and Fardanesh, 1991). Readers interested in more detailed description of the method are referred to (Rastegar and Fardanesh, 1991; Fardanesh and Rastegar, 1992; Tu and Rastegar, 1993; Tu , et al., 1994).

Basically, the method can be said to be a generalization of the computed torque method. The desired motion is synthesized by the selection of appropriate trajectory parameters to satisfy the desired end conditions and/or tracking requirement.

As a simple example, for a one dimensional system with the following dynamic equations:

$$u(t) = m\ddot{x}(t) + c\dot{x}(t) + kx(t) \qquad (1)$$

where $u(t)$ is the control action at time t, x(t), $\dot{x}$(t), $\ddot{x}$(t) are the position, velocity and acceleration and m, c, k are constants.

The controller's task is to identify the control signals $u(t)$ for a given initial and end position, velocity and acceleration. The dynamic equation (Equation 1) is either assumed to be unknown or having inaccurate parameters. In the Trajectory Pattern Method, we synthesize the desired motion as composed of superposition of sinusoidals with a fundamental frequency and its higher harmonics. Furthermore, the parameters of the trajectory are synthesized such that the

initial and end conditions as specified by the problem are met.

For this simple example, given that the initial ($t=0$) and final (t= t_f) position, velocity and acceleration are:

$$x(0) = \dot{x}(0) = \ddot{x}(0) = 0$$
$$x(t_f) = 1.0, \dot{x}(t_f) = \ddot{x}(t_f) = 0 \qquad (2)$$

We can synthesize the desired trajectory as:

$$x(t) = k_0 + k_1(\cos\omega t - \cos 3\omega t / 9)$$
$$\dot{x}(t) = -k_1\omega(\sin\omega t - \sin 3\omega t / 3) \qquad (3)$$
$$\ddot{x}(t) = -k_1\omega^2(\cos\omega t - \cos 3\omega t)$$

where $k_0 = 0.5$ and $k_1 = -9/16$ are chosen to meet the initial and end conditions (Equation 2) requirements with the condition that $\omega t_f = \pi$. That is, when the fundamental frequency is small, t_f is large and the motion is slow. Since the desired motion is synthesized to compose of sinusoidals with a fundamental frequency ω and its third harmonics 3ω, the desired control signal is of the form:

$$u(t) = A\cos\omega t + B\sin\omega t + C\cos 3\omega t + D\sin 3\omega t + E \qquad (4)$$

where A, B, C, D, E are constants analogous to the Fourier coefficients of the control signals. Once this observation is made, the controller's task is now simplified to identifying these constant coefficients. Because of this simplification, HDP, the simple form of ACDs can now be effectively utilized. Another advantage of the Trajectory Pattern Method is that the desired motion is always composed of superpositions of sinusoidal signals with a fundamental frequency and its higher harmonics. These motions are smooth functions similar to that of natural motions. Furthermore, when the fundamental frequency ω is small, the motion is slow. By gradually increasing ω, the motion becomes faster.

Once the Trajectory Pattern Method has been incorporated into the control problem, in the next section, we discuss how the original HDP algorithm can be used to learn the coefficients A,

B, C, D, E first for a slow motion and then for a faster one.

It should be noted also that in the above example, we synthesized the desired motion to contain the fundamental frequency and its third harmonics (ω, 3ω). In general, however, the synthesized motion may contain the fundamental frequency ω and any of its higher harmonics. Readers interested in more advanced Trajectory Pattern Method are referred to (Rastegar and Fardanesh, 1991; Fardanesh and Rastegar, 1992; Tu and Rastegar, 1993; Tu, *et al.*, 1994).

2.2 *Natural Heuristic Dynamic Programming (NHDP) Components*

The key element of the HDP design is the "Critic" network. The function of the critic network is to learn the cost-to-go *J* function in the Bellman equation of dynamic programming (Werbos, 1992; Prokhorov and Wunsch, 1997) :

$$J(t) = \sum_{k=0}^{\infty} \gamma^k U(t+k) \qquad (5)$$

where γ is a discount factor and *U(.)* is the utility function. The basic idea is that the current action should be optimized to minimize or maximize not just current utility but future utility. This goal is achieved through minimization or maximization of *J*, since it is a sum of current and future utility. The multilayered backpropagation neural network is a convenient tool for such optimization as the gradient of *J* can be obtained through backward propagation. However, during our practice with HDP, we found that the algorithm does not perform well for problems involving multiple variables. This perhaps is the major reason that ACD practitioners generally opted for the more advanced design such as DHP.

From a seminar in neurobiology, we recently learned that in the human brain, various functions are highly localized (Damasio, 1997). Each subset of neurons is only responsible for a specific local task. This combined with the approach on Trajectory Pattern Method, gives rise to a new form of HDP that mimics the natural learning behavior of humans, hence the name Natural HDP.

From the discussions in Trajectory Pattern Method, the original control problem is transformed to the problem of identifying the coefficients, *A, B, C, D, E* as in the example of Section 2.1. To localize the effect of each parameter, we use four critic networks, each learning how changes in each coefficient affects the cost-to-go *J* function (the constant *E* does not change with ω). The end result is schematically depicted in Figure 4. From Figure 4, the neuro-controller contains four simple, critic networks, each can perform backpropagation in parallel. Such a neuro-architecture is consistent with the biological model that a subset of neurons is responsible for a local task, learning *J* as a function of a specific coefficient. Such learning is accomplished in parallel, taking the full advantages of the Neural Network properties.

Furthermore, due to the gradient decent nature of backpropagation, the initial values of the coefficients *A, B, ..., E* is important. To have an intelligent guess of the initial coefficients, we first use Backpropagation of Utility (Werbos, 1990; Tang and Pingle, 1996) to find the coefficients for a very slow motion with a small value for the fundamental frequency ω and large value of t_f.

Once a good initial guess is obtained, Natural Heuristic Dynamic Programming with multiple critic networks is used to control the system for a faster motion (larger value for the fundamental frequency ω). Learning is achieved in two phases. In phase I, each critic network learns the cost-to-go function *J* associated with each coefficient through standard backpropagation. Then in Phase II, through backpropagation, the gradient of *J* with respect to the coefficients is computed and a new coefficient is obtained through gradient decent. As an example, for the first critic network, we obtain $\partial J_A / \partial A$ through backward propagation. New values of the coefficient *A* is then computed as:

$$\text{New } A = \text{Old } A - \alpha \, \partial J_A / \partial A \qquad (6)$$

where $0 < \alpha < 1$ is a constant learning rate. Once a new coefficient is obtained the two phases repeated for another iteration until the current utility is less than or bigger than a prescribed value if we are minimizing or maximizing the utility function.

3. A SIMPLE DYNAMIC SYSTEM

To illustrate Natural HDP, we used the simple example outlined in Section 2.1. The dynamic equation is described by Equation 1 with the constants assuming the following values: $m=10$, $c=2$, $k=5.5$. The desired initial and final position, velocity, and acceleration are specified by Equation 2. First, we use backpropagation of utility (Werbos, 1990; Tang and Pingle, 1996) to learn a very slow motion, $\omega=0.01\ \pi$, $t_f=100$. The desired coefficients are:

$$A = -3.088,\ B=3.534e\text{-}2,\ C=3.382e\text{-}1,$$
$$D=-1.178e\text{-}2,\ E=2.75 \qquad (7)$$

Using Backpropagation of Utility, the coefficients are identified as:

$$A = -3.097,\ B = 4.080e\text{-}3,\ C = 3.456e\text{-}1,$$
$$D = -1.017e\text{-}2,\ E = 2.774 \qquad (8)$$

Next we use these as initial values to control a faster motion, $\omega=0.1\pi$, $t_f=10$. Since the coefficients, with the exception of E are functions of the fundamental frequency, ω. The desired coefficients for this faster motion are:

$$A = -2.539,\ B = 3.534e\text{-}1,\ C = -2.114e\text{-}1,$$
$$D = -1.178e\text{-}1,\ E = 2.75 \qquad (9)$$

For each of the coefficient, with the exception of E, we created a critic network that inputs a coefficient and outputs its corresponding cost-to-go function J as defined in Equation 5. The discount factor and the learning rate in Equations 5 and 6 are $\gamma=\alpha=0.1$ and the Utility function is:

$$U = \int_0^{t_f} (x(t) - x_d(t))^2\, dt \qquad (10)$$

where $x(t)$ is the actual position as a result the action signals $u(t)$ and $x_d(t)$ is the desired position as described in the synthesized motion (Equation 3). Obviously, the neuro-controller's task is to find the coefficients that can minimize this utility function over time.

Once the coefficients are determined by multiple critic networks, the control signals are obtained by Equation 4. Figure 2 is a comparison of the control signal $u(t)$ at various iterations versus time ($t=0$, ..., $t_f=10$). The result at $Iter=0$ corresponds to the initial coefficients (Equation 8 with $t_f=100$). As the iteration progresses, the control signal approaches that of the desired value. Figure 3 plots the absolute error between the obtained and the desired control signals.

Figure 4 is a comparison for the position trajectory at various iterations. Again, the position of the dynamic system gradually approaches that of the desired values. Note that at all iterations, the trajectory is a smooth function. This is a direct consequence of the Trajectory Pattern Method. The trajectory profile for velocity and acceleration depicted similar behavior and is not shown here. Figure 5 plots the absolute trajectory error between the actual and desired position, velocity and acceleration after 5,000 iterations. In all cases, the absolute error is less than $3.0e\text{-}2$, which means that the trajectory generated by the control signal is very close to their desired values.

4. SUMMARY

In this paper, a new version of HDP, the simplest kind of ACDs, is presented. The main features of this new version include: (i) use of Trajectory Pattern Method to guarantee smoothness of trajectory and control signals; and simplification of the control problem; (ii) use multiple critic networks to localize effect of each parameter mimicking the natural biological model of human brain; and (iii) allow the controller to learn from slow to fast motion, analogous to the natural learning behavior of humans. In other words, this new version of the HDP algorithm uses the natural model of human learning as a reference, hence the name Natural HDP.

As a preliminary result, we applied the algorithm for a simple, one-dimensional, linear dynamic system. The results are encouraging. The NHDP controller is able to control the system from slow to fast motion. It is also important to note that the strength of NHDP algorithm will be more apparent for complex, non-linear system as the Trajectory Pattern Method provides a greater degree of simplifications for these systems. We are currently working on the application of NHDP to a nonlinear dynamic system.

REFERENCE

Damasio, A. (1997). The ghost in the Machine: Exploration on the Minded Brain, Neurobiology seminar, SUNY Stony Brook.

Fardanesh, B. and J. Rastegar. (1992). A New Model Based Tracking Controller for Robot Manipulators Using the Trajectory Pattern Inverse Dynamics. *IEEE Transactions on Robotics and Automation*, **8(2)**:279--285.

Prokhorov, D.V. and D.C. Wunsch II. (1997). Adaptive Critic Design. *IEEE Transactions on Neural Networks*, **8(5)**:1997—1007.

Rastegar, J. and B. Fardanesh. (1991). Inverse Dynamic Models of Robot Manipulator Using Trajectory Patterns - With Application to Learning Controllers. *In Proceedings of the 8th World Congress on the Theory of Machines and Mechanisms*, Czechoslovakia.

Tang K.W., and G. Pingle. (1996). Neuro-Remodeling via Backpropagation of Utility.' *Journal of Mathematical Modeling and Scientific Computing*, Accepted for Publication.

Tu, Q. , J. Rastegar, and R.J. Singh. (1994). Trajectory Synthesis and Inverse Dynamic Model Formulation and Control of Tip Motion of a High Performance Flexible Positioning System. *Mechanism and Machine Theory*, **29(7)**:959--968.

Tu, Q. and J. Rastegar. (1993). Manipulator Trajectory Synthesis for Minimal Vibrational Excitation Due to Payload. *Transactions of Canadian Society of Mechanical Engineers*, **17(4)**:557--566.

Werbos, P.J. (1990). Backpropagation Through Time: What It Does and How to Do It. *Proceedings of the IEEE*, **78(10)**:1550--1560.

Werbos, P.J. (1992). Approximate Dynamic Programming for Real-Time Control and Neural Modeling. In *Handbook of Intelligent Control*, (White, D.A. and Sofge, D.A. (Ed)), pages 493--525. Van Nostrand Reinhold, New York.

ACKNOWLEDGEMENT

The authors acknowledge and appreciate discussions with Paul Werbos. This research was supported by the National Science Foundation under Grant ECS-9626655.

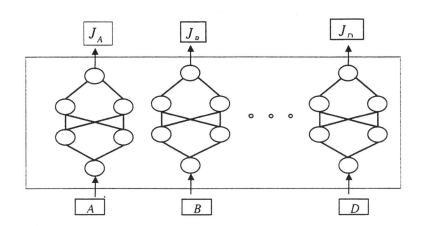

Figure 1: Schematic Representation of NHDP Architecture

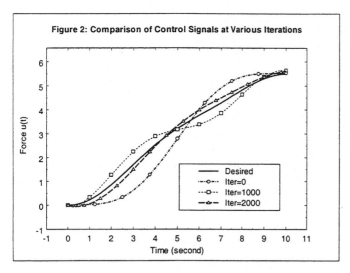

Figure 2: Comparison of Control Signals at Various Iterations

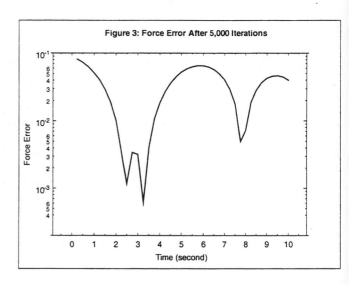

Figure 3: Force Error After 5,000 Iterations

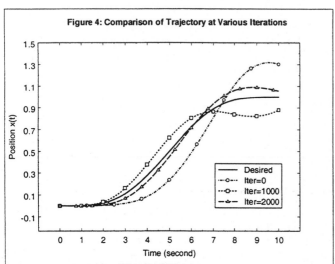

Figure 4: Comparison of Trajectory at Various Iterations

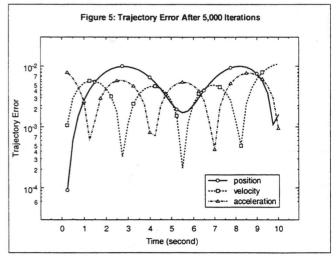

Figure 5: Trajectory Error After 5,000 Iterations

VERIFICATION OF CONTROL LAWS USING PHASE-SPACE GEOMETRIC MODELING OF DYNAMICAL SYSTEMS

Jeff May * Feng Zhao **

* Department of Computer and Information Science
Ohio State University
Columbus, OH 43210
Email: may-j@cis.ohio-state.edu
** Xerox Palo Alto Research Center
3333 Coyote Hill Road
Palo Alto, CA 94304
Email: zhao@parc.xerox.com

Abstract: This paper presents an algorithm for verifying control laws using phase-space geometric modeling of dynamical systems. The algorithm evolves a hierarchically-refined bound of system nonlinear dynamics and can address practical concerns such as sensor, actuator, and modeling uncertainties in a systematic manner. The algorithm has been applied to verifying a control law for a magnetic levitation system, and the computational results are compared against the performance of the actual physical system. Copyright © 1998 IFAC

Keywords: Control systems, Verification, Phase space, Geometric approaches, Computational methods.

1. INTRODUCTION

Control verification ensures correct behaviors for controlled physical systems. Important applications range from safety-critical systems such as aircraft controllers, where improper behavior can result in loss of life, to cost-critical systems such as factory controllers, where a faulty controller can result in costly inefficiency. Unfortunately, obtaining a closed-form analytic solution to the verification problem is often impractical. The nonlinear nature of many man-made systems requires that approximations be made to apply most verification techniques. Uncertainties such as modeling and sensing error make it difficult to express the range of possible behaviors of the system in a tractable form. Thus, verifying a controlled system frequently requires that linear approximations be made, and that considerations for factors such as modeling error and sensing error be omitted.

This paper describes a computational verification algorithm that relies on evolving phase-space geometric models of system dynamics. The contributions of this paper are threefold: (1) It introduces a novel hierarchical refinement of bounds on system dynamics to avoid unnecessary over-approximation of nonlinear behavior; (2) The phase-space representation it uses can model nonlinearity and uncertainty in a systematic, intuitive manner; (3) The algorithm has been applied to verify a nonlinear maglev control system prototyped in our laboratory. Although the phase space of the example used in the paper is two dimensional, the algorithm is applicable to higher-order control systems as well as those whose dynamics does not admit a closed-form analytic description but whose states can be fully observed via experimental means.

A number of approaches to verification exist for simple linear systems. Recent work has focused on developing methodology for more complex systems such as hybrid systems (see e.g. (Alur *et al.*, 1996; Bouajjani *et al.*, 1993)). The approaches in (Henzinger *et al.*, 1995; Puri *et al.*, 1996; Greenstreet and Mitchell, 1998; Dang and Maler, 1998) use a polygonal or grid-based representation for phase-space regions of a hybrid system during verification, and share several similar concepts with the algorithm presented here. Our algorithm evolves from earlier work in phase-space control synthesis (Bradley and Zhao, 1993; Zhao *et al.*, 1997).

2. PHASE-SPACE VERIFICATION ALGORITHM

The phase-space verification algorithm is used to verify proper regions of operation that have the desired limit behaviors for stabilization control systems (i.e. control systems that are designed to stabilize the plant in a goal region). Other properties such as overshoot or convergence rate can also be verified with only minor changes to the algorithm. The algorithm is applicable to discrete-time control systems with a fixed sampling frequency. The underlying dynamics of the plant may be continuous, discrete, or hybrid. It is assumed that the system is designed to operate within a bounded region of the phase space. Let the system dynamics be described by: $\dot{x} = F(x, u)$, where x is the system state, and u is the control input. Since the system is discretely sampled, the dynamics of the controlled system can be written as $x_{n+1} = f(x_n)$, where x_i denotes the system state at time t_i and $f(x_i) = \int F(x_i, o(x_i))$ after one time period ($o(x_i)$ is the controller output at x_i).

The algorithm proceeds as follows:

(1) Partition the phase-space region of interest into a finite set of cells C.
(2) Determine the initial controllable region R_{cont}.
(3) For each cell c in $C - R_{cont}$,
  (a) Find a polytope p_c bounding the image of c under f.
  (b) Compute the "escape polytope" $e_c = p_c - c$.
  (c) If e_c is contained within R_{cont}, and f generates no cycles within c, mark c as verified, and set $R_{cont} = R_{cont} \cup c$.
(4) If any new cells were added to R_{cont} in step 3, repeat step 3.
(5) If the region of interest has been verified, or if a pre-specified number of steps has been

taken, quit. Otherwise, form a new set C' by subdividing the unmarked cells in C. Set $C = C'$, and go to step 3.

The partitioning of the phase space in step 1 is arbitrary; however, regular partitions are often used, and certain types of control suggest preferred partitions. For example, control based on cell maps suggests an initial partition identical to the one used to generate the cell maps (Hsu, 1987).

Determination of the initial controllable region requires a bit more effort. There are two basic approaches. If the controller has already been verified for a certain region R_1 (e.g. using analytic techniques), and the verification algorithm is being used to extend that region, R_{cont} is set to R_1, and all cells that are fully contained within R_1 are marked. This approach is taken, for example, when a local controller (e.g. one based on linear techniques) is being augmented by a global controller.

If no "pre-verified" region is available, R_{cont} cannot just be set to the goal region, because it is possible that for the given controller, the plant can start out in the goal region, but later exit it and never return. Thus, in this approach, a set R_{cont} of "core cells" must be found. The "core cells" have the following two properties:

(1) Every $c \in R_{cont}$ is in the goal region.
(2) For every $c \in R_{cont}$, the image of c under f is contained in R_{cont}.

A maximal set of core cells (for a given phase space partition) can be generated by selecting all cells contained in the goal region, and then iteratively eliminating cells whose image bound lies outside the set of selected cells.

Finding a polytope p_c bounding $f(c)$ can be achieved in many ways. One of the simplest and most efficient ways to find a suitable p_c is to compute the minimum and maximum values for each component of F over the cell c and form a bounding box for $f(c)$ using these values. In hybrid systems terms (see e.g. (Branicky, 1995)), each cell can be thought of as a discrete state, and the bounding polytope is determined by approximating system dynamics with a rectangular differential inclusion.

In step 3, we are checking for two properties:

(1) $\forall x \in c : f^n(x) \notin c$ for some $n \in \mathbf{N}$. That is, all trajectories of the system starting within c eventually exit c.
(2) $\forall x \in c, n \in \mathbf{N} : f_n(x) \notin c \land f_{n-1}(x) \in c \Rightarrow f_n(x) \in R_{cont}$. That is, when a trajectory exits c, it reaches a cell that has already been marked as verified.

Checking the second property is straightforward. To check the first property we intersect $f(c)$ with c to form a polytope p'_c. This process is repeated with p'_c until the intersection is empty, or a pre-specified number of iterations, i, is exceeded. Thus, we replace the first property with a stronger condition—that all trajectories leave c within i time steps.

Assuming that a regular, rectangular initial partition is used, and that subdivision is done uniformly, the space requirement of the algorithm is $O((2^d)^s n)$ where d is the dimension of the phase space, s is the level of subdivision, and n is the number of cells in the initial partition. Thus, the memory requirements depend linearly on the size of the initial partition, and exponentially on the level of subdivision.

Step 1 typically has time complexity linear in n, although a complex partition may require more time. Step 2 requires at most $O(n^2)$ time if a core set is being determined. Each iteration of step 3 takes $O(n)$ time (for the comparison with R_{cont} in 3c), so the entire algorithm requires $O(((2^d)^s n)^3)$. As with the space complexity, the time complexity has an exponential dependence on the level of subdivision, and a polynomial dependence on the size of the initial partition. With a regular partition, the comparison in step 3c depends only on the cells intersected by the image bound, which is typically far less than n. Similarly, the order of selection in step 3 can have a large impact on the efficiency of the algorithm. In practice, the verification of a cell far from the initial controllable region depends on the verification of cells nearer the region. An intelligent ordering, that starts from the cells nearest the controllable region and works outward, often results in more cells being verified for each iteration of step 3, thus reducing the number of times step 3 must be iterated to a number far less than n.

2.1 *Proof of Soundness*

In this section, it is shown that if the system starts in a cell that has been marked as verified, the system will eventually progress to a state within the goal region.

Proof: Number the cells of the phase space partition in the order that they are marked, with all initially verified cells (or core cells) numbered zero. Consider cell number i, with $i > 0$. Since the cells are numbered by order of marking, all trajectories starting in cell i eventually flow into a lower numbered cell (since cell i will be marked only if its image bound lies in cells that have already been marked). Thus, by induction, all trajectories starting in a cell with a positive number will eventually flow into a cell numbered zero. By assumption (or by the definition of core cell), no trajectories starting from a zero numbered cell will leave the set of zero numbered cells, so all trajectories will eventually progress to a state within the goal region.

Note that this only guarantees that marked cells exhibit proper behavior—there is no guarantee that all cells that exhibit proper behavior will be marked (i.e. the algorithm is sound but not necessarily complete).

2.2 *Enhancements and Optimizations*

In the above description, several practical issues, such as measurement error, controller output error, and modeling error are not mentioned. However, because of the geometric nature of the computation, such considerations can be incorporated in a straightforward fashion. Measurement error can be accounted for by expanding the cell when the bounding polytope is determined. Controller output error and model error can be dealt with (assuming the error is bounded) by expanding the image bounding polytope corresponding to the potential range of the function f describing the dynamics.

Continuous time systems can be verified by selecting a time period, and examining the evolution of the system over this time period (i.e. treating the system as a discretely-sampled system, and using the base verification algorithm). If necessary, this time period can be iteratively reduced to provide a less conservative bound on system behavior (as is done with the phase-space partition in the base algorithm).

Certain other properties of a system can be verified with minor modifications to the base algorithm. For example, suppose it is necessary to verify not only that the system reaches the goal region, but also that the percent overshoot is limited. In this case, when a cell is being checked, its image bound must fall within marked cells and each of these intersected cells must be annotated as having trajectories with a tolerable maximum distance to the goal region. The newly marked cell is then annotated with its maximum distance as well (computed as the maximum of the annotated values of the intersected cells and the distance of the cell itself).

In the base algorithm, only the behavior after one sampling period is considered. This is because the bounding polytope of the image of a cell increases in size exponentially with time, thus making the bound less accurate the longer the time period considered. However, when the system dynamics are "slow" in comparison to the partition gran-

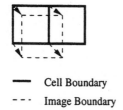

— Cell Boundary

- - - Image Boundary

Fig. 1. Slow dynamics with respect to the partition granularity and the sampling period results in dependence on an adjacent cell.

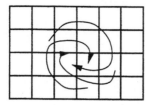

Fig. 2. "Spiraling" trajectories in phase space.

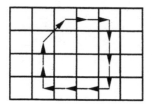

Fig. 3. Cyclic dependence between cells created by spiral trajectories above.

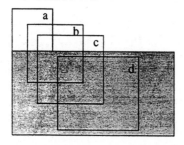

Fig. 4. Sequence of images with decreasing unverified area. The shaded area represents the previous verified region. The transparent boxes represent a sequence of image bounds (b is the image bound of a, c is the image bound of $b - R_{cont}$, et cetera).

ularity and the sampling period, a cell's image bound will often overlap with a neighbor, resulting in a dependence from the cell to its neighbor (Since the neighbor must be verified before the cell in consideration can be; see figure 1). If the system has "spiraling" trajectories (figure 2), a cycle of dependence (figure 3) between a set of unverified cells can occur. In this case, the partition must be subdivided several times to properly verify the system. This subdivision is costly—both memory usage and computation time scale exponentially with the level of subdivision in the worst case. In these cases, the algorithm can be optimized by continuing to iterate the function f when doing so is beneficial. If the escape polytope e_c does

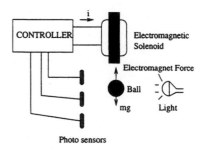

Photo sensors

Fig. 5. Diagram of magnetic levitation system.

not lie entirely within verified cells, the polytope is clipped against those unverified cells it intersects, and the resulting polytopes are recursively checked. This process continues until either the cell is verified (i.e. the iteration produces an image bound that is contained within the verified region), or no further "progress" is made. The current implementation considers "progress" to be made as long as the unverified area (volume) occupied by the bounding polytope is decreasing in size (see figure 4). Other criteria may also be useful.

Algorithmically, this optimization replaces steps 3a - c with a call to the following function on cell c:

bool *CheckRegion*(region r)

(1) Find a polytope p_r bounding the image of r under f.
(2) Compute the "escape polytope" $e_r = p_r - r$.
(3) Set $r_{unverified} = e_r - R_{cont}$.
(4) If $r_{unverified} = \phi$ return true.
(5) Otherwise, if $volume(r_{unverified}) \leq volume(r)$, return *CheckRegion*($r_{unverified}$), else return false.

3. RESULTS AND ANALYSIS

The verification procedure has been tested on the controller for a magnetic levitation system (figures 5 and 6). This system is a useful testbed since it is inherently unstable and nonlinear (due to the inverse square law of magnetic attraction).

The initial verifiable region is obtained using a local controller, generated using a linearized model of the original nonlinear system. This linear controller is augmented by a nonlinear, phase space based global controller (as in (Loh, 1997; Zhao *et al.*, 1997)). The equilibrium point was set to 11.6 millimeters from the bottom of the solenoid to the center of mass of the steel ball. The coordinate system is chosen such that the displacement vector points downwards from the solenoid to the steel ball. The local controller was used when the ball was within one millimeter of the equilibrium point with a velocity having absolute value less than 0.05 meters per second.

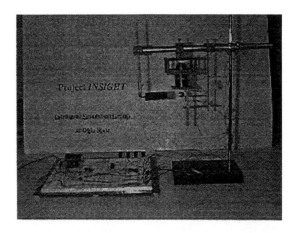

Fig. 6. Photo of the actual magnetic levitation testbed prototyped at Ohio State University.

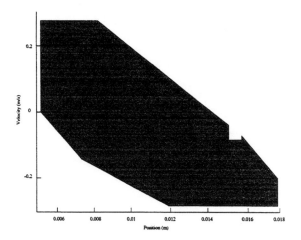

Fig. 7. Region of the magnetic levitation system's phase space that is marked as verified by the verification algorithm.

Outside this region, the global controller was invoked.

The optimized algorithm was used to generate a predicted region of stability for the system — no considerations were made for uncertainties due to the difficultly in measuring these on the simple hardware used. The theoretical region of stability produced by the verification algorithm is shown in figure 7.

The performance of the control law on the real physical system was also measured. This allows for the comparison of the predicted region of stability with the region of stability measured on the actual physical system. However, due to hardware limitations, only a limited portion of the phase space can be explored. In particular, the stability region of the actual system is measured by introducing short disturbances in the form of an increased or decreased input current to the system. This causes the ball to achieve a positive velocity and drop below the desired equilibrium point (reduced input current), or to achieve a negative velocity and rise above the equilibrium point. The region of stability can be measured by observing the system behavior after the disturbance.

Unfortunately, this method of collecting experimental data on the actual system has the drawback that certain regions of the phase space cannot be explored. In particular, regions above the equilibrium point with positive velocity, and regions below the equilibrium point with negative velocity cannot be reached. Another difficulty that occurs is that when an increased input current is applied, the steel ball will often bounce off the solenoid into a region of phase space near the equilibrium point. Thus, for these experiments, only a decreased current type of disturbance is used. Finally, this method enables only coarse grain control of the regions of phase space explored, so often the data obtained will contain "holes" for areas where insufficient samples were obtained.

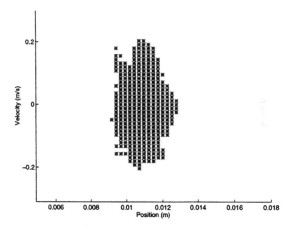

Fig. 8. Measurable region of stability for the actual magnetic levitation system. An asterisk is plotted at the center of each cell that is marked as verified.

The measurable region of stability for the actual system is shown in figure 8. These results were obtained by applying a variety of disturbances, and recording the trajectories of the system when it recovered correctly from the disturbance. Outliers were removed by filtering out data points with few nearby samples.

The measurable region obtained is quite a bit smaller than the predicted region of stability; however, the primary reason for this is the limited area of phase space the hardware enables one to explore, as discussed above. Notice that the top of the measured region corresponds closely with the boundary of the theoretical controllable region. This suggests that a larger disturbance would cause the system to enter an uncontrollable region, and fail to return to the equilibrium point. Thus, for the limited region of phase space explored, the theoretical and actual results are similar.

4. RELATED WORK

The HyTech system (Alur *et al.*, 1996) uses convex polyhedra to represent regions of a hybrid system and shares the concept of verification through exploration of the pre-images of the desired goal states. HyTech allows the expression of properties to be verified as explicit mathematical formulas. For many properties, representation as a simple formula is more flexible than the implicit representation of properties through cell annotation. HyTech is designed to easily represent hybrid systems, which must be represented by an augmented phase space in our approach. However, the geometric approach used here allows for a more straightforward representation of real-world uncertainties. Furthermore, our algorithm is guaranteed to terminate, and allows for a systematic refinement of approximation of dynamics.

Recently, several researchers suggested verification techniques based on projecting phase-space regions. In (Greenstreet and Mitchell, 1998), the authors described a method for computing projection polyhedra obtained from integrating initial regions. While an efficient and exact algorithm exists for two-dimensional linear systems with convex spaces, a general higher-dimensional polyhedra have to be reconstructed from a series of two-dimensional subspace projections. Because the paper did not provide implementation details and computational results for the higher-dimensional case, it is difficult to evaluate the effectiveness of the algorithm. The approaches in (Puri *et al.*, 1996; Dang and Maler, 1998) share a grid-based representation of phase space with the algorithm developed in this paper. In comparison, our algorithm introduces a hierarchically refined bound of system dynamics to avoid unnecessary over-approximation.

5. CONCLUSIONS

An algorithm for verification of control laws using phase-space geometric modeling was presented. This algorithm is applicable to a wide range of control systems that are continuous, discrete, or hybrid, and can be used with a variety of forms of control laws. Once a bound for measurement, controller output, and modeling uncertainty is obtained, considerations for these types of uncertainties can easily be incorporated into the verification algorithm. The algorithm was applied to a nonlinear controller for a magnetic levitation system, and the resulting region of stability was compared to that of the actual physical system.

Future avenues of research include exploring the applicability of the algorithm to other real systems, and investigating optimal initial partitions for different control laws. Furthermore, more accurate and flexible techniques for measuring the region of stability of the physical maglev system need to be developed so that the results of the verification algorithm can more readily be compared to those of the actual system.

6. ACKNOWLEDGMENT

This work is conducted at Ohio State University, supported in part by ONR YI grant N00014-97-1-0599, NSF NYI grant CCR-9457802, and a grant from the Xerox Foundation. We thank S. Loh for designing and prototyping the maglev control testbed.

7. REFERENCES

Alur, R., T. Henzinger and P.H. Ho (1996). Automatic symbolic verification of embedded systems. *IEEE Trans. on Software Engineering.*

Bouajjani, A., R. Echahed and J. Sifakis (1993). On model checking for real-time properties with durations. *Proc. 8th Annual Symposium on Logic in Computer Science.*

Bradley, E. and F. Zhao (1993). Phase-space control system design. *IEEE Control Systems* **13**(2), 39–47.

Branicky, M. (1995). Studies in Hybrid Systems: Modeling, Analysis, and Control. PhD thesis. Massachusetts Institute of Technology.

Dang, T. and O. Maler (1998). Reachability analysis via face lifting. *Hybrid Systems: computation and control, Springer-Verlag Lecture Notes in Computer Science.*

Greenstreet, M.R. and I. Mitchell (1998). Integrating projections. *Hybrid Systems: computation and control, Springer-Verlag Lecture Notes in Computer Science.*

Henzinger, T., P.H. Ho and H. Wong-Toi (1995). Hytech: the next generation. *Proc. 16th Annual Real-time Systems Symposium.*

Hsu, C. S. (1987). *Cell-to-Cell Mapping.* Springer-Verlag, New York.

Loh, S. (1997). An experimental environment for designing and evaluating nonlinear phase space based control laws with application to magnetic levitation. Master's thesis. The Ohio State University.

Puri, A., V. Borkar and P. Varaiya (1996). ε-approximation of differential inclusions. *Hybrid Systems III, Springer-Verlag Lecture Notes in Computer Science.*

Zhao, F., S. Loh and J. May (1997). Phase-space nonlinear control toolbox: The maglev experience. *Hybrid Systems V, Springer-Verlag Lecture Notes in Computer Science.*

BEHAVIORAL PROGRAMMING: ENABLING A "MIDDLE-OUT" APPROACH TO LEARNING AND INTELLIGENT SYSTEMS

Michael S. Branicky [*,1]

* *EE Program*
Case Western Reserve University

Abstract. Motivated by biology and a study of complex systems, intelligent behavior is typically associated with a hierarchical structure, where the lowest level is usually characterized by continuous-variable dynamics and the highest by a logical decision-making mechanism. Consistent with biological findings, we instantiate a particular version of such a hierarchy: a "Middle-out Architecture." Significant there is the plasticity of circuits on both layers, as well as plasticity in their interactions. Intelligent systems must accomplish two broad tasks within this architecture. Based on experience, they must tune their lower-level, continuous, sensorimotor circuits to produce local behaviors that are viable and robust to slight environmental changes. Predicated on focus of attention on sensory input, they must modulate and coordinate these local behaviors to maintain situational relevance and accomplish global goals. Hence, if we are to understand learning and intelligent systems, we must develop ways of *behavioral programming* (taking symbolic descriptions of tasks and predictably translating them into dynamic descriptions that can be composed out of lower-level controllers) and *co-modulation* (simultaneously tuning continuous lower-level and symbolic higher-level circuits and their interactions). Herein, we begin the study of the first of these problems. *Copyright © 1998 IFAC*

Keywords. Learning, behavioral programming, dynamic programming, hybrid systems, intelligent control

1. INTRODUCTION

Though manipulators have gotten more dextrous and sensors more accurate, though processors have gotten faster, memory cheaper, and software easier to write and maintain, truly agile engineering systems that learn and exhibit intelligent behavior have not been demonstrated. The substrate is not lacking; a theoretical approach yielding engineering principles is needed.

Both top-down (cognitive) and bottom-up (reactive) approaches to learning and intelligent systems (LIS) have yielded successes, but only for highly-complex programs performing high-level tasks in highly-structured environments and for simple agents performing low-level tasks in mildly-changing environments, respectively. In contrast, humans and animals possess a middle-level competence between reactive behavior and cognitive skills. They are complex, constrained, and must work in environments whose structure changes.

Thus, the problem of producing LIS confronts us directly with building a theory that bridges the gap between the traditional "top down" and "bottom up" approaches. The missing "middle out" theory we propose requires the development of models and mathematical tools that bridge the gap between "top-down" and "bottom-up" methods.

Motivated by biology and a study of complex systems, intelligent behavior is typically associated with a hierarchical structure. Such a hierarchy exhibits an increase in reaction time and abstraction with increasing level. In both natural and engineered systems the lowest level is usually characterized by continuous-variable dynamics (acting upon and producing continuous signals) and the highest by a logical decision-making mechanism (acting upon and producing discrete symbols). Consistent with

---

[1] Supported by the National Science Foundation, under grant number DMI97-20309. Correspondence: 10900 Euclid Ave., Glennan 515B, Cleveland, OH 44142-7221. `branicky@alum.mit.edu`

biological findings, we instantiate a particular version of such a hierarchy: a "Middle-out Architecture." Significant there—and below—is the plasticity of circuits on both layers, as well as plasticity in their interactions.

Intelligent systems must accomplish two broad tasks within this architecture. Based on experience, they must tune their lower-level, continuous, sensorimotor circuits to produce local behaviors that are viable and robust to slight environmental changes. Predicated on focus of attention on sensory input, they must modulate and coordinate these local behaviors to maintain situational relevance and accomplish global goals. Hence, if we are to understand learning and intelligent systems, we must develop ways of

- Behavioral programming. Taking symbolic descriptions of tasks and predictably translating them into dynamic descriptions that can be composed out of lower-level controllers.
- Co-modulation. Simultaneously tuning continuous lower-level and symbolic higher-level circuits and their interactions.

In control theory and computer science, a mathematics of hybrid systems—those combining continuous and discrete states, inputs, and outputs, e.g., differential equations coupled with finite automata—is emerging that enables the careful study of such questions. Most importantly, we have pioneered a theory and algorithms for optimal control of hybrid systems (Branicky et al., 1998; Branicky, 1995) that are foundational in solving the above mixed symbolic-reactive interdependent optimization problems of LIS. Herein, we study their application to the problem of behavioral programming.

Outline of Paper. In the next section, we motivate our "Middle-out Approach" to learning and intelligent systems. In Section 3, we summarize a formal model framework, *hybrid dynamical systems*, that can be used to model hierarchical systems, from sensor-actuator dynamics to logical decision-making. In Section 4, we review some results on optimal control of hybrid systems that is applicable to solving problems of behavioral programming. Section 5 discusses the behavioral programming idea in more detail and makes the connection with hybrid systems theory explicit.

2. "TOP-DOWN" VS. "BOTTOM-UP" VS. "MIDDLE-OUT"

Currently, there are two unreconciled approaches to learning and intelligent systems:

- the top-down, cognitive, "good old-fashioned AI" approach;

- the bottom-up, reactive, "emergence of behavior" approach. [2]

The first approach has shown successes in structured environments, primarily where search dominates, such as theorem proving or chess playing. In short, "think globally." The second is newer and has shown successes in less-structured environments, primarily where sensorimotor control dominates, such as robotic gait generation or simulated organisms. [3] In short, "(re)act locally."

Our experimental, theoretical, and engineering studies of learning and intelligent systems suggests that neither of these approaches is fully adequate. To illustrate, we describe the problems in applying them to a real-world "intelligent" engineering system and problem, that is error-recovery in an agile manufacturing workcell, in which random errors are inevitable because of statistical mechanical variation:

- **Top-down approach:** Develop a detailed specification and plan for the entire process; run the process, keep an error log, categorize errors and define specific response algorithms; incorporate these as choice points in the robots' control. This is an ineffective approach because task domains change relatively quickly, and the number of possible error conditions is potentially infinite.
- **Bottom-up approach:** Incorporate a range of local reflexes that allow the robots to respond rapidly and automatically to small perturbations in the task. However, this approach does not deal with the larger problem of defining sequences of appropriate actions in the face of a novel problem, optimizing these responses as a function of experience, and adjusting them flexibly to unforeseen contingencies.

The top-down approach tacitly assumes that locally relevant behavior will be produced by thinking globally—and exhaustively. The bottom-up approach is predicated on the belief that global behavior will "emerge" by combining many small behavioral loops. More generally, these two approaches have yielded successes, but to date only for highly-complex programs performing high-level tasks in highly-structured environments (top-down) or simple agents performing low-level tasks in mildly-changing environments (bottom-up).

However, humans and animals occupy a middle-level of competence, lying between the domains of success-

---

[2] Examples of each are Cyc [D.B. Lenat, CYC: A large-scale investment in knowledge infrastructure. *Comm. of the ACM*, 38:11, 1995] and Cog [R.A. Brooks and L.A. Stein. Building brains for bodies, MIT AI Lab Memo #1439, August 1993].

[3] D. Terzopoulos et al., Artificial fishes: Autonomous locomotion, perception, behavior and learning in a simulated physical world, *Artificial Life*, 1:327–351, 1995.

ful top-down and bottom-up systems. They are *responsive*; they must close sensorimotor loops to deal with changing environments. Also, procedural tasks and procedural learning require *sequences* of actions, not just simple behavioral maps. Finally, they require *quick* responses, achieved by *modulating* previously learned behaviors and the current focus of attention. Waiting for globally optimal plans or exhaustively analyzed sensor data from higher-levels means failure. We do not see much effort "in the middle," either from top-down researchers building down, or bottom-up researchers building up. One might argue that such research is unnecessary because future developments of top-down and bottom-up theory will lead to seamless matching between levels. Our group's working hypothesis is that there is no *a priori* reason to believe that a purely symbolic and purely reactive approach can be seamlessly matched without the development of "middle-out" theory and practice. For example, there is no (nontrivial) finite automata representation of certain simple two-dimensional, quasi-periodic dynamical systems (Di Gennaro *et al.*, 1994). Therefore, the *goal* of middle-out theory is to achieve a better understanding of how natural LIS function and how artificial LIS can be effectively implemented by filling the gap between top-down and bottom-up knowledge. Our *approach* to developing this new theory supplies the following missing elements of the two more traditional approaches:

Middle-Out Approach. Develop ways of taking symbolic descriptions of tasks and *predictably* translating them into dynamic descriptions that can be composed out of sensorimotor controllers.

Recapitulating, top-down and bottom-up approaches have contributed in method and principles, each in their own domain of applicability. Humans and animals possess a middle-level competence between reactive behavior and cognitive skills. A "middle-out" approach is necessary, but currently lacking. Recent advances in hybrid systems tehory, summarized below, have us poised—theoretically and experimentally—to pursue such middle ground.

3. WHAT ARE HYBRID SYSTEMS?

Hybrid systems involve both continuous-valued and discrete variables. Their evolution is given by equations of motion that generally depend on all variables. In turn these equations contain mixtures of logic, discrete-valued or **digital** dynamics, and continuous-variable or **analog** dynamics. The continuous dynamics of such systems may be continuous-time, discrete-time, or mixed (sampled-data), but is generally given by differential equations. The discrete-variable dynamics of hybrid systems is generally governed by a **digital automaton**, or input-output transition system with a countable number of states. The continuous and discrete dynamics interact at "event" or "trigger" times when the continuous state hits certain prescribed sets in the continuous state space. See Fig. 1.

Hybrid control systems are control systems that involve both continuous and discrete dynamics and continuous and discrete controls. The continuous dynamics of such a system is usually modeled by a controlled vector field or difference equation. Its hybrid nature is expressed by a dependence on some discrete phenomena, corresponding to discrete states, dynamics, and controls.

For the remainder of this section, we concentrate on modeling of hybrid systems. In particular, we introduce **general hybrid dynamical systems** as interacting collections of dynamical systems, each evolving on continuous state spaces, and subject to continuous and discrete controls, and some other discrete phenomena. The reader is referred to (Branicky, 1995) for more details.

4. THE BASIS: HYBRID DYNAMICAL SYSTEMS

4.1 *Dynamical Systems*

The notion of dynamical system has a long history as an important conceptual tool in science and engineering. It is the foundation of our formulation of hybrid dynamical systems. Briefly, a **dynamical system** is a system $\Sigma = [X, \Gamma, \phi]$, where X is an arbitrary topological space, the **state space** of Σ. The **transition semigroup** Γ is a topological semigroup with identity. The **(extended) transition map** $\phi : X \times \Gamma \to X$ is a continuous function satisfying the identity and semigroup properties (Sontag, 1990). A **transition system** is a dynamical system as above, except that ϕ need not be continuous.

Examples of dynamical systems abound, including autonomous ODEs, autonomous difference equations, finite automata, pushdown automata, Turing machines, Petri nets, etc. As seen from these examples, both digital and analog systems can be viewed in this formalism. The utility of this has been noted since the earliest days of control theory.

We also denote by "dynamical system" the system $\Sigma = [X, \Gamma, f]$, where X and Γ are as above, but the **transition function** f is the **generator** of the extended transition function ϕ.[4] We may also refine the above

---

[4] In the case of $\Gamma = Z$, $f : X \to X$ is given by $f \equiv \phi(\cdot, 1)$. In the case of $\Gamma = R$, $f : X \to TX$ is given by the vector fields

concept by introducing dynamical systems with initial and final states, input and output, and timing maps.[5]

4.2 On to Hybrid ...

Briefly, a hybrid dynamical system is an indexed collection of dynamical systems along with some map for "jumping" among them (switching dynamical system and/or resetting the state). This jumping occurs whenever the state satisfies certain conditions, given by its membership in a specified subset of the state space. Hence, the entire system can be thought of as a sequential patching together of dynamical systems with initial and final states, the jumps performing a reset to a (generally different) initial state of a (generally different) dynamical system whenever a final state is reached.

More formally, a **general hybrid dynamical system (GHDS)** is a system $H = [Q, \Sigma, A, G]$, with its constituent parts defined as follows.

- Q is the set of **index states**, also referred to as **discrete states**.
- $\Sigma = \{\Sigma_q\}_{q \in Q}$ is the collection of **constituent** dynamical systems, where each $\Sigma_q = [X_q, \Gamma_q, \phi_q]$ (or $\Sigma_q = [X_q, \Gamma_q, f_q]$) is a dynamical system as above.
  Here, the X_q are the **continuous state spaces** and ϕ_q (or f_q) are called the **continuous dynamics**.
- $A = \{A_q\}_{q \in Q}$, $A_q \subset X_q$ for each $q \in Q$, is the collection of **autonomous jump sets**.
- $G = \{G_q\}_{q \in Q}$, $G_q : A_q \to \bigcup_{q \in Q} X_q \times \{q\}$, is the collection of **(autonomous) jump transition maps**.
  These are also said to represent the **discrete dynamics** of the HDS.

Thus, $S = \bigcup_{q \in Q} X_q \times \{q\}$ is the **hybrid state space** of H. For convenience, we use the following shorthand. $S_q = X_q \times \{q\}$ and $A = \bigcup_{q \in Q} A_q \times \{q\}$ is *the* autonomous jump set. $G : A \to S$ is *the* autonomous jump transition map, constructed componentwise in the obvious way. The **jump destination sets** $D = \{D_q\}_{q \in Q}$ are given by $D_q = \pi_1[G(A) \cap S_q]$, where π_i is projection onto the ith coordinate. The **switching** or **transition manifolds**, $M_{q,p} \subset A_q$ are given by $M_{q,p} = G_q^{-1}(p, D_p)$, i.e., the set of states from which transitions from index q to index p can occur.

Roughly,[6] the dynamics of the GHDS H are as follows. The system is assumed to start in some hybrid state in

$S \setminus A$, say $s_0 = (x_0, q_0)$. It evolves according to $\phi_{q_0}(x_0, \cdot)$ until the state enters—if ever—A_{q_0} at the point $s_1^- = (x_1^-, q_0)$. At this time it is instantly transferred according to transition map to $G_{q_0}(x_1^-) = (x_1, q_1) \equiv s_1$, from which the process continues. See Fig. 3.

Dynamical Systems. The case $|Q| = 1$ and $A = \emptyset$ *recovers all dynamical systems.*

Hybrid Systems. The case $|Q|$ finite, each X_q a subset of R^n, and each $\Gamma_q = R$ largely corresponds to the *usual* notion of a hybrid system, viz. a coupling of finite automata and differential equations. Herein, a **hybrid system** is a GHDS with Q countable, and with $\Gamma_q \equiv R$ (or R_+) and $X_q \subset R^{d_q}$, $d_q \in Z_+$, for all $q \in Q$: $[Q, [\{X_q\}_{q \in Q}, R_+, \{f_q\}_{q \in Q}], A, G]$, where f_q is a vector field on $X_q \subset R^{d_q}$.[7]

Changing State Space. The state space may change. This is useful in modeling component failures or changes in dynamical description based on autonomous—and later, controlled—events which change it. Examples include the collision of two inelastic particles or an aircraft mode transition that changes variables to be controlled. We also allow the X_q to overlap and the inclusion of multiple copies of the same space. This may be used, for example, to take into account overlapping local coordinate systems on a manifold.

Hierarchies. We may iteratively combine hybrid systems H_q in the same manner, yielding a powerful model for describing the behavior of hierarchical systems (cf. Harel's statecharts).

4.3 ... And to Hybrid Control

A **controlled general hybrid dynamical system (GCHDS)** is a system $H_c = [Q, \Sigma, A, G, V, C, F]$, with its constituent parts defined as follows.

- Q, A, and S are defined as above.
- $\Sigma = \{\Sigma_q\}_{q \in Q}$ is the collection of controlled dynamical systems, where each $\Sigma_q = [X_q, \Gamma_q, f_q, U_q]$ (or $\Sigma_q = [X_q, \Gamma_q, \phi_q, U_q]$) is a controlled dynamical system as above with (extended) transition map parameterized by **control set** U_q.
- $G = \{G_q\}_{q \in Q}$, where $G_q : A_q \times V_q \to S$ is the **autonomous jump transition map**, parameterized by the **transition control set** V_q, a subset of the collection $V = \{V_q\}_{q \in Q}$.

$f(x) = d\,\phi(x, t)/dt|_{t=0}$.

[5] *Timing maps* provide a mechanism for reconciling different "time scales," by giving a uniform meaning to different transition semigroups in a hybrid system.

[6] We make more precise statements in (Branicky, 1995).

[7] Here, we may take the view that the system evolves on the state space $R^* \times Q$, where R^* denotes the set of finite, but variable-length real-valued vectors. For example, Q may be the set of labels of a computer program and $x \in R^*$ the values of all currently-allocated variables. This then includes Smale's tame machines.

- $C = \{C_q\}_{q \in Q}$, $C_q \subset X_q$, is the collection of **controlled jump sets**.
- $F = \{F_q\}_{q \in Q}$, where $F_q : C_q \to 2^S$, is the collection of **controlled jump destination maps**.

As shorthand, G, C, F, V may be defined as above. Likewise, jump destination sets D_a and D_c may be defined. In this case, $D \equiv D_a \cup D_c$.

Roughly, the dynamics of H_c are as follows. The system is assumed to start in some hybrid state in $S \backslash A$, say $s_0 = (x_0, q_0)$. It evolves according to $\phi_{q_0}(\cdot, \cdot, u)$ until the state enters—if ever—either A_{q_0} or C_{q_0} at the point $s_1^- = (x_1^-, q_0)$. If it enters A_{q_0}, then it *must* be transferred according to transition map $G_{q_0}(x_1^-, v)$ for some chosen $v \in V_{q_0}$. If it enters C_{q_0}, then we *may* choose to jump and, if so, we may choose the destination to be any point in $F_{q_0}(x_1^-)$. In either case, we arrive at a point $s_1 = (x_1, q_1)$ from which the process continues. See Fig. 4.

Control results for this model are derived in (Branicky *et al.*, 1998); they are summarized in Section 4.

Definition 4.1. The **admissible control actions** available are the **continous controls** $u \in U_q$, exercised in each constituent regime; the **discrete controls** $v \in V_q$, exercised at the autonomous jump times (i.e., on hitting set A); and the **intervention times** and **destinations** of controlled jumps (when the state is in C).

5. HYBRID CONTROL

Theoretical Results. We consider the following optimal control problem for controlled hybrid systems. Let $a > 0$ be a **discount factor**. We add to our model the following known maps:

- **Running cost**, $k : S \times U \to R_+$.
- **Autonomous jump cost** and **delay**, $c_a : A \times V \to R_+$ and $\Delta_a : A \times V \to R_+$.
- **Controlled jump (or impulse) cost** and **delay**, $c_c : C \times D_c \to R_+$ and $\Delta_c : C \times D_c \to R_+$.

The total discounted cost is defined as

$$\int_T e^{-at} k(x(t), u(t)) \, dt + \sum_i e^{-a\sigma_i} c_a(x(\sigma_i), v_i)$$

$$+ \sum_i e^{-a\zeta_i} c_c(x(\zeta_i), x(\zeta_i')) \tag{1}$$

where $T = R_+ \backslash (\bigcup_i [\tau_i, \Gamma_i])$, $\{\sigma_i\}$ (resp. $\{\zeta_i\}$) are the successive pre-jump times for autonomous (resp. impulsive) jumps and ζ_j' is the post-jump time (after the delay) for the jth impulsive jump. The **decision** or **control** variables over which Eq. (1) is to be minimized are

the *admissible controls* of our controlled hybrid system (see Def. 4.1). Under some assumptions (the necessity of which are shown via examples) we have the following results (Branicky *et al.*, 1998):

- A finite optimal cost exists for any initial condition. Furthermore, there are only finitely many autonomous jumps in finite time.
- Using the relaxed control framework, an optimal trajectory exists for any initial condition.
- For every $\epsilon > 0$ an ϵ-optimal control policy exists wherein $u(\cdot)$ is precise, i.e., a Dirac measure.
- The value function, V, associated with the optimal control problem is continuous on $S \backslash (\partial A \cup \partial C)$ and satisfies the **generalized quasi-variational inequalities (GQVIs)**, which are formally derived in (Branicky *et al.*, 1998).

Algorithms and Examples. We have outlined four approaches to solving the generalized quasi-variational inequalities (GQVIs) associated with optimal hybrid control problems (Branicky, 1995). Our algorithmic basis for solving these GQVIs is the generalized Bellman Equation: $V^*(x) = \min_{p \in \Pi} \{g(x, p) + V^*(x'(x, p))\}$, where Π is a generalized set of actions. The three classes of actions available in our hybrid systems framework at each x are the admissible control actions from Def. 4.1. From this viewpoint, generalized policy and value iteration become solution tools (Branicky, 1995).

The key to *efficient* algorithms for solving optimal control problems for hybrid systems lies in noticing their strong connection to the models of impulse control and piecewise-deterministic processes. Making this explicit, we have developed algorithms similar to those for impulse control (Costa and Davis, 1989) and one based on linear programming (Costa, 1991; Ross, 1992) (see (Branicky, 1995)). Three illustrative examples are solved in (Branicky, 1995). In each example, the synthesized optimal controllers verify engineering intuition.

6. BEHAVIORAL PROGRAMMING

Others have noticed the powerful possibility of solving complex learning problem by composing predictable local behaviors to achieve predictable global goals. For example, (Krishnaswamy and Newman, 1992) presents a means to string goals together in order to accomplish global planning (with local obstacle avoidance guaranteed by the reflex and servo layers). Related work in procedural tasks such as juggling has proved successful (Burridge *et al.*, 1995). Our behavioral programming method starts by constructing families of controllers with guaranteed stability and performance properties. These constituent controllers can then be automatically combined on-line using graph-theoretic algorithms so that higher-level dynamic tasks can be accomplished without regard to lower-level dynamics or safety constraints. The process is analogous to building sensible speech using an adaptable, but predictable, phoneme generator. Such problems can be cast as optimal hybrid control problems as follows:

- Make N copies of the continuous state-space, one corresponding to each behavior, including a copy of the goal region in each constituent space.
- Using as dynamics of the ith copy the equations of motion resulting when the ith behavior is in force.
- Allow the controlled jump set to be the whole state space; disable autonomous jumps. Impose a small switching cost.
- Solve an associated optimal hybrid control problem, e.g., penalize distance from the goal.

The result (if the goal is reachable) is a switching between behaviors achieving that achieves goal.

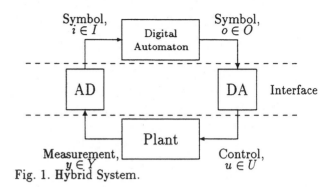

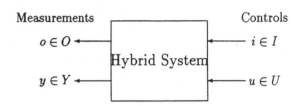

Fig. 1. Hybrid System.

Fig. 2. Hybrid Control System.

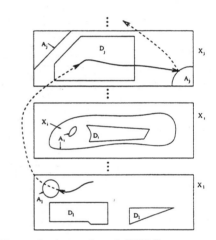

Fig. 3. Example dynamics of GHDS.

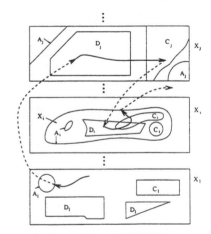

Fig. 4. Example dynamics of GCHDS.

7. REFERENCES

Branicky, M. S. (1995). Studies in Hybrid Systems: Modeling, Analysis, and Control. ScD thesis. Massachusetts Institute of Technology. Dept. of Electrical Engineering and Computer Science.

Branicky, M. S., V.S. Borkar, and S.K. Mitter. (1998). A unified framework for hybrid control: Model and optimal control theory. *IEEE Trans. Automatic Control* **43**(1), 31–45.

Burridge, R.R., A.A. Rizzi, and D.E. Koditschek. (1995). Toward a dynamical pick and place. In: *RSJ/IEEE Intl. Conf. Intell. Robots and Systems*.

Costa, O. L. V. (1991). Impulse control of piecewise-deterministic processes via linear programming. *IEEE Trans. Automatic Control* **36**(3), 371–375.

Costa, O. L. V. and M. H. A. Davis (1989). Impulse control of piecewise deterministic processes. *Math. Control Signals Syst.* **2**, 187–206.

Di Gennaro, S., C. Horn, S. R. Kulkarni, and P. J. Ramadge. (1994). Reduction of timed hybrid systems. In: *Proc. 33rd IEEE Conf. Decision and Control*, pp. 4215–4220 Lake Buena Vista, FL.

Krishnaswamy, V.K. and W.S. Newman. (1992). On-line motion planning using critical point graphs in two-dimensional configuration space. In: *Proc. IEEE Intl. Conf. Robotics and Automation*, pp. 2334–2340, Nice, France.

Ross, S. M. (1992). *Applied Probability Models with Optimization Applications*. Dover. New York.

Sontag, E. D. (1990). *Mathematical Control Theory*. Springer. New York.

COCOLOG LOGIC CONTROL FOR
HIERARCHICAL AND HYBRID SYSTEMS[1]

Peter E. Caines* Tom Mackling
Carlos Martínez-Mascarúa***

* *Department of Electrical Engineering, McGill University, 3480
University Street, Montreal, Quebec, Canada H3A 2A7 and
Department of Mechanical and Automation Engineering, The
Chinese University of Hong Kong, Shatin, N.T.Hong Kong.
peterc@cim.mcgill.edu*
** *Optimal-Robotics Corp., Montreal, Quebec, Canada,
tomm@optimal-robotics.com*
*** *Cegelec ESCA , 11120 NE 8th Place, Bellevue, WA 98004,
carlos.mascarua@esca.com*

Abstract: COCOLOG (Conditional Observer and Controller Logic) (Caines and Wang
1990, 1995) is a framework for the design and implementation in first order logic of
controllers for finite deterministic input-state-output machines. A COCOLOG control
system consists of a partially ordered family of first order logical theories which
describe and control the evolution of the state of a given partially observed finite
machine $\mathcal{M}$. In order to increase the efficiency of COCOLOG we introduce the so-
called systems of *Markovian fragments* which, in a precise sense, have the same control
power as standard COCOLOG systems but for which the amount of information to be
stored does not increase with time. Furthermore, we present an enhancement of the
basic COCOLOG system involving, first, Macro (COCOLOG) Languages and, second,
Macro (COCOLOG) Actions, which, together, consititute *Macro COCOLOG*. These
extensions and enhancements yield a COCOLOG system which, for certain classes
of logic control problems, is (i) syntatically and logically general and flexible, (ii)
efficient in its problem expression and solution, and (iii) efficient in the execution of
feedback control. Applications of COCOLOG include robotics and elevator control. An
implemented COCOLOG controller utilized by a team at McGill University, Montreal,
for part of the control system of a physical mobile robot called the MMRA (McGill
Mobile Robotics Architecture) is briefly described. *Copyright © 1998 IFAC*

Keywords: logic control, artificial intelligence, automated theorem proving,
intelligent control, hierarchical systems, hybrid systems

---

[1] Work partially supported by NSERC grant number
OGP 0001329, NSERC-FCAR-Nortel grant number CRD
180190, NCE-IRIS II IS-5 and NASA-Ames Research Center grant number NAG-2-1040.

1. BASIC COCOLOG LOGIC CONTROL

COCOLOG (Conditional Observer and Controller
Logic) (Caines and Wang 1990, 1995) is a framework for the design and implementation in first

order logic of controllers for finite deterministic input-state-output machines. A COCOLOG control system consists of a partially ordered family of first order logical theories expressed in the typed first order languages $\{L_k; k \geq 0\}$ describing and enabling the controlled evolution of the state of a given partially observered finite machine $\mathcal{M}$. The initial theory of the system, denoted Th_0, gives the theory of $\mathcal{M}$ without data being given on the initial state. Later theories, $\{Th(o_1^k); k \geq 1\}$, depend upon the (partially ordered lists of) observed input-output trajectories, where new data is accepted in the form of the *new axioms* $AXM^{obs}(L_k), k \geq 1$. A feedback control input $U(k)$ is determined via the solution of control problems posed in the form of a set of conditional control rules $CCR(L_k)$, which is paired with the theory $Th(o_1^k)$. A CCR itself is composed of two parts: the *Conditional Control Formula* (CCF) and the *Control Action* (CA).

The basic operational cycle of a COCOLOG system begins with the transition of a system into a new state (for instance a mobile robot comes to a halt at a certain location). A set of observation axioms is generated from the observations on the most recent inputs and outputs of the system (i.e. the most recent partial observations of the system state); in addition, the system generates the state estimation axioms which permit the logical computation of the estimates of the new state. The combination of these new axioms with the previous COCOLOG axioms generates a set of axioms for the new COCOLOG theory, denoted Σ_k. Next, the set of extra-logical conditional control rules $CCR(L_k)$ - possibly independent of k - gives a set of (mutually exclusive and exhaustive) formulas asserting the achievability of a set of tasks (such as moving to one of a set of different corners of a room) starting with a different specified control. At the instant k the current set of CCFs are tested for deducibility from the current axiom set; by their mutually exclusive and exhaustive property one and only one will in fact be deducible.

Operationally this step may be implemented by either: (1) a model verification (MV) facility of the system, or (2) an automated theorem proving (ATP) facility of the system, or (3) combination of (1) and (2).

In cases (2) and (3), the system will typically attempt a group of parallel proofs of the CCF formulas and, by the completenes of each COCOLOG theory, finds a proof of the only statement that is true. The system then acts by implementing the associated control input and the cycle repeats.

Some refinements and enhancements of the general COCOLOG formal framework are described next.

2. MARKOVIAN FRAGMENTS OF COCOLOG

In order to increase the efficiency of COCOLOG, a class of restricted versions is introduced called (systems of) *Markovian fragments* (Wei and Caines, 1994, 1996); in this class of COCOLOG systems a smaller amount of information is communicated from one theory to the next than for full COCOLOG systems. This formulation is associated with a restricted set of candidate control problems, denoted $CCR(L_k^m), k \geq 1$. Under certain conditions, a Markovian fragment theory $MTh(o_1^k)$ contains a large subset of $Th(o_1^k)$, which includes, in particular, the state estimation theorems of the corresponding full COCOLOG system, and, for the set of control rules $CCR(L_k^m)$, has what is termed the same control reasoning power. This provides a theoretical basis for the increased theorem proving efficiency of the fragment systems versus the full COCOLOG systems.

3. MACRO COCOLOG

The logical language of Th^k, for any $k, k \geq 1$, is so basic in its nature that the expression of relatively simple control conditions may become extremely complicated. Hence a theory and methodology has been introduced (Martínez-Mascarúa and Caines, 1996a, 1996b, and Martínez-Mascarúa, 1997) for the introduction of new function and predicate symbols together with their definitional formulas. The resulting logical languages are known as *Macro (COCOLOG) Languages* and they permit the efficient treatment of complex sets of formulas in basic COCOLOG. A crucial feature of Macro (COCOLOG) structures is that they can be recursively extended to create further such structures with concomitant reductions of expression complexity at each stage. In order to express more complex (i.e. less elementary) actions, the CCAs themselves are embedded in a language-like structure consisting of sets of CCRs. This enhancement requires the definition of an execution model for the resulting set of what are called *Macro CCRs*.

The resulting enhancement of the basic COCOLOG system involving first, Macro (COCOLOG) Languages and, second, Macro (COCOLOG) Actions (taken together with the associated execution model) is called *Macro COCOLOG*.

4. HIERARCHICAL AND HYBRID SYSTEMS.

Macro (Markovian Fragment) COCOLOG may be organized into heirarchical control architectures for systems which may, or may not, be initially designed with such structures. This is a consequence of the *dynamical consistency* notion of state aggregation for hierarchical control described in (Wei and Caines, 1994, Caines and Wei 1995); in this work, a theory of state aggregation is introduced which permits the construction of hierarchically layered COCOLOG controllers. Furthermore, any such controller may itself contain a hierarchically layered set of linguistic terms and control actions formulated within the framework of Macro CO-COLOG described above.

Finally we note that the base systems subject to control themselves may not necessarily be discrete systems, and, as described in (Caines and Wei, 1998), an initial continuous state aggregation may be used to obtain a system whose base level controllers are continuous time control systems and whose higher level controllers are discrete (possibly logic) controllers.

The structures mentioned in this section may be used as a conceptual model for the organization of the multi-level controller of the mobile robot which is briefly described in the next section, although it is to be stressed that apart from the top layer of COCOLOG logic control, the mobile robot was designed using other methodologies.

5. COCOLOG LOGIC CONTROL APPLICATION TO A MOBILE ROBOT

Applications of COCOLOG include robotics and elevator control (Dyck and Caines, 1995) and this paper outlines an application to mobile robotics.

An implemented COCOLOG controller was developed and utilized for a part of the control system of a physical mobile robot, called MMRA (McGill Mobile Robotics Architecture), by a team at McGill University, Montreal; this work was carried out with the support of the Canadian National Centres of Excellence, Institute of Robotics and Intelligent Systems (IRIS-II, IS-5 Project).

The overall objective of the (COCOLOG) logic control work within the IS-5 project was to develop high level flexible logic control algorithms and for the software to be applied (among other objectives) to path planning for mobile robot systems. The operation of the COCOLOG controller was intrinsically flexible in the following sense:

during the robot's operation, in case there was either (i) an observed evolution of the environment (obstacles, doors, corridors, rooms), or (ii) an evolution of the robot's knowledge base of the environment, or (iii) a change in the robot's objectives, the COCOLOG logic control system was designed to produce a new planned route for the robot.

The theoretical basis for this work was the existing COCOLOG theory, sketched above, together with contributions to automated theorem proving theory presented in (Mackling, 1997). The programming of the COCOLOG logic controller, and the interface of the logic controller with the other lower level control systems of the mobile robot, was carried out by T. Mackling, C. Martínez-Mascarúa and T. Le Blanc. Many experiments of its operation were carried out by the McGill MMRA team. In addition, among the public exhibitions of the system, one notable sequence of public demonstrations took place at the Queen Elizabeth Hotel, Montreal, at the IRIS-PRECARN meeting in June, 1996.

6. OUTLINE OF THE OPERATION OF THE MOBILE ROBOT COCOLOG CONTROL SYSTEM

The COCOLOG software, which was integrated with the mobile robot communications interface software, was used to realize the controller from a COCOLOG (logical state transition) network graph model and control specification input files. It was processed in real time, in an off-board workstation computer, and interfaced with lower-level control software. The low level control computations were also performed in real time on off-board workstation computers connected to the robot by radio link. The COCOLOG controller modelled the system in terms of rooms, doors and corridors, and the robot's location at the level of granularity of a particular room or corridor. During the robot's motion along a planned path it would learn and whether doors were either blocked (obstructed), open or closed, and similarly whether (inter-door sections of) corridors were either blocked or passable. Blockages within corridors were modeled as newly introduced constraints, which essentially indicated that: "the traversal from a given door D_i to a given door D_j through a given corridor C_k is infeasible". Using such a model, the COCOLOG controller steered the robot, in terms of the doors to be sequentially traversed, to given specified target rooms. The robot would dynamically update its COCOLOG model with the discovery of untraversable (blocked or closed) doors and blocked corridors. This project successfully demonstrated the application of real-time COCOLOG logical specification and logic based control to a sophisticated mobile robot system.

A more detailed discussion of the work of the McGill MMRA team on the mobile robot system may be found at http://www.cim.mcgill.ca/ dudek, and references at http://www.cim.mcgill.ca/ dudek /mobile/mobilebib.htm.

7. REFERENCES

Caines, P.E. and S. Wang (1990). COCOLOG: A conditional observer and controller logic for finite machines. In: *Proc. of 29th IEEE CDC*. Honolulu, HA. pp. 2845–2850.

Caines, P.E. and S. Wang (1995). COCOLOG: A conditional observer and controller logic for finite machines. *SIAM J. Cont. and Opt.* **33**(6), 1687–1715.

Caines, P.E. and Y-J. Wei (1998). Hierarchical hybrid control systems: A lattice theoretic formulation. *IEEE Transactions on Automatic Control Special Issue on Hybrid Systems* **43**(4), 1–8.

Caines, P.E. and Y.J. Wei (1995). The hierarchical lattices of a finite machine. *Systems and Control Letters* **25**, 257–263.

Dyck, D.N. and P.E. Caines (1995). The logical control of an elevator. *IEEE Transactions on Automatic Control* **40**(3), 480–486.

Mackling, T. (1997). Contributions to Automated Theorem-Proving and Formal Methods with Applicatons to Control Systems. PhD thesis. Dept. of Electrical Engineering, McGill University.

Martínez-Mascarúa, C. (1997). Syntactic and Semantic Structures in COCOLOG logic control. PhD thesis. Dept. of Electrical Engineering, McGill University.

Martínez-Mascarúa, C. and P.E. Caines (1996a). Macro COCOLOG with an application to a discrete event tank system. In: *Proc. of the WODES'96: Workshop on Discrete Event Systems*. IEE. Edinburgh, UK. pp. 272–277.

Martínez-Mascarúa, C. and P.E. Caines (1996b). Macro control languages and actions for COCOLOG.
Submitted for publication in the *SIAM Journal of Control and Optimization*. Available at http://www.cim.mcgill.ca/~mascarva/papers.html.

Wei, Y.J. and P.E. Caines (1994). Hierarchical COCOLOG for finite machines. In: *11th IN-*

RIA International Conference on the Analysis and Optimization of Systems (G. Cohen and J-P. Quadrat, Eds.). Vol. 199. INRIA-Sophia Antipolis, France. pp. 29–38.

Wei, Y.J. and P.E. Caines (1996). On Markovian Fragments of COCOLOG for logic control systems. *SIAM Journal of Control and Optimization* **34**(5), 1707–1733. Conference version: *Proc. of the 31st IEEE CDC*, pages 2967–2972, Tucson, AZ, 1992.

EVOLVING A JUGGLER: USING THE INVERTED PENDULUM/CART CONTROL TASK TO INVESTIGATE THE ISSUE OF HOW TO MODULATE REFLEXES

*Yoh-Han Pao and Baofu Duan

Case Western Reserve University
Cleveland, OH 44106
Email: yhp@po.cwru.edu

Abstract: Paper reports on an investigation of how adaptive intelligent reflex controllers can be trained through an evolutionary process using recurrent coupled neural nets acting in direct control manner. The motivation is to explore how reflex actions are learned and how reflex action might be modulated adaptively. Two tasks are discussed. Results indicate that 'intelligent' reflex behavior can be implemented through coupling of dynamical systems. This mode of control seems to be eminently suitable for exercising adaptive real-time control. *Copyright © 1998 IFAC*

Keywords: Dynamical Equations, Neural-Net Control, Evolution, Model-based Control, Adaptation, Recurrent Nets, Hybrid Systems.

1. Introduction

The results of investigations in many disparate fields of research suggest that intelligent behavior is associated with hierarchical structure.

In such systems, reaction times and extent of abstraction increase with increase in level. In addition, the lower levels might be characterized by continuous-variable dynamics and the high levels by logical decision-making discrete-valued mechanisms. The interaction of these different levels, with their types of information, leads to a 'hybrid' system.

In [1], Branicky addresses the issue of developing formal modeling, analysis and control methodologies for 'hybrid ' systems. That effort does not include the topic of design which is much less susceptible to formal treatment.

This present paper is concerned with an aspect of design, namely the question of learning how to implement reflex-mode control actions and how these reflex responses might be modulated to suit circumstances and task requirements. The motivation is to investigate how to avoid or delay having to resort, prematurely, to the higher logic-driven entities in a hierarchical system.

The context of the present work is that of formulating and implementing fast, real-time, adaptive intelligent control and how such reflex responses and control strategies can be learned in changing environments.

Two tasks are considered. In the first task, a controller is asked to learn how to balance an inverted pendulum and at the same time control the position of the cart on which the pendulum is mounted. This is a well-known task. Different control strategies have been used including linearization [2], neural networks [3,4] as well as fuzzy logic controllers [5,6].

In contrast to the model-based perspective of neural-net control [7-10], this present study casts the task into the form of direct-control with recurrent networks [11]. In that approach, in contrast to the model-based neural-net approach, there is no need to learn a detailed model of the pendulum and of the cart. What the controller learns is the appropriate reflex response, with the nature of the reflex modulated by consideration of various circumstances that might include changes in the environment. In particular, in the present study, the objective is to investigate how the control might be refined to place more emphasis on cart displacement or angle deviation depending on the circumstances.

In another task, a juggler wonders if he is capable of adding yet another ball to those in the air. He acquires the judgement capability through evolution but the procedure is not simple because there may be non-linearity and hysteresis in his behavior. We have the intertwined issues of coupling of dynamical systems, interaction with environment and embedded systems, learning and modulation of reflexes.

The studies are motivated by a perceived need for ability to implement neural net control systems which are truly adaptive and are also able to switch modes of control in reflex mode rather than through use of additional high level finite state machines or automata. The latter is feasible but usually slower.

Results of the investigations suggest that efficient adaptive real-time controllers might be implemented in the form of coupled neural networks that are recurrent in an indirect manner. In such systems, changes in emphasis or attention are manifested in a special kind of modulated reflex manner. These matters are discussed briefly in this preliminary report.

The approach of this work is influenced by the work of researchers interested in the behavior of artificial intelligent insects or in the performance of intelligent robots [12-15], by those who emphasize the special aspects of embedded systems[16, 17] and by studies of coupled dynamical systems [18].

2. The Mode-Adapting Inverted Pendulum /Cart Controller

If this task were to be handled in hybrid system manner with discrete valued decision-making at the higher levels, there might be need for two modes of operation. In one extreme there would be no consideration of the position of the table, all attention would be devoted to keeping the pendulum at the desired angle. In the other extreme, all attention would be given to bringing the table reasonably close to the center of the platform and maintaining it in that condition. In practice what is required is a strategy and an ability to vary the emphasis depending on the circumstances, so as to maintain the pendulum in upright position or at a feasible angle and with the cart on the platform for as long a period of time as possible.

The control task is illustrated schematically in Figure 1. In the present direct-control method, there is no need for learning a neural net model of the pendulum or of the cart. There is only a neural net controller which when presented with the system state vector as input, returns a suggested value of the impulse I, the control action. In addition there is a neural net which generates the value of α, an emphasis parameter used in generating the value of the impulse I. Those matters are illustrated schematically in Figure 1.

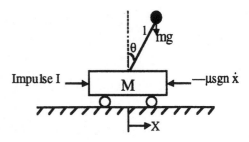

The environment :

$$\theta(k+1) = 2\theta(k) - \theta(k-1) + \frac{h^2 g}{l}\sin\theta(k) - \frac{Ih}{Ml}\cos\theta(k)$$

$$x(k+1) = 2x(k) - x(k-1) + \frac{Ih}{M+m} - \frac{\mu h \operatorname{sgn}\dot{x}}{M+m}$$

l = length of the pendulum

μ = friction coefficient

Figure 1: The combined Inverted-Pendulum/Cart control task

The relative importance of the angular and position deviations is determined by the value of a parameter 'alpha'. Given the values of these deviations a net generates an appropriate value of alpha to provide guidance on the level of emphasis to be given to one deviation or the other. That value of alpha is passed on to the controller itself for use in learning the value of the optimal impulse. The output of the controller is simply the direction and magnitude of the impulse to be applied to the cart. That control action is exercised and the environment returns the values of the results of that action. Those updated values of the pendulum/cart configuration are then supplied to the emphasis network and that supplies a value for alpha. But alpha is a parameter in the control action generator net and so the next impulse value is generated using the updated value of alpha. These matters are illustrated schematically in Figure 2.

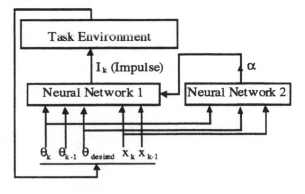

Figure 2: Structure of coupled adaptive reflexes acting in recurrent manner

This control configuration is learned through evolution, it is of the nature of direct-control, it is recurrent but in an indirect manner, and it is coupled in that the output from one net is used as an adaptation parameter in another network. The two networks operate in reflex manner but the output of one modulates the behavior of the other. The nature of the learning has characteristics of reinforcement learning.

In learning phase, at every juncture, the system learns through evolution what emphasis to place on angle or cart position and what the impulse magnitude and direction should be. Some of the trajectories synthesized in the training process are exhibited in Figure 3.

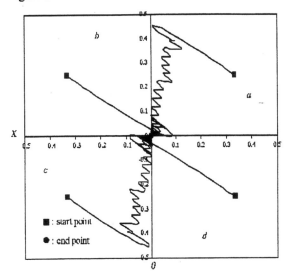

Figure 3: Some Trajectories Synthesized in the Training Process

The task is to bring the inverted pendulum to the upright position and the cart to the center of the platform from various starting positions. These training trajectories illustrate how the system converges onto the desired end configurations from various initial states. The manner in which the impulse changes during the course of one of these trajectories is depicted in Figure 4.

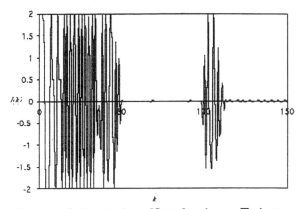

Figure 4: Optimal value of Impulses in one Trajectory Used for Learning Reflex Response.

A great deal of additional interesting details can be said about this intelligent reflex control action generator. For the present, it suffices to demonstrate that an intelligent reflex controller can indeed be evolved readily in this manner as shown in Figure 5, but it is both interesting and troublesome that for certain quadrants (a and c) of initial positions, the target final configuration cannot be reached. The system gets trapped in stable local minima, near the target but not quite there.

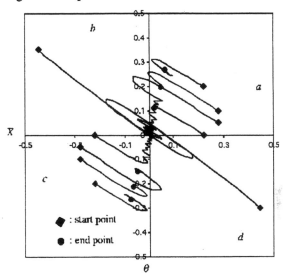

Figure 5: Controlled Trajectories of Pendulum/Cart

3. The Ambitious Cautious Juggler

The second control task was and is motivated by a desire to explore further what can and what cannot be accomplished by the coupling of lower level dynamical systems, acting in reflex mode.

Abstracting the juggling task, the basic act is that a juggler catches a ball, transfers it from one hand to the other hand and then to launches it into the air again. After having put N balls into play, he wonders if he could put yet another ball into play by increasing the launch velocity. An increase in launch velocity results in an increase in round trip time for each ball meaning that more balls could be squeezed into the time frame between launch and catch. However there is a problem. The mean and variance in the catch-transfer-launch time also increase with increase in launch velocity, in a nonlinear manner. At any one juncture, in steady state mode, the controller estimates and adjusts the value of the launch velocity so as use up all the time available. In buildup mode, the controller tests to see if yet one more ball is feasible. In wind-down mode, the controller estimates the launch velocity for a decreased number of balls. In this case there is indeed a higher level which does not do anything besides specifying the status of the mode of operation, steady-state, build-up or wind-down. A plausible coupled control architecture essentially of adaptive reflex manner is illustrated in Figure 6.

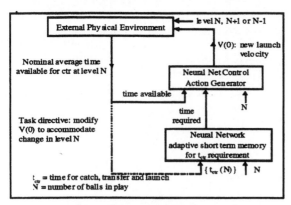

Figure 6: Schematic Illustration of Elements of the Juggler

References:

[1] Branicky, M.S., 1995. Studies in Hybrid Systems: Modeling, analysis and control, MIT D.SC. thesis.

[2] Bryson, A.E. and D.G. Leuenberger, 1970. The synthesis of regulator logic using state-variable concepts, Proceedings of the IEEE, vol.58, pp.1803-1811.

[3] Widrow, B, 1987. The original adaptive net broom-balancer, IEEE Int. Symposium on Circuits and Systems, pp.351-357.

[4] Williams, V. and K. Matsuoka, 1991. Learning to balance the Inverted pendulum using neural networks. 1991 IEEE International Joint Conference on Neural Networks, Singapore, pp.214-219.

[5] Kosko, B., 1991. Neural networks and fuzzy systems: a dynamical systems approach to machine intelligence, Prentice Hall, Englewood Cliffs, NJ.

[6] Jang, R.,, 1992.Self-learning fuzzy controllers based on temporal back propagation, IEEE Trans. Neural Networks, vol. 3, pp.714-723.

[7] Narendra, K.S. and K.Pathasarthy, 1990. Identification and control of dynamical systems using neural networks, IEEE Transactions on Neural Networks, Vol.1, pp.4-27.

[8] Nguyen, D.H. and B. Widrow, 1990. Neural networks for self-learning control systems, IEEE Control Systems Magazine, Vol.4, pp.18-23.

[9] Pao, Y.H., S. Phillips, and D.J.Sobajic, 1992. Neural-net computing and the intelligent control of systems, Special Intelligent Control Issue of International Journal of Control, Vol. 56, pp.263-290, Taylor and Francis, Ltd., London, U.K.

[10] Kawato, M, Y. Uno, M. Isobe, and R. Suzuki, 1988. Hierarchical neural network model for voluntary movement with applications to robotics, IEEE Control Systems Magazine, Vol.8, pp.8-16.

[11] Yip, P.P.C. and Y.H. Pao, 1994. A recurrent neural-net approach for 1-step ahead control problems, IEEE Transactions on Systems, Man and Cybernetics, vol.24, pp.678-683.

[12] Brooks, R.A., 1989. A robot that walks: emergent behaviors from a carefully evolved network, Neural Computation, Vol.1, pp. 253-262.

[13] Brooks, R.A., 1991. Intelligence without representation, Artificial Intelligence, Vol. 47, pp.139-159.

[14] Beer, R.D., H.J. Chiel, R.D. Quinn, K.S. Espenchied and P. Larsson, 1992, A distributed neural network architecture for hexapod robot locomotion, Neural Computation, Vol.4, pp. 356-365.

[15] Beer, R.D., 1995. A dynamical systems perspective on agent-environment interaction, Artificial Intelligence , Vol.72, pp.173-215.

[16] Clark, A., 1997. Being There: Putting brain, body, and world together again, MIT Press, Cambridge, MA.

[17] Clancey, W.J., 1997. Situated Cognition: On human knowledge and computer representations, Cambridge University Press, Cambridge, U.K.

[18] Yao, Y. and W.J. Freeman, 1990. Model of biological pattern recognition with spatially chaotic dynamics, Neural Networks, Vol.3, pp. 153-170.

HYBRID AUTOMATA FOR MODELING DISCRETE TRANSITIONS IN COMPLEX DYNAMIC SYSTEMS

Pieter J. Mosterman [*,1] **Gautam Biswas** [**,2]

* *Institute of Robotics and System Dynamics, DLR
Oberpfaffenhofen, D-82230 Wessling. Germany.*
** *Knowledge Systems Laboratory, Stanford University, Stanford,
CA 94305. U.S.A.*

Abstract: Hybrid system models combine continuous behavior evolution with discrete mode transitions. These transitions may cause discontinuous changes in the field that defines continuous system behavior and the variable values associated with the continuous state vector. In reality, these discontinuous changes are fast continuous transients. To simplify the analysis of these transients *time scale* and *parameter* abstractions are applied to system models with very different impacts on the analysis of system behavior. We have developed a systematic modeling approach based on hybrid automata which combines *a priori* and *a posteriori* switching values to formally implement switching semantics associated with the abstraction events.
Copyright © 1998 IFAC

Keywords: hybrid systems, compositional modeling, modeling abstractions.

1. INTRODUCTION

The pressure to achieve more optimal and reliable performance on complex systems such as aircraft and nuclear plants, while meeting rigorous safety constraints is leading to more detailed analysis of the embedded controllers for these systems. In embedded systems, the continuous physical process interaction with digital control signals requires modeling schemes that facilitate the analysis of mixed continuous and discrete, i.e., *hybrid* behavior. Discrete phenomena may also occur when modeling abstractions are applied to simplify fast nonlinear continuous process behavior.

Consider the primary aerodynamic control surfaces of an airplane in Fig. 1 (Seebeck, 1998). Modern avionics systems employ electronic signals generated by a digital computer, which are

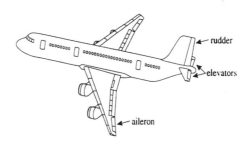

Fig. 1. **Aerodynamic control surfaces.**

transformed into the power domain by electro-hydraulic actuators. The primary flight control system exemplifies the need for hybrid modeling in embedded control systems. At the lowest level in the control hierarchy, positioning of the rudder, elevators, and ailerons are achieved by continuous PID control. Desired set point values are generated directly by the pilot or by a supervising control algorithm implemented on a digital processor. Digital control may mandate *mode* changes at different stages of a flight plan (e.g., *take-off*, *cruise*, and *go-around*). Detection of failed components may lead to discrete changes

[1] Supported by a grant from the DFG Schwerpunktprogramm KONDISK.
[2] On leave from Dept. of Computer Science, Vanderbilt University, Nashville, TN. Partially supported by a grant from Hewlett Packard Company.

in system configuration. Model simplification by discretizing fast nonlinear transients also results in discontinuous variable changes.

We have developed a hybrid modeling paradigm that encompasses analysis of embedded systems and modeling abstractions in physical systems. In this paper we present our formalisms for abstracting complex transients into hybrid automata models, and discuss formal semantics for computing the discontinuous changes in the system state vector. The methodology is applied to the elevator positioning subsystem of the primary flight control system of aircraft to demonstrate the correspondence between the model semantics and physical system behavior.

2. HYBRID DYNAMIC SYSTEMS

Hybrid modeling paradigms (Alur *et al.*, 1993; Guckenheimer and Johnson, 1995; Mosterman *et al.*, 1998*b*) supplement continuous system description by mechanisms that model discrete state changes resulting in discontinuities in the field description and the continuous state variables. In previous work we have established an ontology of discrete transition types in physical system behavior (Mosterman *et al.*, 1998*a*).

Differential equations form a common representation of continuous system behavior. The system is described by a *state vector*, x, and other variables called *signals*, s, are derived algebraically, $s = h(x)$. Behavior over time is specified by field f. Interaction with the environment is specified by *input* and *output* signals, u and y.

Discrete systems, modeled by a state machine representation, consist of a set of discrete modes, α. Mode changes caused by events, σ, are specified by the *state transition function* ϕ, i.e., $\alpha_{i+1} = \phi(\alpha_i)$. A transition may produce additional discrete events, causing further transitions.

In hybrid dynamic systems, a mode change from α_i to α_{i+1}, may result in a field definition change from f_{α_i} to $f_{\alpha_{i+1}}$. Discontinuous changes in the state vector are governed by an algebraic function g, $x^+ = g_{\alpha_i}^{\alpha_{i+1}}(x)$. Discrete mode changes are caused by an *event generation function* γ associated with the current active mode, α_i, $\gamma_{\alpha_i}(x) \leq 0 \rightarrow \sigma_j$.

The resultant general architecture for hybrid models of embedded control systems appears in Fig. 2 (Mosterman and Biswas, 1997*a*) Signal value changes (s_p) and closed-loop control active in mode α (s_c) may cause discontinuous changes. The corresponding physical events, σ_p and σ_c, or open loop control generated discrete events σ_x cause mode transitions defined by ϕ.

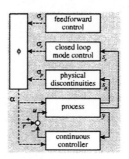

Fig. 2. **Hybrid control.**

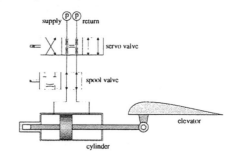

Fig. 3. **Hydraulics of one actuator.**

3. THE ELEVATOR SYSTEM

Attitude control in an aircraft is achieved by the elevator control subsystem (Seebeck, 1998). This system may consist of two mechanical elevators (Fig. 1) which are positioned by two electrohydraulic actuators. When a failure occurs, redundancy management switches the actuators and oil supply systems to ensure maximum control. Fig. 3 shows the operation of an actuator. The elevator positioning is controlled by servo valve, which is implemented by a continuous feedback mechanism. When the actuator is *active*, the spool valve is in its *supply* mode and the control signal generated by the servo valve is transferred to the cylinder that positions the elevator. When the actuator is *passive*, the spool valve is in its *loading* mode, and control signals cannot be transferred to the cylinder. Oil flows between the chambers through a loading passageway, otherwise the cylinder would block movement of the elevator, canceling control signals from the redundant *active* actuator.

Consider a scenario where a sudden pressure drop is detected on one of the left elevator actuators. Redundancy control moves the spool valve this actuator from *supply* to *loading* and the spool valve of the other actuator from *loading* to *supply*. This causes transients that are studied in greater detail below.

4. MODELING THE ELEVATOR SYSTEM

We employ *parameter* and *time scale* abstractions (Mosterman and Biswas, 1997*b*) to design a simpler but adequate model of the elevator

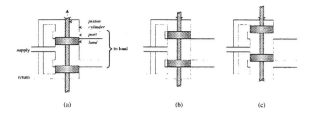

Fig. 4. **A typical spool valve.**

subsystem for control purposes. Parameter abstraction removes small and large parameter values (parasitic dissipative and capacitive elements) from the model. Time scale abstractions collapse the end effect of phenomena associated with very fast time constants to a point in time. We show how these different abstraction types relate back to physical parameters in the real system.

4.1 *The Spool Valve*

A typical spool valve shown in Fig. 4, consists of a piston that moves in a cylinder. A number of cylinder ports connect the supply and return part of the hydraulic system with the load. Fig. 4(a) and (c) show two possible oil flow configurations. When moving from one to the other, the spool valve passes through the configuration where the ports are closed by lands (Fig. 4(b)).

Mode changes in the actuators are facilitated by the spool valve. To enable analysis of behavior during mode changes, four modes of operation are modeled:

(α_0) *loading*: The spool valve operates as a load. Pressure changes generated by the servo valve are blocked. Oil flow between chambers of the elevator positioning cylinder is possible through a loading passageway.

(α_1) *closed*: The spool valve is closed. Pressure changes generated by the servo valve are blocked. Oil flow between the chambers of the elevator positioning cylinder is not possible. This is a transitional mode between α_0 and α_3.

(α_2) *opening*: The valve is opening. While its lands move past the ports, fluid inertial effects may become active. Depending on the physical construction of the valve, these may have significant effects on transient behavior.

(α_3) *supply*: The spool valve is opened and supplies control power. Pressure changes generated by the servo valve are transferred to the cylinder that positions the elevator. Flow of oil into and out of this cylinder is possible.

Mode changes of the spool valve are controlled by a redundancy management module which monitors a number of critical system variables. In the fault scenario, a sensor reading in actuator1 generates the failure event, σ_f. In response, the redundancy management reconfigures control by generating a sequence of discrete control signals that cause a switch of actuators. The resulting state α_{ij} indicates the state of actuator2, α_i, and actuator1, α_j.

(1) A control event (σ_c) is generated that causes the piston in the *left* spool valve to move from its *supply* to *loading* position at a constant rate of change. Along the trajectory, a number of physical events (σ_p) occur:
(i) $\Delta x > -\epsilon \rightarrow \sigma_{close} \Rightarrow \alpha_1$, the overall system mode becomes α_{01} (actuator2 - loading, actuator1 - closed).
(ii) $\Delta x > \epsilon \rightarrow \sigma_{open} \Rightarrow \alpha_2$, the overall system mode becomes α_{02} (actuator2 - loading, actuator1 - opening).
(iii) $\Delta x > x_{th} \rightarrow \sigma_{load} \Rightarrow \alpha_0$, the overall system mode becomes α_{00} (actuator2 - loading, actuator1 - loading).

(2) A second control event is generated to move the piston in the *right* spool valve to move from its *loading* to *supply* position with a constant rate of change, causing the overall system to switch through modes $\alpha_{00} \rightarrow \alpha_{10} \rightarrow \alpha_{20} \rightarrow \alpha_{30}$.

The values of ϵ and x_{th} are based on physical parameters of the valve, e.g., the shape of ports and lands (Merritt, 1967). For a *critical center* type valve $\epsilon = 0$, and for a *closed center* valve ϵ has a small nonzero value. The value of x_{th} and ϵ determine when inertial effects become active, $\Delta x > \epsilon$ and for how long $\Delta x > x_{th}$.

4.2 *Model Assumptions*

When an actuator moves to its *closed* mode, oil flow into and out of the cylinder that positions the elevator is blocked. This implies that the cylinder piston that controls elevator position cannot move, and, the elevator stops moving as well. In reality, internal dissipation and small elasticity parameters of the oil cause the elevator velocity to change continuously during the transition. The behavior in the continuous transient mode between *supply* and *closed* is shown in Fig. 5. How quickly the system reaches the 0 velocity state in the closed mode depends on the elasticity and internal dissipation parameters chosen for the oil. After a short time in the *closed* mode, the actuator moves to the *opening* mode, and the inertial effects become active. Fig. 6 illustrates the continuous transients involved in the transition. The inertial parameter determines the final elevator velocity, v_e. In the *opening* mode, the inertial effect decreases as the clearance between port and land increases. After some time its value becomes negligible, and the actuator operates as a simple load (*loading* mode). This is shown in Fig. 7 for an inertial parameter with two different values.

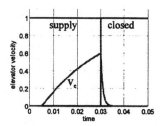

Fig. 5. **Continuous transients:** *closed* **mode.**

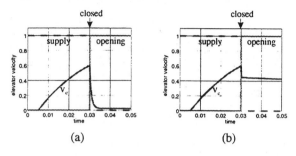

(a)   (b)

Fig. 6. **small Continuous transients:** *supply* $\rightarrow$ *closed* $\rightarrow$ *opening* **for spool valves - (a)** $I = 1$ **and (b)** $I = 100$.

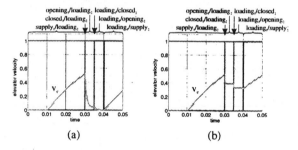

(a)   (b)

Fig. 7. **Continuous transients:** *loading* **mode -** **(a)** $I = 1$ **and (b)** $I = 100$.

The continuous transients described above are not of much interest to the modeler for analysis and control (see Fig. 7 where the transients in the opening mode are still clearly visible but the continuous transients in the closed mode are not). Model simplification results in removal of small elasticity and inertial effects but Fig. 7 illustrates that depending on their magnitude, they may have a distinct impact on the overall system behavior.

4.3 Abstraction Types

We apply previous work on model simplification by abstraction (Mosterman and Biswas, 1997b) to analyze the elevator control subsystem.

Time Scale Abstraction. In the *opening* mode, fluid inertia and dissipative effects in the clearance between land and port, cause a second order build-up of fluid flow. Though the fluid flow velocity and its time derivative are 0 initially, the velocity of the elevator and the driving piston are not. This results in a pressure build-up in the cylinder governed by the elasticity coefficient of the oil

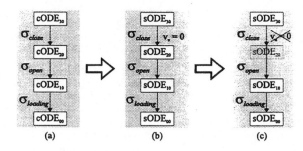

(a)   (b)   (c)

Fig. 8. **Hybrid automata specifying the actuator1 elevator subsystem.**

which causes a rapid increase of fluid flow through the land/port opening. The pressure also causes the elevator velocity to decrease rapidly resulting in the transient in Fig. 6(a). The initial transient from moving into the *closed* mode is replaced by the transient moving into the *opening* mode. The difference is best seen by comparing Fig. 5 with Fig. 6(a). The final value of the velocity after this transient depends on the dissipative effects and starting point and duration of the *opening* mode.

If the elastic and inertial effects are abstracted away, the *closed* and *opening* modes are traversed instantaneously in sequence into the *loading* mode. However, the inertial element has a distinct derivative effect on system behavior, and the influence occurs over a small time interval. This is an example of time scale abstraction, where mode change phenomena is expressed at *a point* in time. An important implication is that the state vector has to be modified through the sequence of mode changes. An algebraic relation is derived to compute the elevator velocity to correspond to the fast transient behavior in the mode transitions (Fig. 5 and Fig. 6(a)).

Parameter Abstraction. When dissipation in the land/port clearance dominates the inertial effect, a much faster response in fluid flow velocity occurs because dissipation does not introduce a time derivative effect. The flow of oil into and out of the cylinder is fast, and the pressure build-up in the cylinder is small. As a result, elevator velocity remains almost unchanged as the model switches from *closed* to *opening* (Fig. 6(b)). Small parameter values are abstracted away, and the transitions through the *closed* and *opening* modes are instantaneous (no time derivative effects are present). For small parameter values (Fig. 5 and Fig. 6(b)), the transients to *opening* (Fig. 6) may result in very different behavior from transients into *closed* (Fig. 5). When a discontinuous jump occurs, the eventual elevator velocity is not computed by first executing the jump to *closed* and then to *opening*, but immediately to *opening*. Otherwise, *closed* would have set the velocity to 0, which would also be the value in the *opening* mode. For parameter abstractions the intermediate steps are completely abstracted away.

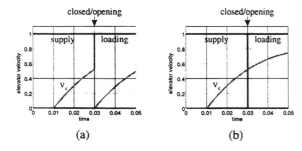

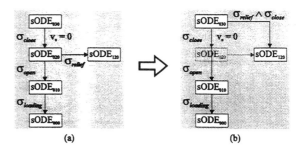

Fig. 9. **Effect of a (a) time scale abstraction and (b) parameter abstraction.**

Fig. 10. **Mode switching with pressure relief valve.**

5. HYBRID AUTOMATA FOR MODELING COMPLEX SYSTEMS

Application of time scale and parameter abstractions while producing simpler models for analysis requires the explication of the type of abstraction applied to the model, so that formal semantics can be applied to ensure correct behavior generation. Hybrid automata provide a powerful formalism for specifying hybrid dynamic systems.

5.1 Modeling with Hybrid Automata

Fig. 8(a) illustrates the hybrid automata implementation of the switching of an actuator from its *active* to *loading* mode in the elevator subsystem. The behavior models include the fast continuous transients, therefore, they are numerically complex ODEs (cODE). Time scale and parameter abstractions produce numerically simpler ODEs (sODE), but require the specification of discrete transition functions, ϕ, γ, and g. In the first case, transition conditions were based on spool valve position, x. However, in the latter model, the detailed continuous behavior of the system around $x = 0$ is abstracted away, so the corresponding events $\{\sigma_{supply}, \sigma_{close}, \sigma_{open}, \sigma_{load}\}$ have to be generated by explicit discrete control. Our analysis shows that transients into *closed* results in $v_e = 0$. If these transients are abstracted away, a discontinuous jump specified by the hybrid automaton transition sets $v_e = 0$ (Fig. 8(b)).

Time scale abstraction applied to the actuator model produces correct behaviors (Fig. 9(a)). Parameter abstraction, however, produces incorrect behavior (Fig. 9(b)) because the underlying continuous transient that changes $v_e = 0$ in *closed* is aborted by the *opening* mode becoming active, therefore, v_e remains unchanged. In effect, this means that mode α_{20} is never active, i.e., it is *mythical* (Mosterman and Biswas, 1997a), and the change in continuous state never occurred. An intuitive solution would be to remove mode α_{20} from the hybrid automata model, as shown in Fig. 8(c). However, this model reduction requires global applicability conditions about the

mode change behavior of the elevator system. Therefore, this approach makes it hard to develop complex system models by composing constituent elements. Additional transitions require the real system to be re-evaluated (in a sense re-modeled) to establish the correct discrete state transition structure for the extended model.

Consider the situation where a pressure relief valve becomes active when the pressure in the cylinder exceeds a threshold value. When the spool valve is closed, a rapid pressure build up occurs induced by the fast change in v_e. If the continuous transients are abstracted into a discontinuous change, this pressure is modeled as a Dirac impulse function whose area is determined by the v_e values immediately before *closed* and initial values in the *closed* mode ($v_e = 0$). If this area exceeds a critical value, the pressure relief valve opens up to prevent excessive pressures in the cylinder. The energy transient undergoes a continuous trajectory, and mode α_{20} should not be removed.

The hybrid automata model for this mode change behavior is shown in Fig. 10(a). To prevent the pressure build-up, the pressure relief valve prevents $v_e = 0$. Therefore, when the pressure relief valve comes on (indicated by a 1 in the left most index of the sODE subscripts) $v_e = 0$ does not apply. However, this information is not available in the hybrid automata when the straightforward extension with σ_{relief} and corresponding $sODE_{120}$ mode is applied (Fig. 10(a)). An exhaustive analysis of the real system is required to reduce the hybrid automata to the one in Fig. 10(b) but now the transition to the relief mode is invoked when σ_{relief} **and** σ_{close} is generated. This results in a non intuitive and complex state transition structure that has little relation to the actual transition behavior of the real system. Detailed pre-analysis of the discrete transition behavior of the complete system is required before hand to generate the correct model, and this complicates the development of compositional modeling techniques in the hybrid automata framework.

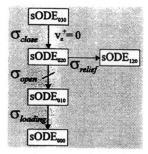

Fig. 11. **Compositional hybrid automata.**

5.2 Structure Preserving Hybrid Automata

To enable compositional modeling additional transition semantics in the form of discontinuous changes from the *a priori* state vector, x, to the *a posteriori* values, x^+, have been developed (Mosterman *et al.*, 1998b). In combination with recognition of parameter and time scale abstraction events this has proved to be a powerful mechanism for modeling complex systems (Mosterman and Biswas, 1997a). Applications to the secondary sodium cooling system of a fast breeder reactor appear in (Mosterman *et al.*, 1998b). The hybrid automata formalism makes state vector assignments to x^+. Transients enabled by events caused by parameter abstraction are traversed instantaneously and the *a priori* state vector is unchanged. Time scale abstraction generated events cause an update of the *a priori* state to the current *a posteriori* values, $x = x^+$. State transitions with time scale abstraction events are marked by a sloped stroke line.

For the pressure relief valve, the event σ_{relief} is a function of the change in v_e between α_{030} and α_{020}. The v_e value in α_{020} is assigned to the *a posteriori* value v_e^+, and the event generation can be specified as $v_e^+ - v_e > v_{th} \rightarrow \sigma_{relief}$. The illustration of the hybrid automata in Fig. 11 clearly shows that the state transition structure and the corresponding discontinuous jumps in state vector values are preserved while still generating correct behavior. The description is complete, and does not require modifications when new transitions are added to the overall system.

6. CONCLUSIONS

Hybrid automata combine discrete transitions with continuous behavior evolution to provide a powerful formalism for modeling hybrid systems. Discrete transitions cause changes in the system behavior model, but discontinuous changes in the state vector values may also occur. These changes are specified as transitions. We have incorporated the two abstraction types (i) time scale abstraction and (ii) parameter abstraction and the associated semantics that govern discontinuous

changes in behavior specification into our hybrid automaton framework. An important feature of our work is that these abstractions relate back to physical parameters of the physical system that cause fast continuous transients. We have also shown how the use of *a priori* and *a posteriori* switching values help specify the formal semantics for the two types of abstractions. Analysis of the hydraulic cylinder with the pressure relief valve demonstrates the usefulness of this method for compositional modeling of complex systems.

7. REFERENCES

Alur, R., C. Courcoubetis, T.A. Henzinger and P. Ho (1993). Hybrid automata: An algorithmic approach to the specification and verification of hybrid systems. In: *Lecture Notes in Computer Science* (R.L. Grossman, A. Nerode, A.P. Ravn and H. Rischel, Eds.). Vol. 736. pp. 209-229. Springer-Verlag.

Guckenheimer, J. and S. Johnson (1995). Planar hybrid systems. In: *Lecture Notes in Computer Science* (P. Antsaklis, W. Kohn, A. Nerode and S. Sastry, Eds.). Vol. 999. Springer-Verlag. pp. 202-225.

Merritt, H.E. (1967). *Hydraulic Control Systems*. John Wiley and Sons. New York.

Mosterman, P.J. and G. Biswas (1997a). Formal Specifications for Hybrid Dynamical Systems. In: *IJCAI-97*. Nagoya, Japan. pp. 568-573.

Mosterman, P.J. and G. Biswas (1997b). Principles for Modeling, Verification, and Simulation of Hybrid Dynamic Systems. In: *Fifth Intl. Conf. on Hybrid Systems*. Notre Dame, Indiana. pp. 21-27.

Mosterman, P.J., F. Zhao and G. Biswas (1998a). An Ontology for Transitions in Dynamic Physical Systems. In: *AAAI98*, Madison, WI, pp. 219-224.

Mosterman, P.J., G. Biswas and J. Sztipanovits (1998b). A hybrid modeling and verification paradigm for embedded control systems. *Control Engineering Practice*.

Seebeck, J. (1998). *Modellierung der Redundanzverwaltung von Flugzeugen am Beispiel des ATD durch Petrinetze und Umsetzung der Schaltlogik in C-Code zur Simulationssteuerung*. Diplomarbeit. Arbeitsbereich Flugzeugsystemtechnik. Technische Universität Hamburg-Harburg.

PRELIMINARY THOUGHTS ON THE APPLICATION OF REAL-TIME AI GAME-TREE SEARCH TO CONTROL *

Todd W. Neller

*Knowledge Systems Laboratory, Gates Building 2A, Stanford
University, Stanford CA 94305-9020*
email: neller@ksl.stanford.edu

Abstract: We introduce a formal definition of a general hybrid system control game,
and a decision procedure for an agent which uses simulation to inform game-playing
decisions in real-time. We discuss heuristics informing the estimation of expected
utilities of actions, describe metalevel reasoning criteria which direct game-tree
expansion according to decision relevance, and point out challenges in applying ideas
from AI real-time game-tree search to control. *Copyright © 1998 IFAC*

Keywords: real-time AI, game theory, hybrid systems, model-based control, artificial
intelligence, agents

1. INTRODUCTION

The fundamental question of this research is:
"How can a rational agent use simulation to
choose intelligent actions for hybrid system control in real-time?" In this paper, we introduce a
formal framework which views the reasoning of
such an agent as real-time game-playing. The *general hybrid system control game* formalism is more
general than that of discrete- and continuous-
time dynamic games (Başar and Olsder, 1995), in
that it allows general hybrid system behavior and
strategies involving both discrete and continuous
actions.

As problem frameworks become more general,
constraints which are helpful in problem solving become fewer, so we must elucidate trade-
offs and architectural design decisions which make
such a framework practical for application. This
paper describes a spectrum of simulation-based
approaches which range from weak, uninformed
game-tree search to strong, informed (heuristic)
approaches which can benefit from other forms of
analysis. Since the expected utility of the agent's
external game actions is time-dependent, the expected utility of performing internal simulation
actions in order to deliberate about such external actions is time-dependent as well. We later
provide a general decision procedure for such reasoning with simulation.

Defining a rational agent as that which chooses
actions which maximize expected utility, our real-
time limitations force us to examine the expected
utility of the act of simulation itself. For example,
imagine you are driving on a highway through
a city with a prior notion of a best route. Suddenly, you see a sign for an impending exit which
presents the possibility of a much better route.
Another exit with similar possibilities follows.
You're faced with a decision on how to deliberate
as you quickly grab your map: Should you examine the route from the immediately approaching
exit, or is your deliberation better focussed on
examining the route from the exit beyond? Such
deliberation about deliberation, or *metalevel rea-*

* This work was supported by the Defense Advanced
Research Projects Agency and the National Institute of
Standards and Technology under Cooperative Agreement
70NANB6H0075, "Model-Based Support of Distributed
Collaborative Design".

soning, is necessary to focus computational deliberation in promising directions.

Of course, metalevel deliberation is itself deliberation and one could regress infinitely were it not for the fact that such metalevel deliberation also needs to be justified in terms of maximization of expected utility. Metalevel reasoning must use less runtime than the reasoning it obviates. Encouraging research by Russell and Wefald (Russell and Wefald, 1991) indicates that, provided good heuristics for focusing search, metalevel reasoning in game-tree search justifies its cost. Their research is the inspiration for our simulation approach to game-playing. Of course, many different types of analysis can better inform play of a hybrid system control game, and we shall see that such analysis is beneficial for forming the heuristics to productively focus metalevel deliberation. We will suggest domain-independent and -dependent heuristics for informing the metalevel of game-tree search.

Before delving into formalism and discussion, we should first discuss something of the motivation of this research. Why have we chosen to pursue our particular line of inquiry? First, controllers are being called on to meet more complex objectives than ever before. Particularly, systems which are highly reconfigurable may present the control engineer with such complexity as cannot be humanly or analytically overcome. We believe that enabling controllers to improvise control with online models will push the envelope of applicability of both artificial intelligence and control engineering. Second, we observe that larger-scale real-time constraints may play a role in the design and testing of controllers. Such techniques could spawn a new generation of design tools which deal with real-time design constraints explicitly within their reasoning framework. In short, we see an opportunity for the fruitful application of real-time AI game-tree search to practical control problems.

The paper is structured as follows: In Section 2, we formally define a hybrid system control game. The remainder of the paper informally introduces the concepts of AI real-time game-tree search and discusses the challenges in applying such ideas to hybrid system control game play. In Section 3, we discuss different means of evaluating game-trees. In Section 4, we discuss principles for expanding such game-trees in ways most relevant to rational decision-making. In Section 5, we bring real-time constraints into relevance consideration and present a general decision procedure for simulation-based hybrid system control game play. Finally, in Section 6 we describe ways of heuristically informing the calculation of estimates of expected utilities in the game-tree.

2. HYBRID SYSTEM CONTROL GAME

A common example of a hybrid system is a physical system controlled through mode-switching. Such a system may evolve through a sequence of continuous phases with instantaneous[1], discrete changes of control modes in between. If we assume that mode switching is under control of one or more agents (an agent being a controller, a disturbing influence, etc.), then the control of such a system may be viewed as a game where each agent is a player seeking to maximize some utility measure as their game score.

Consider this simple illustrative example of a hybrid system: A car is traveling along a road with a stop light which is known to have one of a constrained set of possible states and behaviors. The driver knows bounds on the different delay times associated with the different states of the light. From the driver's initial observation of the traffic light at a distance, the driver has a collection of utility influences: travel progress, fuel efficiency, safe speed, avoiding speeding tickets, and staying out of the intersection when the light is red. Suppose the driving agent sets its driving policy (accelerate, cruise, or decelerate) safely according to worst-case possible traffic light timings and delays. This then can be viewed as a sort of game where the driving agent is a player seeking to optimize its score (the utility of its performance), and the traffic light controller is an adversary seeking to reduce the driver's score by turning the light red when the car is in the intersection.

Informally, our hybrid system control game evolves with continuous and discrete (instantaneous) behaviors. For all non-terminal game states, the continuous behaviors define a unique trajectory until such time as a terminal state is reached. For any given state, some discrete transitions (also called actions) may be either forced by the system or enabled for player(s) to enact. We call a player-enacted discrete transition a *move*. Given an initial state, this then defines a possibly infinitely-branching tree of trajectories, where branches are caused by player moves. Given scoring functions over trajectories, and termination conditions for the game, we encode the desirability of each possible system trajectory for each player.

We seek to most generally and formally express what it means to play a control game with a hybrid system. Our formalism is based on that of Branicky's controlled hybrid dynamical system (CHDS) (Branicky, 1995). With a desire to both (1) preserve the expressive power of a CHDS, and (2) create a formal language suited to expressing hybrid system game play, we have found it most natural to describe transitions in terms of actions,

---

[1] for practical purposes of modeling and simulation

and allow for the possibility of multiple actions being taken simultaneously. We formalize a *general hybrid system control game* G as follows:

$$G = [Q, \Sigma, S_I, A, V, p, W, F, E, R, s, \omega]$$

where

- Q is a set of discrete (or index) states.
- $\Sigma = \{\Sigma_q\}_{q \in Q}$ is the collection of constituent dynamical systems indexed by Q, where each dynamical system $\Sigma_q = [X_q, \Gamma_q, \phi_q, U_q]$ (or $\Sigma_q = [X_q, \Gamma_q, f_q, U_q]$), where X_q is an arbitrary topological space (the state space of Σ_q), transition semigroup Γ is a topological semigroup, U_q is the control set, transition map $\phi_q : X_q \times \Gamma_q \times U_q \to X_q$ satisfies identity, semigroup, and continuity properties, and transition function f_q is the generator of extended transition function ϕ_q. $U = \{U_q\}_{q \in Q}$ is the total control set. In short, each constituent dynamical system is defined as a standard controlled dynamical system.

  We define the total state space $S = \bigcup_{q \in Q} (q \times X_q)$.
- $S_I \subseteq S$ is the set of possible initial states of the system.
- A is the set of discrete actions $\alpha_1, \alpha_2, \dots$.
- V is the transition control set.
- $p \in \mathbb{Z}^+$ is the number of players (or agents). We define player index set $P = \{1, \dots, p\}$.
- $W : U \cup V \to P$ maps the control sets to the players that control them.
- $F : S \to A \cap \emptyset$, the forcing function, maps each system state to a single forced (required) action or to an empty set indicating that no action is forced.
- $E : S \to 2^{P \times A}$, the enabling function, maps each system state to a set of pairs of players and actions which the players are enabled to enact.
- $R : S \times 2^A \times V \to S$, the result function, maps a system state and actions taken in that state to a set of possible resulting system states. Since forced actions take precedence over enabled actions, then for all A' and s such that $A' \subseteq A$, $s \in S$, $F(s) \neq \emptyset$, and $F(s) \in A'$, we require that $R(s, A') = R(s, F(s))$.

We define a trajectory $\tau = [s_0, t_0, h, t_{n+1}]$ to be a unique evolution of the game where

- $s_0 \in S$ is the start state (with q_0 and x_0 the discrete and continuous start states, respectively),
- $t_0 \in \Gamma_{q_0}$ is the initial time,
- $h = [[p_1, \alpha_1, t_1, s_1], \dots, [p_n, \alpha_n, t_n, s_n]]$ is the move history, a list of $n \geq 0$ moves, $[p_i, \alpha_i, s_i, t_i]$ indicating that player p_i took enabled action α_i at time t_i to reach state s_i (for $1 \leq i \leq n$, $p_i \in P$, $\alpha_i \in A$, $t_i \in \Gamma_{q_i}$, $t_i \geq t_{i-1}$, and $s_i \in S$).

- $t_{n+1} \in \Gamma_{q_{n+1}}$, $(t_{n+1} \geq t_n)$ is the final time, and
- the system evolves from the initial state and time according to system dynamics (including forced discrete actions) except for enabled actions taken in the order listed until the final time.

$\mathcal{T}$ denotes the set of all legal trajectories allowed by the system. Complete trajectories begin with an initial state and end with a single terminal state. Other legal trajectories are called partial trajectories. Trajectories beginning with an initial state are called initial trajectories.

- $s : P \times \mathcal{T} \to \mathbb{R}$ is the scoring function, mapping players and trajectories to scores (also called utilities),
- $\omega \subset S \times \mathbb{R}$ is the termination set, the set of states and times for which the game ends. We require $\exists t \forall s (t' \geq t \to [s, t'] \in \omega)$, i.e. a game must eventually end.

Within this paper, we will restrict our attention to a subset of general hybrid system control games, here called *hybrid system control games*, for which the set of discrete states Q is countable, each $X_q = \mathbb{R}^{d_q}$ (i.e. continuous state spaces are real-valued), $d_q \in \mathbb{Z}^+$, $\Gamma_q = \mathbb{R}^+$ (the system evolves forward in real-valued time), S_I is a singleton (contains a single initial state), R maps to a singleton (action results are deterministic), $U = V = \emptyset$ (no control sets), and behavior is non-Zeno (players must allow time to progress). Note that nondeterministic system behavior may be modeled by introducing a "chance" player controlling nondeterministic choices.

In a given state and time, game-playing consists of each player choosing between possible moves and inaction in order to maximize its score. Formally, a player maps initial trajectories and possible moves for the player in the final state of the trajectory to one such possible move or $\emptyset$ (inaction). If players' scores sum to zero for each trajectory, the game is zero-sum adversarial. If players' scores are equal for each trajectory, the game is identically cooperative. Each game represents a player's beliefs about how the world evolves, how the players can affect change in the world, and the intentions of each player.

3. IDEAL AND BOUNDED RATIONAL PLAY

Ideally a game-playing agent would be omniscient of all possible system behaviors and would play optimally with respect to models of the other agents' decision procedures. Simulation alone cannot give us such omniscience. However, we can use simulation to sample the infinitely branching game-tree and judge between alternatives.

Given a sampled game-tree, decision-making consists of propagating trajectory scores bottom-up through the tree according to the expected behavior of each player given choices at each branch. We could simply randomly sample possible trajectories, triggering random hypothetical moves for each player and hoping that the trajectories most relevant to each player's decision is contained in the sampling. However, we are not committed to randomly sampling the game-tree. This idea is introduced as a baseline of comparison for better approaches to come. At heart, we are attempting a complex type of optimization on two-levels. First, we wish to optimize our agent's choice of simulation according to relevance to decision making. Second, we wish to optimize our agent's choice of action according to the game-tree.

Although our game has termination conditions such that all trajectories are finite, our computational resources are limited and we generally cannot simulate complete trajectories and expect to gain an adequate sampling of possible behaviors. For this reason, we allow partial simulation, i.e. simulation which can terminate before the game terminates. Let us call the final state of each trajectory in the game-tree a *leaf*. The utility of reaching a terminal leaf is computed with a trajectory scoring function, but what of the expected utility of reaching non-terminal leaves? Ideally, the expected utility of a non-terminal leaf would be the score expected from taking ideal actions until game termination. However, we cannot expect to compute this ideal value. How can we estimate the expected utility of reaching a non-terminal node?

We could assume that the current trajectory score is a good indication of the relative utility of following that trajectory. This greedy sort of game-playing approach is vulnerable to short-term gain in score at the expense of greater long-term loss, but may be preferable for simplicity and effectiveness when short-term ramifications of actions are key. Another approach would be to limit our search to an arbitrary time horizon, restricting the agent's deliberation to only consider trajectories up to a given time limit. Knowledge of one's problem domain is necessary to set a good time horizon.

When longer-term ramifications are important, or choice of a time horizon is not obvious, we then need to have some form of heuristic estimate of the expected score from a non-terminal leaf until the game's end. Such heuristic estimates can possibly be efficiently derived from analysis of the simulation model, from previous experience, or prior expectations of the designer. We revisit this topic in Section 6. Note that if we uniformly overestimate the score, we can expect shorter branches to look preferable. If we uniformly underestimate the score, we can expect longer branches to look preferable. In fact, not only do we need such an estimate, we also need a way to express our uncertainty about the estimate as well.

In acting rationally under uncertainty, the agent must deliberate between actions of varying estimated expected utility with possibly differing certainty for such estimations. A densely sampled space should give a more certain estimate than a sparsely sampled one. The certainty of the heuristic estimate expected utility of an action informed by sound mathematical analysis should be greater than the certainty of estimates based on general rules-of-thumb. In providing both the estimate and a measure of certainty of the estimate, the agent can decide which estimates are worth performing simulation to reduce uncertainty. An uncertain estimate showing great promise may be worth following up if it has a high probability of being better than the current best action. Two very certain and similar estimates may not be worth further deliberation. These thoughts are expressed mathematically in (Russell and Wefald, 1991), and point to the importance of reasoning about the relevance of computation to an agent's decision.

In summary, designing a game-playing agent with bounded computational resources challenges us to be clever in our heuristic estimation of expected utilities, and in the choice of additional simulations to reduce the uncertainty of such estimations.

4. CHOOSING RELEVANT SIMULATIONS

Assume we are given a finite game-tree of trajectories where each of the infinite states represented has associated with it a probabilistic estimation of the expected utility of reaching that state. Except in degenerate cases, an infinite subset of these states enable actions which can be taken and offer possible simulations the agent may choose to perform to further inform its decision on how and when to act next. Ideally, the agent should choose to compute the simulation which is most expected to add value to its decision. Given probabilistic estimates of expected utilities, how can the agent choose simulations most relevant to maximizing utility?

For brevity, we will summarize some main points of (Russell and Wefald, 1991): At any point in time, the game-playing agent faces a decision between a set of internal computational actions and a commitment to a current default external game action. We make the simplifying assumption that our algorithm is *meta-greedy*, that is, our algorithm chooses the next action maximizing expected utility. (Such meta-greediness can mislead us if we place too much confidence in (i.e. underestimate uncertainty of) myopic utility estimates.) We then wish to greedily choose a simulation with greatest expected informational value for our decision. *The information a simulation is expected to gain has value to the extent that it will cause a change of the agent's belief in the best external action, and to the extent that the new best external action has significantly greater utility than the old.* Therefore, a simulation expected to increase expected utility of the agent's current best action, or decrease expected utilities of alternative actions has no direct, immediate informational value for the agent's decision.

Unlike the problem domain in (Russell and Wefald, 1991), we cannot assume the property of *subtree independence*, where expanding the game-tree affects the utility estimate of only one action at the root of the expanded subtree. For instance, in updating the player utility estimates of an action taken at a given time, we must also update the utility estimates of the same action taken at nearby times. As another example, discovery of a particularly (un)desirable set of states could affect the expected utilities of subtrees which possibly reach such states. In the worst case, each simulation could cause the agent to update all of its utility estimates. For example, consider a car crossing a bridge which suddenly detects that a section of the bridge is missing. One of the challenges of applying these ideas to real-time control is to develop an efficient means of propagating utility estimate updates through the tree. In fact, one does not need to propagate updates where they are provably irrelevant to the decision at hand.

The lack of subtree independence which presents a challenge for updating estimates of expected utilities is also cause for optimism that such game-playing approaches can be feasible. Such interdependence between the infinite possible subtrees gives us hope that our finite sampled game-tree can adequately inform our decision over the infinite possibilities which remain unsimulated.

5. DELIBERATION VERSUS COMMITMENT TO ACTION

Were it not for real-time computational constraints, one might think that one would simply perform simulations with greatest informational value until such a value went below a certain threshold. However, the utility of committing to an action at a specific time is meaningless once that time is in the past. Since the utility of such a commitment is time dependent, the utility of deliberating about that commitment is time dependent as well. Recall our highway exit deliberation example. Time spent deliberating whether or not to take the first exit is time not available to deliberate whether or not to take the next exit. Using the traffic light example, one would attach far less value to a search for an action yielding a marginally greater fuel efficiency than a search for an action which would successfully stop at a red light. Simply put, we must work with an explicit notion of computational value in our decision making framework. Let us now sketch the basic decision procedure we have been discussing:

GT := init_simulation(default_move(S_I))
while game not terminated
      and no adversary action **do**
  u := u_est(GT)
  M := generate_moves(GT)
  choose $m{\in}$**M** maximizing exp_net_gain(GT, m)
  if exp_net_gain(GT, m) > 0
    add_simulation(GT, m)
  else
    commit(best_action(**u**))
    GT := subtree after best_action(**u**)
    add_simulation(GT, default_move(root(GT)))
  end if
end while

We initialize the game-tree to contain a simulation of a trajectory reflecting the default next move of the agent. While there is a reason to continue working with this game-tree, we update estimated expected utilities of the agent's possible next moves in the game-tree. We generate a set of possible moves (of any player) and pick the one which maximizes expected net gain from the simulation (the expected improvement in the action utility after the computation minus the computational cost). If the net gain is positive, we perform the simulation and add it to the game-tree. Otherwise, we commit to our current best action, prune the tree accordingly, and simulate the next default move.

6. HEURISTICS FOR INTELLIGENT SIMULATION

In this section, we discuss means by which one might heuristically compute expected utilities for states in the game-tree and expected information gain of trajectories not yet simulated. We discuss the difficulty of dealing with adversarial moves in real-time, and give an example of how complementary techniques can inform game-tree search.

In classical game-playing, the expected utility of reaching a leaf state is computed using a heuristic evaluation function, and the expected utility of reaching a non-leaf state (or node) is computed as a function of its immediate children. If the node has too many children to fully expand, a sampling can inform a probabilistic expectation of the utility of reaching that node. This evaluation of non-leaf states relies on the property of subtree independence, which we do not have. However, it is still the case that evaluations of the leaves of the subtree beneath a state form the primary source of information in estimating the expected utility of non-leaf states in our game-tree.

Similar or symmetric trajectories are a second source of information. Just as similarity or symmetry play a role in reducing branching factors in AI search, we may also expect similar subtrees to inform each other's estimates of expected utility, possibly reducing the amount of simulation necessary. This is most easily seen in examining an action which may be taken with different timing. For example, consider a game such as baseball. If I swing too early, I foul to one side. Too late, I foul to the other side. With minimal domain knowledge, I can use such observations (or simulations) to inform proper timing of my swing. With each enabled action, we seek to optimize timing to provide the most informational value. Information-based optimization (Neller, 1998) is particularly suited to optimization problems where the objective function is computationally expensive to compute.

Another means of informing our estimation of expected utilities is through learning. If the game involves repetitive tasks, then past experiences may inform future experiences. A person learns to balance a pole through trial and error. In practice, we may not have the luxury of such learning where one error means the game is over for good. Still this is a source of knowledge which, applied skillfully and efficiently, may significantly inform utility expectations.

Finally, through analysis of one's model, one may acquire domain knowledge useful for estimating or bounding one's utility expectations. The work of Tomlin, Lygeros, and Sastry (Tomlin *et al.*, 1998) is complementary to game-tree search in this respect. Through analysis of a nonlinear hybrid game, the state space is divided into safe and unsafe states from which the game can be guaranteed to be won or lost, respectively. While presenting computational difficulties, such an algorithm working in parallel with a game-tree search can yield better utility estimates over the run of the system. If such analysis is done a priori, we reap a large benefit in advance. Other techniques computing topological knowledge of the dynamical system or qualitatively abstracting the

hybrid system model may also inform our utility expectations.

One final point involves a complication in game play which has been glossed over until now: When an adversary makes a move, it's likely that almost all of the agent's game-tree consists of trajectories which will never be traveled. Is all of an agent's deliberation nullified whenever an adversary moves? Yes and no. Yes, in that a new tree must be grown reflecting change in the real world. No, in that the previous game-tree may inform the new game-tree as discussed above. If the agent has examined trajectories where the adversary has taken the same action at a different time, then that subtree may provide much valuable information relevant to the growth of the new game-tree. Similarly, if such a game-playing control system is implemented, then over time we can expect that the model will deviate significantly from the world. The reconstruction of a revised game-tree could be similarly informed from the previous one.

With all of these sources of heuristic information, it is of course essential to underscore the point that a heuristic should always pay for its computational cost in improved performance. Finding efficient ways to draw from these and other sources of information are key to the feasibility of this approach.

7. REFERENCES

Başar, Tamar and Geert Jan Olsder (1995). *Dynamic Noncooperative Game Theory, 2nd Ed.*. Academic Press. London.

Branicky, Michael S. (1995). Studies in Hybrid Systems: modeling, analysis, and control. PhD thesis. Massachusetts Institute of Technology. Cambridge, MA, USA.

Henzinger, Thomas A. and Sastry, Shankar, Eds.) (1998). *LNCS 1386: Hybrid Systems: computation and control, First International Workshop, HSCC'98, Proceedings*. Springer. Berlin.

Neller, Todd W. (1998). Information-based optimization approaches to dynamical system safety verification. pp. 346–359. In: Henzinger and Sastry (1998).

Russell, Stuart and Eric Wefald (1991). *Do the Right Thing: studies in limited rationality*. MIT Press. Cambridge, MA, USA.

Tomlin, Claire. John Lygeros and Shankar Sastry (1998). Synthesizing controllers for nonlinear hybrid systems. pp. 360–373. In: Henzinger and Sastry (1998).

MODELING EVOLUTION

Thomas L. Vincent

University of Arizona, Tucson, AZ, USA

Abstract: A brief review of the G-function method for modeling the evolutionary
process is presented. This includes some basic definitions used in evolutionary game
theory and an introduction to the ESS maximum principle used to determine
evolutionarily stable strategies (ESS). How a species actually arrives at an ESS is an
AI problem nature solves by means of genetic processes. Here we solve this problem
though the introduction of strategy dynamics. The strategy dynamics used here is
based on selective redistribution of phenotypes so that it mimics an evolutionary
process through time, however, it does not require any genetic modeling. Strategy
dynamics provides a new dimension to the study of evolution. Here we will briefly
look at the "speed" at which strategies evolve and the role it plays in coexistence.
Copyright © 1998 IFAC

Keywords: Evolutionarily Stable Strategies, Strategy Dynamics, Evolutionary
Games

1. INTRODUCTION

Since the fitness of each individual organism in
a biological community may be affected by the
strategies of all other individuals, the essential element of a 'game' exists. This game is an evolutionary game where the individual organisms (players)
inherit their strategies from a continuous play of
the game through time. Here, the strategies are
assumed to be parameters which can adapt (such
as sunlight conversion efficiency for plants or body
length in animals) in a set of differential equations
which describe the population dynamics of the
community. By means of natural selection, these
parameters will evolve to a set of strategy values
that natural selection, by itself, can no longer
modify, i.e., an **evolutionarily stable strategy** (ESS). An extensive literature translates the
ESS concept into a mathematical setting, see for
example (Maynard-Smith, 1982), (Riechert and
Hammerstein, 1983), and (Hines, 1987). Previous
work related to the development of an **ESS maximum principle** includes (Vincent, 1985), (Brown
and Vincent, 1987), (Vincent and Fisher, 1988),
(Vincent and Brown, 1987), and (Vincent *et al.*,
1996).

For a given class of models, it is possible to
predict the outcome of evolution by finding the
ESS (assuming it exists) using the ESS maximum principle. This principle states that a special
function, called the **G-function**, associated with
the fitness of every individual in the community
must take on a maximum with respect to ESS
strategies. Evolutionarily stable strategies may be
determined directly from this principle for a large
class of models. However our focus here will be
on examining a dynamical evolutionary process
which will take the system to an ESS by means
of **strategy dynamics.** See (Vincent, 1990) and
(Vincent *et al.*, 1993) for the development of
the method used here. See (Charlesworth, 1990),
(Abrams *et al.*, 1993*b*), and (Abrams *et al.*, 1993*a*)
for related results.

2. POPULATION DYNAMICS

Assume there are a finite but arbitrary number r
of different strategy values $\mathbf{u} = [\mathbf{u}_1, \ldots, \mathbf{u}_r]^T$ in a
population of individuals. The strategies $\mathbf{u}_i$ may
be scalars u_i or vectors $\mathbf{u}_i = [u_{i1}, \ldots, u_{is}]^T$ with
s components. Each strategy $\mathbf{u}_i$ is required to lie

in the same subset $\mathcal{U} \subseteq \mathcal{R}^s$. We will shorten this latter requirement to read $\mathbf{u} \in \mathcal{U}$ (every component of $\mathbf{u}$ must lie in $\mathcal{U}$). While it is often assumed that $\mathcal{U} = \mathcal{R}^s$ more generally the strategy set must be bounded for any real system.

Let x_i be the density (i.e. the number of individuals) of strategy type $\mathbf{u}_i$ at time t for $i = 1, \ldots, r$ and let $\mathbf{x} = [x_1, \ldots, x_r]^T$ be the vector of all such densities. Assume that changes in population densities are given by

$$\dot{x}_i = x_i H_i[\mathbf{u}, \mathbf{x}], \quad i = 1, \ldots, r \qquad (1)$$

where H_i is the fitness of individuals x_i using strategies $\mathbf{u}_i$. The individual fitness function $H_i[\mathbf{u}, \mathbf{x}]$ is assumed to be continuous (in practice, it is usually continuously differentiable) in $\mathbf{x}$ in the non-negative orthant

$$\mathcal{O}^r = \{\mathbf{x} \in \mathcal{R}^r \mid \mathbf{x} = [x_1, \ldots, x_r], x_i \geq 0\}$$

and also continuous in $\mathbf{u}$ if $\mathcal{U}$ has a given topology.

In what follows it is assumed that for any $\mathbf{u} \in \mathcal{U}$ there exists an equilibrium solution $\mathbf{x}^* \in \mathcal{O}^r$ satisfying $x_i^* H_i[\mathbf{u}, \mathbf{x}^*] = 0$ for $i = 1, \ldots, r$. If for some i, the equilibrium condition is satisfied by the non-trivial solution $H_i[\mathbf{u}, \mathbf{x}^*] = 0$ with $x_i^* > 0$ then the equilibrium solution is called an **ecological equilibrium point**. Reorder the i index, if necessary, so that all non-trivial solutions are listed first. Then, at an ecological equilibrium point there exists a $\sigma \geq 1$ such that the equilibrium solution is given by

$$\begin{aligned} H_i[\mathbf{u}, \mathbf{x}^*] &= 0, x_i^* > 0 \text{ for } i = 1, \ldots, \sigma \\ x_i^* &= 0 \qquad\qquad\quad \text{for } i = \sigma + 1, \ldots, r \end{aligned} \quad (2)$$

The strategies corresponding to the non-trivial solutions are designated by

$$\mathbf{u}_c = [\mathbf{u}_1, \ldots, \mathbf{u}_\sigma]^T$$

with the remaining strategies designated by

$$\mathbf{u}_m = [\mathbf{u}_{\sigma+1}, \ldots, \mathbf{u}_r]^T.$$

For the definition which follows we will require certain properties that are similar to asymptotic stability for the ecological equilibrium point. We cannot assume that $\mathbf{x}^*$ is asymptotically stable at the onset since $\mathbf{x}^*$ will, in general, lie on the boundary of $\mathcal{O}^r$ and trajectories starting at neighboring points not in $\mathcal{O}^r$ are of no interest and need not approach $\mathbf{x}^*$. Rather, we require that every trajectory starting in $\mathcal{O}^r \cap \mathcal{B}$ where $\mathcal{B}$ is an open ball about $\mathbf{x}^*$ remains in $\mathcal{O}^r$ for all time and converges to $\mathbf{x}^*$ as $t \to \infty$. An ecological equilibrium point which satisfies these properties is called an **ecologically stable equilibrium point** (ESE). If $\mathcal{B}$ can be made arbitrarily large

then $\mathbf{x}^*$ is said to be a global ESE otherwise it is a local ESE. These considerations ensure that the ESE consists of positive population densities.

Definition 1. The vector $\mathbf{u}_c \in \mathcal{U}$ is an **ESS** for the ecological equilibrium point $\mathbf{x}^* \in \mathcal{O}^r$ if, for any $r > \sigma$ and any $\mathbf{u}_m \in \mathcal{U}$, $\mathbf{x}^*$ is an ecologically stable equilibrium point.

Definition 1 incorporates the intuitive concept that an ESS cannot be invaded by rare mutants. In fact, the ESS coalition $\mathbf{u}_c$ is stable against simultaneous invasion by any possible set of mutant strategies $\mathbf{u}_m$. The mutant strategy types need not use strategies near those used by members of the coalition. In this sense, the ESS is always global with respect to $\mathcal{U}$. As an ESS, the ESE becomes a stable equilibrium point for any $\mathbf{x}$ near $\mathbf{x}^*$ regardless of the increased dimension of $\mathbf{x}$ and $\mathbf{u} = [\mathbf{u}_c, \mathbf{u}_m]$. If $\mathbf{x}^*$ is a local ESE then the mutants must be rare. If $\mathbf{x}^*$ is a global ESE then the mutants need not be rare. The ESS Maximum Principle (Theorem 1 below) is based on the existence of a fitness generating function G defined as follows.

Definition 2. A function $G(\cdot): \mathcal{U} \times \mathcal{U}^r \times \mathcal{O}^r \to \mathcal{R}$ is a G-function for (1) if, for every $i = 1, \ldots, r$,

$$G(\mathbf{u}_i, \mathbf{u}, \mathbf{x}) \equiv H_i[\mathbf{u}, \mathbf{x}].$$

We will write the G-function in terms of the dummy variable $\mathbf{v}$ as $G(\mathbf{v}, \mathbf{u}, \mathbf{x})$. Replacing $\mathbf{v}$ by the strategy used by individuals of strategy type $\mathbf{u}_i$ results in the fitness function for these individuals. Thus, $\mathbf{v}$ represents the strategy of an arbitrary individual while $\mathbf{u}, \mathbf{x}$ defines the current biotic environment.

Theorem 1. (ESS Maximum Principle). Let there exist a G-function for (1). If $\mathbf{u}_c = [\mathbf{u}_1^*, \ldots, \mathbf{u}_\sigma^*]$ is an ESS for the ecological equilibrium point $\mathbf{x}^* \in \mathcal{O}^r$, then $G(\mathbf{v}, \mathbf{u}^*, \mathbf{x}^*)$ must take on its maximum value as a function of $\mathbf{v} \in \mathcal{U}$ at $\mathbf{v} = \mathbf{u}_1^*, \ldots, \mathbf{u}_\sigma^*$. Furthermore $\max G(\mathbf{v}, \mathbf{u}^*, \mathbf{x}^*) = 0$.

The ESS maximum principle is a necessary condition for an ESS since it was derived by assuming that $\mathbf{x}^*$ is a local ESE. It is a constructive principle in that it may be used to directly solve for ESS strategies either analytically or numerically. In complex problems, the numerical approach is easy to implement. For example, MATHEMATICA was programed in (Vincent *et al.*, 1993) to find the ESS using a model in which, by varying a fixed parameter, the ESS could be a coalition of one, two, or more strategies. This relatively simple numerical procedure allowed for a demonstration of how coexistence could come about and be maintained in a rather complex model.

The ESS maximum principle has a simple geometric interpretation when the strategies are scalars. We will make this assumption in what follows. Let an adaptive landscape be defined as a plot of $G(v, \mathbf{u}^*, \mathbf{x}^*)$ with respect to v. According to the ESS Maximum Principle, each strategy of an ESS must be a global (with respect to $v \in \mathcal{U}$) maximum of the adaptive landscape. For example, suppose that $\mathbf{u}_c$ is an ESS coalition of two with the strategies u_1^* and u_2^*. If we substitute $\mathbf{u}^*$ and $\mathbf{x}^*$ into G and plot G as a function of v, its global maximum value must occur at u_1^* and u_2^* with a value of zero. An ESS must be invasion resistant. The above conditions are the necessary conditions for insuring that $\mathbf{u}^*$ is resistant to invasion by any $\mathbf{u} \in \mathcal{U}$.

3. STRATEGY DYNAMICS

Adaptive dynamics consider how the strategy of the i-th population, u_i, changes in response to the direction and magnitude of the fitness gradient on the adaptive landscape. Under the above assumptions regarding population dynamics and some assumptions regarding the distribution and redistribution of heritable strategies about the population's mean strategy, the mean strategy value of the i-th population evolves according to

$$\dot{u}_i = h\sigma^2 \left. \frac{\partial G(v, \mathbf{u}, \mathbf{x})}{\partial v} \right|_{v=u_i} \quad (3)$$

where $h\epsilon(0,1)$ is a heritability coefficient, σ^2 is some measure of genetic variance, and u_i is the mean strategy value. To simplify notation let

$$\varepsilon = h\sigma^2$$

be defined as the **speed** associated with the strategy dynamics. The dynamics of the strategy depends on the gradient of G. All of the strategies of $\mathbf{u}$ with positive population sizes evolve by climbing the adaptive landscape. As they evolve, the landscape changes and a strategy ceases to evolve after it reaches a point of zero gradient which can be a minimum, a maximum, or an inflection point on the adaptive landscape.

4. TOTAL SYSTEM DYNAMICS

With the G-function approach we write the coupled dynamics of population density and strategy as

$$\begin{aligned} \dot{x}_i &= x_i G_i(u_i, \mathbf{u}, \mathbf{x}) \\ \dot{u}_i &= \epsilon \left. \frac{\partial G_i(v, \mathbf{u}, \mathbf{x})}{\partial v} \right|_{v=u_i} \end{aligned} \quad (4)$$

for $i = 1, ..., r$. We will examine the total system dynamics for the Lotka-Voltera competition model defined by the G-function

$$G_i(v, \mathbf{u}, \mathbf{x}) = R - \frac{R}{K(v)} \sum_{j=1}^{r} \alpha(v, u_j) x_j.$$

where

$$K(u) = K_m \exp\left\{ -\frac{u^2}{2\sigma_k^2} \right\}$$

$$\alpha(u, u_i) = 1 + \exp\left\{ -\frac{(u - u_i + \beta)^2}{2\sigma_\alpha^2} \right\}$$

$$- \exp\left\{ -\frac{\beta^2}{2\sigma_\alpha^2} \right\}$$

The parameters R, K_m, β, σ_α, and σ_k are all assumed to be fixed parameters not subject to evolution. For this study these fixed values were chosen according to

$$\begin{aligned} R &= 0.25 \\ K_m &= 100 \\ \beta &= 2 \\ \sigma_\alpha &= \sqrt{8} \\ \sigma_k &= 2 \text{ or } \sqrt{12.5} \end{aligned}$$

The evolutionary parameters determine the value for the carrying capacity K and the interaction coefficient α. Using these values, it may be determined from the ESS maximum principle that for $\sigma_k = 2$, an ESS coalition of one exists with $u_1 = 1.2131$ with a corresponding equilibrium population of $x_1^* = 83.198$. For the case of $\sigma_k = \sqrt{12.5}$ an ESS coalition of two exists with $u_1 = -0.23973$, $u_2 = 3.1294$, with the corresponding equilibrium populations $x_1 = 39.284$ and $x_1 = 51.062$.

Here we will examine a very limited situation in which there are only two different types ($r = 2$) of individuals with the following initial conditions

$$\begin{aligned} x_1(0) &= 30 \\ x_2(0) &= 40 \\ u_1(0) &= 4 \\ u_2(0) &= 4.5 \end{aligned}$$

These values do not corresponds to equilibrium conditions, hence we are not starting at an ESS. All of the figures which follow are generated using (4) under the conditions already specified for the fixed parameters and initial conditions. The particular value of ε used is noted in each case. A solid line is use for the x_1 and u_1 trajectories and a dashed line is used for the x_2 and u_2 trajectories. Two sets of figures are produced for each case.

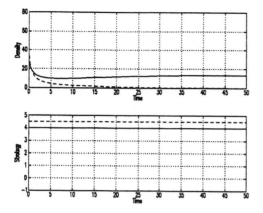

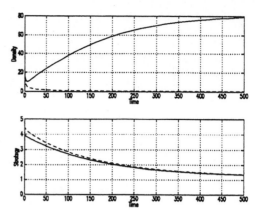

Fig. 1. Total system dynamics with $\sigma_k = 2$ and $\varepsilon = 0$.

Fig. 3. Total system dynamics with $\sigma_k = 2$ and $\varepsilon = 0.1$.

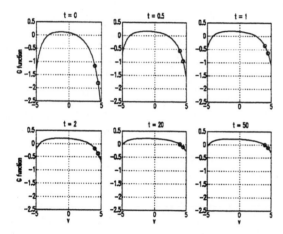

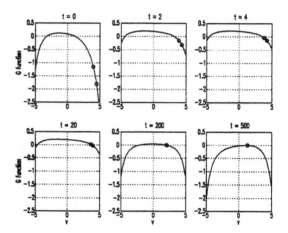

Fig. 2. G-function plots with $\sigma_k = 2$ and $\varepsilon = 0$.

Fig. 4. G-function plots with $\sigma_k = 2$ and $\varepsilon = 0.1$.

The first set of figures illustrate the population densities (x_1 and x_2) as a function of time and the strategies used by these populations (u_1 and u_2) as a function of time. The second set of figures illustrates a series of plots of the G-function over a range of v values with $\mathbf{x}$ and $\mathbf{u}$ fixed at their values corresponding to the time indicated at the top of each plot. The actual values of u_1 and u_2 being used at the time noted are indicated by the circles. Note that the time slices used differ for each case.

The first three cases correspond to the situation in which the ESS is a coalition of one ($\sigma_k = 2$). Figures 1 and 2 illustrates the results obtained then the strategies are not allowed to evolve by setting $\varepsilon = 0$. Initially the densities for each population drop dramatically as shown in Figure 1. This is because the initial values for x_1 and x_2 are well above equilibrium values corresponding to the two the strategies used. However x_1 recovers and asymptotically approaches the equilibrium solution of $x_1^* = 13.534$ while x_2 dies out. Figure 2 illustrates more clearly why we obtain this result. At $t = 0$, the value of the fitness function for each individual is negative. While the value of both fitness functions increases very quickly at

first, the long term result is for u_1 to produce a slightly positive fitness and u_2 to produce a slightly negative fitness. As the system finally goes to equilibrium the fitness of the individuals using u_1 goes to zero while the fitness of individuals using u_2 remains negative. Note that this does not imply that u_1 is an ESS strategy. We see from the last G-function plot that the equilibrium value of u_1 does not satisfy the maximum principle. What this means is that if we were to start this simulation over again with $u_1 = 1.2131$ (the ESS solution) and u_2 equal to any other value, then the individuals using u_1 would always survive and the individuals using u_2 would always die out. The final G-function plot in this case would have u_1 at the maximum point.

Figures 3 and 4 illustrates the results obtained by allowing the strategies to evolve slowly by setting $\varepsilon = 0.1$. Again the densities of both populations drop initially, with the individuals using u_1 recovering quickly and then asymptotically approaching an equilibrium density of $x_1^* = 82.570$. The individuals using u_2 continue to decline approaching an equilibrium density of $x_2^* = 0.62841$. Note that strategies of both population evolve to the ESS solution of $u_1 = u_2 = 1.2131$ and that the sum

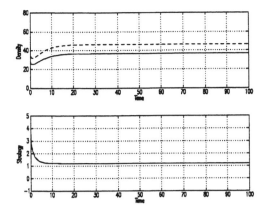

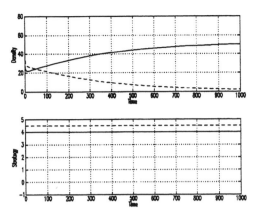

Fig. 5. Total system dynamics with $\sigma_k = 2$ and $\varepsilon = 10$.

Fig. 7. Total system dynamics with $\sigma_k = \sqrt{12.5}$ and $\varepsilon = 0$.

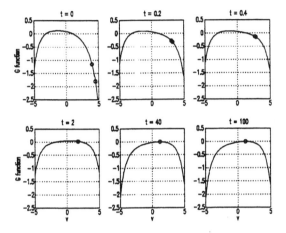

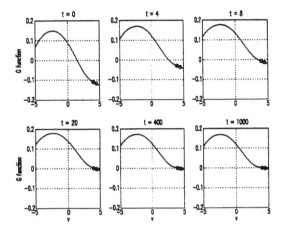

Fig. 6. G-function plots with $\sigma_k = 2$ and $\varepsilon = 10$.

Fig. 8. G-functions with $\sigma_k = \sqrt{12.5}$ and $\varepsilon = 0$.

of the two equilibrium densities $x_1^* + x_2^* = 83.198$ is equal to the density corresponding to the ESS coalition of one. In other words we have a case of **convergent evolution**. At equilibrium, each species has evolved to the same type and they are indistinguishable with respect to the theory since individual types are identified by the strategies that they are using. However if we were possible to keep track of these individuals by some characteristic (color, perhaps) that does not affect fitness and if each type were to breed true, then we would have a population of two types. However it is doubtful that the second type could remain viable at such low population numbers. The G-function plots illustrate what is happening. As the two strategies u_1 and u_2 "climb" up the adaptive landscape they approach one another and eventually merge to the same value at the ESS.

Figures 5 and 6 show the results of speeding up evolution ($\varepsilon = 10$). In this case both types again evolve to the ESS solution ($u_1 = u_2 = 1.2131$) but at reasonable equilibrium densities ($x_1^* = 36.728$, $x_2^* = 46.470$, $x_1^* + x_2^* = 83.198$). In this case one would expect that both populations could be maintained. Note that the final population numbers are a function of both the speed ε and

the size of the initial populations. We would expect that convergent evolution to be much more common in those situations where the speed ε is high.

The next two cases correspond to the situation where an ESS coalition of two exists ($\sigma_k = \sqrt{12.5}$). Figures 7 and 8 correspond to the case where the strategies are not allowed to evolve ($\varepsilon = 0$). The results in this case are not dissimilar to the case illustrated in Figures 1 and 2. The individuals using strategy u_1 survive with an equilibrium population if $x_1^* = 52.729$ and the individuals using strategy u_2 die out. However the G-function plots in this case are significantly different, with the final equilibrium solution a long way from the maximum as required by the ESS maximum principle. The strategy $u_1 = 4$ is clearly not an ESS.

If we now increase the speed to $\varepsilon = 5$, and rerun this same case, we get dramatically different results as illustrated in Figures 9 and 10. In this case the two different types evolve to the ESS coalition of two solution with equilibrium values as noted above. How this is possible is best illustrated using the information contained in the

59

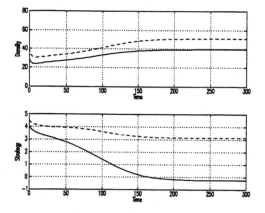

Fig. 9. Total system dynamics with $\sigma_k = \sqrt{12.5}$ and $\varepsilon = 5$.

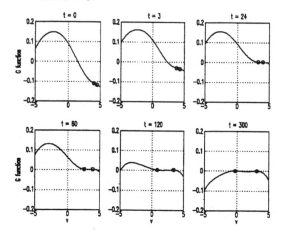

Fig. 10. G-function plots with $\sigma_k = \sqrt{12.5}$ and $\varepsilon = 5$.

G-function plots. As illustrated in the first three plots of Figure 10 the two strategies decrease in value as they climb the adaptive landscape until a valley appears between the two strategies. At this point u_1 continues to climb the hill to the left and u_2 starts to climb the hill to the right. As this happens, the adaptive landscape is also changing, so that as u_2 climbs the hill to the right it increases in value only slightly and then decreases in value again as the hill on the right side of the adaptive landscape gets shifted to the left. All this time both strategies are always climbing the adaptive landscape according to the strategy dynamics, but due to the flexible nature of the adaptive landscape it allows u_1 and u_2 to occupy the two isolated peaks as they finally achieve equilibrium. It follows from the last plot in Figure 10 that the two strategies satisfy the ESS maximum principle and have evolved to an ESS coalition of two. In this case, evolution provides stable coexistence with two distinct strategies. No other different strategy types can coexist along side this ESS solution. This same result will be obtained for any non-zero speed ε. The only difference in varying ε will be in the time it takes to reach the ESS. Note that this result is totally different from the convergent evolution case as the same distinct ESS equilibrium strategies and equilibrium population densities are obtained for all starting conditions.

5. REFERENCES

Abrams, P.A., H. Matsuda and Y. Harada (1993a). Evolutionary unstable fitness maxima and stable fitness minima of continuous traits. *Evolutionary Ecology* **7**, 465–487.

Abrams, P.A., Y. Harada and H. Matsuda (1993b). On the relationships between quantitative genetics and ESS. *Evolution* **47**, 982–985.

Brown, J.S. and T.L. Vincent (1987). A theory for the evolutionary game. *Theoretical Population Biology*.

Charlesworth, B. (1990). Optimization models, quantitative genetics, and mutation. *Evolution* **44**, 520–538.

Hines, W.G.S. (1987). Evolutionary stable strategies. *Theoretical Population Biology*.

Maynard-Smith, J. (1982). *Evolution and the Theory Games*. Cambridge University Press, Cambridge.

Riechert, S.E. and P. Hammerstein (1983). Game theory in the ecological context. *Annual Review of Ecology and Systematics*.

Vincent, T.L. (1985). Evolutionary games. *Journal of Optimization Theory and Applications* **46**(4), 605.

Vincent, T.L. (1990). *Strategy Dynamics and the ESS*. pp. 236–249. Birkhauser, Boston, New York.

Vincent, T.L. and J.S. Brown (1987). Evolution under nonequilibrium dynamics. *Mathematical Modelling*.

Vincent, T.L. and M.E. Fisher (1988). Evolutionarily stable strategies in differential and difference equation models. *Evolutionary Ecology*.

Vincent, T.L., M.V. Van and B.S. Goh (1996). Ecological stability, evolutionary stability and the ess maximum principle. *Evolutionary Ecology*.

Vincent, T.L., Y Cohen and J.S. Brown (1993). Evolution via strategy dynamics. *Theoretical Population Biology*.

EVOLUTIONARY DYNAMICS AND OPTIMAL LEARNING
UNDER BOUNDED RATIONALITY

R. Cressman

Department of Mathematics
Wilfrid Laurier University
Waterloo, Ontario N2L 3C5
Canada

K. H. Schlag

Economic Theory III
University of Bonn
Adenauerallee 24-26
D-53113 Bonn
Germany

Abstract: Individuals repeatedly face a sequential decision-making problem (or multi-decision bandit) in which the basic structure of which decision is faced under what circumstances is known whereas underlying payoff distributions are unknown. They have minimal memory and update their strategy by observing previous play (and not strategy) of someone else. We select behavioral rules that increase average payoffs as often as possible in a large population where all use the same rule. Here imitation generalizes to a pasting procedure. When decisions within the multi-decision bandit are unrelated, individuals eventually learn the efficient strategy but the underlying dynamic is not monotone. However, when choices influence which decisions are subsequently faced in the multi-decision bandit, play may not be efficient in the long run as it approaches a Nash equilibrium of the agent normal form. *Copyright © 1998 IFAC*

Keywords: Behavior dynamic, game theory, stability, frequency-dependent

1 INTRODUCTION

People are constantly faced with choices, that affect their well-being, without a complete knowledge of the consequences of their possible actions. In a society, people in these situations often make their choice through observing (or *sampling*) actions and their consequences of other individuals. As argued by Samuelson (1997), they develop rules of thumb, based on their limited information, which experience suggests will lead to their best choice. We call these *behavioral rules*. The purpose of this paper is to analyze the effect of different behavioral rules on both individual choice and aggregate population evolution.

As an elementary example, consider a single decision that has two possible actions (called a *Two -Armed Bandit* by Rothschild (1974)). Here, an individual's *strategy* consists of specifying which action is chosen in this game against nature. The strategy can be adjusted between rounds of play based on the consequences, that are measured numerically through a single variable called the *realized payoff*, of previous play. The objective is to make adjustments in such a way as to eventually adopt the strategy that

yields the highest *expected payoff* (i.e. to learn the *best strategy*). We are especially interested in such rules the are *improving* (i.e., that never decrease average payoffs when all individuals use the same rule.

Throughout the paper, we assume an extreme form of bounded rationality; namely, an individual's strategy adjustment between rounds can only be based on his own last strategy, play and realized payoff together with the last play and realized payoff of one randomly sampled individual. When there is no uncertainty about the observed payoff consequences of each action in the above example (i.e. when realized payoff always equals expected payoff), the behavioral rule "Imitate if Better"[1] will clearly achieve the objective if there are initially some individuals using the best strategy. It is also the best (or *dominant*) rule in these circumstances in the sense

---

[1] Explicitly, this rule is "switch to the sampled individual's strategy if it yielded a higher payoff than your own in the last round, otherwise maintain own last strategy".

that any other rule that achieves the objective will either take more rounds on average to accomplish this or else will sometimes cause average population payoff to decrease. Notice that such rules as "try each action once and then choose the one that realized the highest payoff" are not allowed due to our limits on memory even though they may quickly achieve the objective in this example.

We will see in Section 2, where behavioral rules for *Multi-Armed Bandits* (i.e. a single decision with two or more possible actions) are analyzed, that Imitate if Better is not good when realized and expected payoffs are not always equal (i.e. payoffs are given by a distribution with positive variance). However, rules that satisfy our objective while being improving must be imitating (i.e. strategies can only be adjusted by switching to sampled play). Theorem 2.2 characterizes all dominant rules for Multi-Armed Bandits. Two of particular importance are PIR and POR (Definition 2.1) that switch with a net probability that depends linearly on own and sampled realized payoffs. Section 2 also shows that, whenever all individuals adopt the same dominant behavioral rule in a Multi-Armed Bandit, population strategy frequency evolves according to a discrete version of the replicator dynamic. As an immediate consequence!, the best strategy(ies) is globally asymptotically stable since its frequency is monotonically increasing (i.e. dominant rules are *best reply monotone*).

Section 3 generalizes the above concepts to *Parallel Bandits* which model situations where, in each round, an individual faces one of m decisions (where $m \geq 2$) that have unrelated consequences. An individual's strategy must now specify which action will be chosen if play reaches decision l. The best strategy is now the one that specifies the action with the highest expected payoff at each decision. Since each play reaches exactly one decision, we picture a Parallel Bandit as m Multi-Armed Bandits connected in parallel. Only the last play is sampled under our assumption of bounded rationality and so imitation of strategies is no longer possible. Although this will imply no behavioral rule is best reply monotone, our objective can still be reached with an improving rule. These rules must use a *playwise imitating* pasting p!rocedure (the only strategy adjustments allowed between rounds are the replacement of planned action at decision l by the sampled action there). We restrict attention to *consistent rules* in that they must simplify to our previous rules selected when information can be used in a way that the present complicated bandit cannot be distinguished from a simpler one analyzed earlier. For instance, if own and sampled last play both occur at decision l, we insist the pasting follow an earlier dominant rule such as PIR for these *within-decision learning* situations. For *cross-decision learning* where own and sampled last play occur at different decision points, improving rules must ignore own last payoff. It then turns out that consistent rules

must use POR here since it is the only dominant rule that treats cross-decision learning like a within-decision learning situation. The resultant strategy frequency dynamic for every consistent rule induces th!e discrete replicator dynamic at each decision from which it quickly follows that the best strategy is again globally asymptotically stable.

Section 4 briefly considers complications that arise in multi-decision bandits where actions chosen at one decision influence the subsequent decision points reached. We do this by analyzing an example, the *Centipede Bandit*, that has an initial decision with two actions, D(own) and A(cross). Action D at decision 1 leads directly to a payoff. Action A leads to a move by nature with two branches, the first leading to a payoff, the second leading to another decision point of the player (decision 2) with actions d and a. For the same reason as given for Parallel Bandits, best reply monotone rules do not exist. We also show that there is no improving behavioral rule and no rule that makes the best strategy globally asymptotically stable. However, consistency and domination conditions lead to a class of playwise imitating rules (the CPIRs of Theorem 4.1) where the only relevant flexibility is what the switching proba!bilities between strategies Dd and Da are when sampled play is Aa or Ad. One of these rules is emphasized due to its simple implementation and desirable dynamic properties (Theorem 4.2). Here, the population evolves to a Nash equilibrium of the two-player game where players one and two are assigned to decision 1 and 2 respectively (this is the *agent normal form* of the Centipede Game).

2. MULTI-ARMED BANDITS

An n-armed bandit ($n \geq 2$) consists of a single decision point with n possible actions. Action i realizes a payoff that is drawn independently from a probability distribution P_i with expected value π_i and finite support in the interval $[\alpha, \omega]$. From our assumptions of bounded rationality, a behavioral rule F is a map $F(i,x,j,y)_k$ which specifies the probability of choosing action k in the next round if one's own last action was i with payoff x and the sampled action was j with payoff y. If all individuals in a large population use F and p_k^t denotes the proportion of individuals using strategy k in round t, the discrete-time population adjustme!nt process is

$$p_k^{t+1} = \sum_{i,j=1}^{n} \sum_{x,y \in [\alpha,\omega]} p_i^t p_j^t F(i,x,j,y)_k P_i(x) P_j(y). \quad (1)$$

Example. Suppose the two-armed bandit mentioned in the Introduction has payoffs in [10,20]. Explicitly, suppose action 1 always realizes payoff 18 whereas action 2 realizes payoff 20 with probability 2/3 and payoff 10 with probability 1/3 for an expected payoff of (2/3)(20)+(1/3)(10)=50/3<18. Thus, the choice of action 1 is the best strategy. Imitate if Better has more individuals switching from action 1 to action 2 than vice versa since action 2 has the higher realized payoff 2/3 of the time that one's own action differs from sampled action. Thus, this behavioral rule is not improving. In fact, in the long run, all individuals will eventually choose the worse action if there are initially some individuals who use action 2.

On the other hand, PIR and POR of Definition 2.1 below are both improving rules that lead the population to the best action. For instance, the only nonzero switching probabilities for PIR are

$$F(1,18,2,20)_2=(1/10)(20-18)=1/5 \text{ and}$$
$$F(2,10,1,18)_1=(1/10)(18-10)=4/5.$$

Thus $(2/3)(1/5)=2/15$ of action 1 players sampling action 2 switch whereas $(1/3)(4/5)=4/15$ of action 2 players sampling action 1 switch for a net switch in favor of action 1 of 2/15. A similar calculation using POR shows that, although switching rates are higher, the net switching rate is the same.

Definition 2.1. The *Proportional Imitation Rule* (PIR) only imitates the sampled action if it realizes a higher payoff and does this with probability equal to $(y-x)/(\omega-\alpha)$. That is, for $i{\neq}j$, $F(i,x,j,y)_j=(y-x)/(\omega-\alpha)$ if $y>x$ and equals 0 otherwise. The *Proportional Observation Rule* (POR) imitates the sampled action with probability equal to $(y-\alpha)/(\omega-\alpha)$. That is, $F(i,x,j,y)_j=(y-!\alpha)/(\omega-\alpha)$ if $i{\neq}j$.

All three behavioral rules mentioned in the above example are imitating since $F(i,x,j,y)_k=0$ whenever $k{\notin}\{i,j\}$. It is straightforward to verify that

$$F(i,x,j,y)_j - F(i,x,j,y)_i=(y-x)/(\omega-\alpha) \qquad (2)$$

are the net switching rates between actions $i{\neq}j$ for PIR and POR but not for Imitate if Better. The following theorem is proved in Schlag (1998) by exploiting any nonlinearity in x and y for the left-hand side of (2) for other rules. Recall that F is *dominant* if it is improving and the expected increase between rounds in any Multi-Armed Bandit is at least as great as for any other improving rule.

Theorem 2.2. If F is an improving rule, then F is imitating. F is dominant if and only if F is improving and (2) holds whenever $i{\neq}j$.

If F is any dominant rule, then (1) and (2) imply

$$p_k^{t+1}=\Sigma_i\Sigma_{x,y} \, p_i^{\,t} \, p_k^{\,t}[(y-\alpha)/(\omega-\alpha)+F(k,y,i,x)_i]P_i(x) \, P_k(y)$$
$$+\Sigma_i\Sigma_{x,y} \, p_i^{\,t} \, p_k^{\,t}[1-F(k,y,i,x)_i]P_i(x) \, P_k(y)$$
$$= p_k^{\,t}+(1/(\omega-\alpha))\Sigma_i p_i^{\,t} \, p_k^{\,t}(\pi_k-\pi_i)$$
$$= p_k^{\,t}+(p_k^{\,t}/(\omega-\alpha))(\pi_k-\pi(\,p^t)).$$

where $\pi(p^t)=\Sigma_i p_i^{\,t}\pi_i$ is the average population payoff when strategy frequencies are $p^t=(p_1^{\,t},\_,p_n^{\,t})$. This discrete dynamic is a version of the replicator dynamic for the game against nature. In particular, $p_k^{\,t+1}{\geq}p_k^{\,t}$ for any k in B=argmax$\{\pi_i\}$. Since this

inequality is strict whenever $p_j^{\,t}p_k^{\,t}>0$ for some $k{\in}B$ and some j in the complement of B, $\Sigma_{k{\in}B}p!_k^{\,t}$ is monotonically increasing and converges to 1 for any initial population where some individuals are using a best strategy. That is, dominant rules are *best reply monotone* and, in particular, the convex hull of the set of best strategies is globally asymptotically stable.

3. PARALLEL BANDITS

As stated in the Introduction, in Sections 3 and 4, we generalize behavioral rules selected for simpler Bandits to more complicated situations. For instance, the simplest Parallel Bandit has two decision points (m=2) with only one possible action at decision 1. Thus, payoffs observed at decision 1 are irrelevant to learning the best strategy. The following theorem results from this fact. The proofs of the results in Sections 3 and 4 are in Cressman and Schlag (1998).

Theorem 3.1. A behavioral rule F for the Simple Parallel Bandit described above is dominant if and only if
(i) F uses a dominant rule (such as PIR) for within-decision learning when own and sampled last play are at decision 2 (i.e. F is consistent with the single decision analysis of Section 2) and
(ii) F uses POR for cross-decision learning when own and sampled last play are at decisions 1 and 2 respectively.

It follows immediately from Theorem 3.1 that all dominant rules for Simple Parallel Bandits are best reply monotone. However, this is no longer true for general Parallel Bandits that have at least two decisions containing two or more actions.

Theorem 3.2. There is no rule that is best reply monotone for all Parallel Bandits. An improving rule must be playwise imitating.

For some intuition behind the above result, consider the Parallel Bandit with m=2 where the possible actions at decisions 1 and 2 are {A,Z} and {a,z} respectively. The fact that an improving rule or a best reply monotone rule must be playwise imitating follows by the same arguments as in Theorem 2.2. However, playwise imitating precludes the possibility of best reply monotone rules given the following observation. Suppose Aa is the unique best strategy. Under any playwise imitating rule, the frequency of Aa cannot strictly increase when the population consists entirely of Aa and Zz individuals (i.e. playwise imitation can only lead to Az or Za strategies). Thus no best reply monotone rule exists in general.

However, increase in the frequency of play of the best action can be implemented. The rules described in the following theorem have this *decision-wise monotone* property. Consistency referred to below encompasses all Simple Parallel Bandits contained in the general Parallel Bandit. Here ${}^{l}p_{k}^{t}$ is the frequency of individuals who plan to use the kth action at decision l in round t, ${}^{l}\pi_{k}$ is the expected payoff of this action, q_{l} is the exogenous probability decision l is reached and $[{}^{l}\alpha, {}^{l}\omega]$ is the finite interval containing the distributions of the actions at decision l.

Theorem 3.3. A behavioral rule F is both consistent and improving if and only if F uses a dominant rule (such as PIR) for within-decision learning and POR for cross-decision learning.

It is not difficult to show that all consistent improving rules induce the dynamic

$${}^{l}p_{k}^{t+1} = {}^{l}p_{k}^{t} + ((q_{l})({}^{l}p_{k}^{t})/({}^{l}\omega - {}^{l}\alpha))[{}^{l}\pi_{k} - {}^{l}\pi ({}^{l}p_{k}^{t})]$$

at each decision l. For the same reason given at the end of Section 2, this dynamic implies the convex hull of the set of best strategies is globally asymptotically stable for any consistent improving rule.

4. CENTIPEDE BANDIT

For the Centipede Bandit, as described in the Introduction, there are four possible outcomes of the game against nature {D, n, d, a} where, for example, n is the outcome that occurs when A is chosen at decision 1 and nature takes its first branch. Let q be the probability nature takes this branch and let π_{D}, π_{n}, π_{d}, π_{a} be the expected payoffs at the respective outcomes. For simplicity, we will assume all four payoff distributions are contained in the same finite interval $[\alpha, \omega]$.

The first question that arises is whether {Dd, Da, Ad, Aa} or {D, Ad, Aa} is the appropriate strategy set. The second component i of strategy Di acts like memory by specifying the action at decision 2 even though this is never reached when an individual uses D at decision 1. Without this memory in the latter strategy set, individuals switching from D when sampled play is A (i.e. the sampled individual either uses strategy Ad or Aa and nature takes its first branch) must decide on either Ad or Aa. Cressman and Schlag (1998) have shown that intuitive means to make this decision (e.g. flip a coin) lead to behavioral rules where D is asymptotically stable in cases where it is not the best strategy. We therefore consider only the model with four strategies from now on.

Unfortunately, there is no improving rule that always makes the best strategy asymptotically stable. To see this, denote the expected payoffs of the four strategies as $\pi_{Dd} = \pi_{Da} = \pi_{D}$, $\pi_{Ad} = q\pi_{n} + (1-q)\pi_{d}$ and $\pi_{Aa} = q\pi_{n} + (1-q)\pi_{a}$. Suppose F is !an improving rule. If $\pi_{D} = \pi_{Ad} > \pi_{Aa}$ and there are no Aa individuals in the population (i.e. $p_{Aa}^{t} = 0$), then $p_{Aa}^{t+1} = 0$. By interchanging the roles of Ad and Aa above, we see that F cannot switch from Di to either Aa or Ad when sampled outcome is D or n. Thus, for q=1 and $\pi_{Ad} = \pi_{Aa} > \pi_{D}$, there are no switches to the best strategy, in particular, the best strategy is not asymptotically stable.[2]

In the following, we consider only consistent behavioral rules, F, which considerably reduces the number of possibilities. For instance, if outcome n is neither experienced nor sampled, F must restrict to a dominant behavior rule for the Three-Armed Bandit with strategies {D, Ad, Aa}. Similarly, if D is neither experienced nor sampled, F must be a dominant rule for the Simple Parallel bandit with outcomes {n, d, a} as given in Theorem 3.1 with POR used for cross-decision learning. A careful analysis of these two consistency requirements plus those of two other simpler bandits shows that, if F is a consistent rule, then

(i) F must be playwise imitating (i.e. strategies can only be adjusted between rounds by pasting sampled actions at some or all of sampled decisions). In fact F cannot switch actions at an earlier decision (i.e. decision 1) on the sampled path without also switching at later decisions (i.e. decision 2 if it is sampled).

(ii) Net switching rates always satisfy (2) where switches of memory in Di when sampling outcome j (i.e. switching to Dj) are not counted.

Theorem 4.1. A behavioral rule F is dominant among the consistent rules if and only if F uses PIR for all switching rates except in the case of the Simple Parallel Bandit where POR is used and in the case of within-decision 2 learning where any dominant Two-Armed Bandit rule (e.g. PIR or POR) can be used. Due to the added prevalence of PIR, we call any such F a *Centipede Proportional Imitation Rule* (CPIR).

The only substantive difference between two CPIRs is the amount of memory switching from Di to Dj when outcome j is sampled. These effects only appear after two rounds in the evolution of population expected payoff. By adding another order to our dominance relation[3] that compares payoffs in two rounds, we are able to eliminate all CPIRs except two and certain convex combinations of them. We will only consider one of them from now on. F^{max} is the CPIR With Maximal Switch that always switches

---

!grid
[2] Taking q=1 produces a Two-Armed Bandit. The argument then resembles an application of our consistency requirement for simpler bandits.

[3] Specifically, if two behavioral rules increase payoffs between consecutive rounds for the same set of Centipede Bandits and by the same amount, we prefer one to the other if it recovers more of any decrease in payoffs within two rounds.

i in Di if Aj is sampled and i≠j (i.e. the individual either adopts Dj of Aj). This rule has many appealing properties - especially those relevant for population evolution that are summarized in the following result.

Theorem 4.2. Consider population strategy evolution under F^{max}. If D is either the best or worst choice (i.e. $\pi_D > \max\{\pi_{Ad}, \pi_{Aa}\}$ or $\pi_D < \min\{\pi_{Ad}, \pi_{Aa}\}$, then the convex hull of the set of best strategies is globally asymptotically stable. If the initial population has some individuals of all four strategy types, then strategy frequency converges to a Nash equilibrium of the agent normal form of the Centipede Game. In particular, if $0_{Aa} < \pi_D < \pi_{Ad}$, either

(i) P_{Ad}^t converges to 1 or

(ii) P_{Ad}^t converges to 0 and p_{Da}^t converges to some number between 0 and $(\pi_D-\pi_{Aa})/(\pi_{Ad}-\pi_{Aa})$.

F^{max} is eventually improving on each path (i.e. for t_0 sufficiently large, the average population payoff increases between rounds for all $t > t_0$).

5. CONCLUSION

By combining the rule F^{max} for Centipede Bandits with the behavioral rules selected in Sections 2 and 3 for Multi-Armed Bandits and Parallel Bandits respectively, we obtain a consistent behavioral rule for general multi-decision bandits where choices at one decision may affect subsequent decision points reached. This rule is easy to describe. The key is to find the "deviation" node at which sampled play differs from own play for the first time. If the deviation node is associated to a decision point, then imitate the sampled action at this node with probability given by PIR and imitate sampled actions at all other nodes with probability 1. If the deviation node is a move by nature, imitate simultaneously all sampled actions with probability given by POR.

REFERENCES

Cressman, R. and K.H. Schlag (1998). Updating strategies through observed play - Optimization under bounded rationality. *SFB Discussion Paper* **B-432**, University of Bonn .

Rothschild. M. (1974). A two-armed bandit theory of market pricing. *J. Econ. Theory*, **9**, 185-202.

Samuelson, L. (1997). *Evolutionary Games and Equilibrium Selection*. MIT Press, Cambridge, Mass.

Schlag, K.H. (1998). Why imitate and if so, how? A bounded rational approach to multi-armed bandits. *J. Econ. Theory*, **78**, 130-156.

ESS AND THE CONFORMIST GAME

Yosef Cohen

Department of Fisheries and Wildlife
University of Minnesota
St. Paul, MN 55108 USA

Abstract: Evolutionary stable strategies (ESS) are such that players who adopt them survive the evolutionary game and do not allow other strategies (mutant strategies) to invade or coexist—all mutant strategies go extinct. An imaginary game among humans, that has the characteristics of evolutionary games is developed. The game and its predictions can be tested in a laboratory set-up. Solutions are demonstrated. An algorithm is proposed such that a player need not know anything about the game (except that it is an evolutionary game) in order to guarantee a win. The algorithm is demonstrated. *Copyright © 1998 IFAC*

Keywords: Game theory, evolution, Differential games

1. INTRODUCTION

Heuristically, evolution works this way: Organisms transfer their traits to their progeny via some inheritance mechanisms. This transfer frequently happens via reproduction. For example DNA and genes carry information from one generation to the next. By information here we mean the set of traits that the parents own. During the life time of an organism, random errors in duplicating these genes occur. Because of these errors, not all progeny look alike. There are both genetic and phenotypic variations. Phenotypes are the individual organisms who express particular genetic traits. Because of genetic variation phenotypes look, behave, etc. differently. During their life-time, these phenotypes are exposed to limitations that cause their death and affect their reproductive capacity. Because phenotypes are different, these limitation act differently on different phenotypes (and therefore genotypes). This process, often called natural selection, leads

in the long run to changes in gene frequencies in the population Those phenotypes that withstand these limitations better than others will wind up producing more progeny and their genes will spread in the population. Over time, organisms therefore become less sensitive to selective pressures and this process is called adaptation by natural selection.

Charles Darwin and his contemporary Alfred Wallace were the first to articulate this theory (Darwin, 1859). More recently, Maynard-Smith (1982) have realized that during the process of adaptation (i.e., the process by which phenotypic characters of organisms change), evolving organisms interact with other organisms that evolve also. This process is called coevolution: all organisms that participate in a biotic community—via processes such as competition, predation, cannibalism—affect each others' values of these adaptive traits. This then calls for game-theoretic mathematical modeling. Games that describe the system outlined above constitute what is now known as evolutionary games. Because of the repeated process of mutation, natural selection, and coevolution, one is likely to observe (in the long run) a biotic community in which the collection of phenotypic traits (coalition) is such that no other trait values can invade and coexist with the resident collection of trait values. The trait values are equated to strategies, and the coalition of such values is called evolutionary stable strategies (ESS). Here these ideas are applied to an experimental game; the so-called the conformist game. One approach to the solution of such games is discussed (very briefly) in Section 2. The game is described and then solved in Section 3. Finally, some possible extensions to the game are discussed in Section 4.

2. THE THEORY

Here is a brief outline of the theory, skipping numerous mathematical details, which are given in Vincent et al. (1996). Definitions and concepts are modified from the traditional use (in evolution) to fit the present game scenario—see Section 3. Assume that there are a finite but arbitrary number r of different strategy values $\mathbf{u} = [\mathbf{u}_1, \dots, \mathbf{u}_r]^T$. The strategies $\mathbf{u}_i$ may be scalars u_i or vectors $\mathbf{u}_i = [u_{i1}, \dots, u_{is}]^T$ with s components. Each strategy $\mathbf{u}_i$ is required to lie in the same subset $\mathcal{U} \subseteq \mathcal{R}^s$.

Let x_i be the fortune of strategy type $\mathbf{u}_i$ at time t for $i = 1, \dots, r$ and $\mathbf{x} = [x_1, \dots, x_r]^T$ be the vector of all fortunes of the r strategy types. A player is identified with a strategy type, not with the number of individuals that play the game. Thus, for each strategy type $\mathbf{u}_i$ there corresponds a fortune x_i. Assume that changes in the dynamics of x_i are given by

$$\dot{x}_i = x_i H_i[\mathbf{u}, \mathbf{x}], \quad i = 1, ..., r. \qquad (1)$$

In the context of evolutionary games H is called the fitness of individuals x_i using strategies $\mathbf{u}_i$ and this convention will be followed here. The fitness function $H_i[\mathbf{u}, \mathbf{x}]$ is assumed to be continuous (and usually continuously differentiable) in $\mathbf{x}$ in the non-negative orthant

$$\mathcal{O}^r = \{\mathbf{x} \in \mathcal{R}^r \mid \mathbf{x} = [x_1, \dots, x_r], x_i \geq 0\}.$$

In other words, a strategy type with fortune 0 is eliminated from the game.

Assume that for any $\mathbf{u} \in \mathcal{U}$ there exists an equilibrium solution $\mathbf{x}^* \in \mathcal{O}^r$ satisfying $x_i^* H_i[\mathbf{u}, \mathbf{x}^*] = 0$ for $i = 1, \dots, r$. It is required

that for some i, the equilibrium condition is satisfied by the non-trivial solution $H_i[\mathbf{u}, \mathbf{x}^*] = 0$ with $x_i^* > 0$—at least one player must participate in the game. This is called the game equilibrium point (GEP). Reorder the i index if necessary so that all non-trivial solutions are listed first. Thus, at the GEP there exists a $\sigma \geq 1$ such that the equilibrium solution is given by

$$\begin{aligned} H_i[\mathbf{u}, \mathbf{x}^*] = 0, \, x_i^* > 0 \quad & i = 1, \dots, \sigma \\ x_i^* = 0 \quad & i = \sigma + 1, \dots, r \end{aligned}$$
$$(2)$$

The strategies corresponding to the non-trivial solutions is designated by

$$\mathbf{u}_c = [\mathbf{u}_1, \dots, \mathbf{u}_\sigma]^T$$

with the remaining strategies designated by

$$\mathbf{u}_m = [\mathbf{u}_{\sigma+1}, \dots, \mathbf{u}_r]^T.$$

It is required that every trajectory starting in $\mathcal{O}^r \cap \mathcal{B}$ where $\mathcal{B}$ is an open ball about $\mathbf{x}^*$ remains in $\mathcal{O}^r$ for all time and converges to $\mathbf{x}^*$ as $t \to \infty$. A GEP which satisfies these properties is called stable game equilibrium point (SGEP). These considerations ensure that the stable equilibrium consists of strategy types with positive fortunes. This equilibrium is called the game equilibrium point (GEP).

Definition 1 *The vector $\mathbf{u}_c \in \mathcal{U}$ is an **ESS** for the game equilibrium point $\mathbf{x}^* \in \mathcal{O}^r$ if, for any $r > \sigma$ and any $\mathbf{u}_m \in \mathcal{U}$, $\mathbf{x}^*$ is a stable game equilibrium point.*

Definition 1 incorporates the intuitive concept that an ESS cannot be invaded by different strategies with small fortune. These different strategies are called mutants. In fact, the ESS coalition of σ strategy types $\mathbf{u}_c$ is stable against simultaneous invasion by any possible set of mutant strategies $\mathbf{u}_m$. As an ESS, the SGEP becomes a stable point for any $\mathbf{x}$ near $\mathbf{x}^*$ regardless of the increased dimension of $\mathbf{x}$ and $\mathbf{u} = [\mathbf{u}_c, \mathbf{u}_m]$. Thus, according to the above definition, an ESS is not only uninvadable but it must also be convergent stable in the sense that mutants that invade with strategy values near the ESS values will either go extinct (their fortune becomes zero) or the strategy value converges to one of the ESS values; in other words, they become indistinguishable from exiting players. To proceed, the following definition is needed.

Definition 2 *A function $G(\cdot): \mathcal{U} \times \mathcal{U}^r \times \mathcal{O}^r \to \mathcal{R}$ is a G-function for (1) if, for every $i = 1, \dots, r$,*

$$G(\mathbf{u}_i, \mathbf{u}, \mathbf{x}) \equiv H_i[\mathbf{u}, \mathbf{x}].$$

G is called the fitness generating function. The G-function is written in terms of the dummy variable $\mathbf{v}$ as $G(\mathbf{v}, \mathbf{u}, \mathbf{x})$. Replacing $\mathbf{v}$ by the strategy type $\mathbf{u}_i$ results in the fitness function for these types. Thus, $\mathbf{v}$ represents an arbitrary strategy value while $\mathbf{u}, \mathbf{x}$ defines the state of the game (the number and values of the various coalition strategies, and the fortune of each of these).

Theorem 1 (ESS Maximum Principle)
Let $G(\mathbf{v}, \mathbf{u}, \mathbf{x})$ be a G-function for (1). If $\mathbf{u}_c = [\mathbf{u}_1^*, \ldots, \mathbf{u}_\sigma^*]$ is an ESS for the game equilibrium point $\mathbf{x}^* \in \mathcal{O}^r$, then $G(\mathbf{v}, \mathbf{u}^*, \mathbf{x}^*)$ must take on its maximum value as a function of $\mathbf{v} \in \mathcal{U}$ at $\mathbf{v} = \mathbf{u}_1^*, \ldots, \mathbf{u}_\sigma^*$. Furthermore $\max G(\mathbf{v}, \mathbf{u}^*, \mathbf{x}^*) = 0$.

The ESS maximum principle is a necessary condition for an ESS. For example, suppose that $\mathbf{u}_c$ is an ESS coalition of two with the strategies u_1^* and u_2^*. If $\mathbf{u}^*$ and $\mathbf{x}^*$ are substituted into G and G is plotted as a function of v, its global maximum value must occur at u_1^* and u_2^* with a value of zero. Furthermore if $\mathcal{U}$ is unbounded, then the ESS maximum principle results in the following necessary conditions

$$\left.\frac{\partial G(v, \mathbf{u}^*, \mathbf{x}^*)}{\partial v}\right|_{v=u_1^*} = \left.\frac{\partial G(v, \mathbf{u}^*, \mathbf{x}^*)}{\partial v}\right|_{v=u_2^*} = 0$$

$$\left.\frac{\partial^2 G(v, \mathbf{u}^*, \mathbf{x}^*)}{\partial v^2}\right|_{v=u_1^*} < 0$$

$$\left.\frac{\partial^2 G(v, \mathbf{u}^*, \mathbf{x}^*)}{\partial v^2}\right|_{v=u_2^*} < 0.$$

3. THE GAME

In this section the theory is applied to a game that can be played by humans. The game could also be played by animals, provided that one can teach them to access food (or reward of any other kind) via some repetitive activity. First the rules of the game are discussed and the necessary model is specified. Next, the game is solved for some parameter set. Finally, some interpretation and predictions based on the theory are provided.

3.1 Set-up

Consider a game with n players. Denote the fortune of player i at time t by $x_i(t)$. Each player faces a lever. Players are asked to press their lever with a certain amount of force. Player's

i strategy, u_i, is the amount of force applied to the lever. When player i presses the lever with force u_i, a reward is given according a function of his fortune, $a(x_i)$ where a is an increasing function of x_i. This reward is adjusted by some penalty. The penalty is proportional to $a(x)$ and depends the force applied by and fortune of all other players $\mathbf{u}$ and all $\mathbf{x}$. This proportion has two attributes, $b(u_i)$ and $c(u_i, \mathbf{u}, \mathbf{x})$, which will be discussed in a moment. Thus,

$$\dot{x}_i = a(x_i) - a(x_i)b(u_i)c(u_i, \mathbf{u}, \mathbf{x}). \quad (3)$$

A player's strategy depends on the amount of wealth he acquired, his own strategy, and the strategy of all other players. At this point, the rules need to be specified more explicitly.

The simplest rule for reward is linear:

$$a(x_i) = \alpha_a x_i.$$

Because u_i is the amount of force applied by a player, and because one wishes to (highly) discourage players from applying excessive force, exponential increase on the penalty is used:

$$b(u_i) = \alpha_b e^{-\beta_b u_i^2}.$$

Next, $c(u_i, \mathbf{u}, \mathbf{x})$ reflects the fact that a player is a "non-conformist" in the sense that the force he applies is as different as possible from the force applied by all other players, weighed by the corresponding fortunes. The rule used for c is then

$$c(u_i, \mathbf{u}, \mathbf{x}) = \sum_{j=1}^{n} \left[1 + e^{-\left(\frac{u_i - u_j + a_c}{\sigma_c}\right)^2}\right] x_j.$$

This rule may be interpreted as follows: The exponent takes a maximum value of $\exp\left[-(a_c/\sigma_c)^2\right]$ when the strategy of u_i equals that of u_j. As the difference between pairs of strategies increases, the value of the exponent decreases. The exponent is adjusted by the value of x_j, and all these values are added for the proportion by which the penalty for player i is adjusted. This means that player i should try to maintain as a large difference in strategy value between himself and all others. Furthermore, the player should play in such a way that the wealth of all other players is as small as possible. Reduction in both of these quantities will drive the penalty—the second term in equation (3)—on player's reward down. The players are allowed to negotiate on the amount of force they wish to apply to their corresponding levers. In this game a unique value of the strategy identify a unique player, not the actual person that applies the force. Thus, if the players negotiate a certain amount of force, and $u_i = u_j$, then the fortunes of both players are combined and they become a single player. At any point of the game, one or two persons that constitute a single player may decide to change u_i. At this instant they become a new player. To avoid an all out "war", the amount of change in force applied by a player is restricted to a nominal (small) percent of the force applied—by that player—in the previous instant. Players can choose a constant strategy throughout the game.

These rules essentially say: "Try to get as much wealth as you can. However, your strategy should be as different as possible from all others. The more force you apply to the lever, the higher the penalty you pay, and the more wealth you have, the faster you will accumulate wealth. But if your strategy is close to all others, the faster you will lose your wealth. Finally, you cannot change your strategy drastically." The game regulator then connects all levers to a computer and computes rewards and penalties in real time.

For a one player game (after very little algebra)

$$\dot{x} = x\left[\alpha_a - \alpha_a\alpha_b e^{-\beta_b u^2}\left(1 + e^{-\left(\frac{\alpha_c}{\sigma_c}\right)^2}\right)x\right]$$

This of course is a veiled Lotka-Volterra competition model. For an arbitrary strategy v, write G thus:

$$G(v, x) = \frac{a(x)}{x} - a(x)b(v)c(v, u, x).$$

and the strategy dynamics becomes

$$\dot{u} = \sigma^2 \left.\frac{\partial G(v, u, x)}{\partial v}\right|_{v=u}$$

where σ^2 scales the rate at which strategy dynamics change. Now that the game is set up, some solutions are examined next.

3.2 Some Solutions

For the parameter values:

$$\alpha_a = 0.25;\ \alpha_b = 0.01;\ \beta_b = 0.025;$$
$$\alpha_c = 1;\ \beta_c = 2;\ \sigma_c = 4;\ \sigma = 0.0016 \quad (4)$$

the trajectories are shown in Figures 1 and 2.

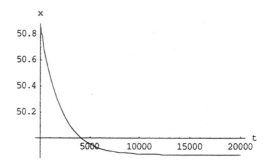

Figure 1. Fortune dynamics.

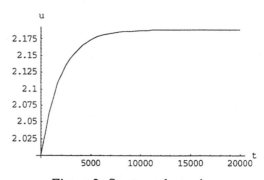

Figure 2. Strategy dynamics.

From these figures one can see that the game does settle to a stable equilibrium. Is this equilibrium an ESS? To answer this question, solve for the necessary conditions

$$G(u^*, u^*, x^*) = 0,$$
$$\left.\frac{\partial G(v, u^*, x^*)}{\partial v}\right|_{v=u^*} = 0$$

to find that $u = 2.19$ and $x = 49.87$. However, a plot of $G(v, u^*, x^*)$ (Figure 3) reveals that an ESS solution may require a coalition of 2.

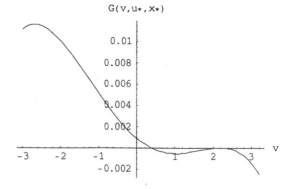

Figure 3. The G-function indicates that the ESS may be a coalition of 2.

To be able to derive G easily and keep the function $a(x)$ in a general form, write the equations thus:

$$\dot{x}_i = x_i \left[\frac{a(x_i)}{x_i} - \frac{a(x_i)}{x_i} b(u_i) \right.$$
$$\left. (c(u_i, u_1) x_1 + c(u_i, u_2) x_2) \right],$$
$$i = 1, 2.$$

Now without loss of generality, we write

$$G(v, \mathbf{u}, x, \mathbf{x}) = \frac{a(x)}{x} -$$
$$\frac{a(x)}{x} b(v)$$
$$(c(v, u_1) x_1 + c(v, u_2) x_2).$$

To retrieve the fitness of player i replace u_i and x_i for v and x. Upon using the necessary conditions

$$G(u_i^*, \mathbf{u}^*, x_i^*, \mathbf{x}^*) = 0,$$
$$\left. \frac{\partial G(v, \mathbf{u}^*, x_i^*, \mathbf{x}^*)}{\partial v} \right|_{v=u_i^*} = 0;$$
$$i = 1, 2$$

and with the parameter values given in (4) one gets

$$u_1^* = -1.85, \; u_2^* = 2.15, \tag{5}$$
$$x_1^* = 4.08, \; x_2^* = 47.55 \tag{6}$$

(Figure 4).

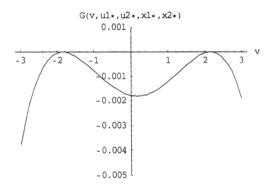

Figure 4. The G-function for a coalition of 2 gives an ESS candidate solution.

The dynamics of the trajectories from a variety of initial conditions end up with the ESS values in (5).

4. DISCUSSION

For the parameter values used in (4) one ends up with 2 players remaining in the game, provided that they play the ESS strategies. If a single player plays the constant strategy $u_1 = 2.19$ and a new player joins the game with a strategy close enough, the single player will win. However, if a new player joins with a strategy far enough from $u_1 = 2.19$, either both players or other players who play the ESS strategies will win the game. This then is the intuitive meaning of local evolutionary stability and the notion of global evolutionary stability. Reminiscent of the outcome of capitalism, one player is rich, the other is poor, and they coexist (contrast this with socialism, where all players end up being poor). Why is that so? Because the way the penalty is enforced. The parameter β_b plays an important role; as its value decreases, so does the penalty on using excessive force. This in turn allows players to play their strategies closer to each other (u_i close to u_j in the function $c(\cdot)$ in equation 3) without increasing their penalty significantly. In short, more unique strategy values are allowed to coexist. This prediction can be tested in an experimental setup of the game.

From an experimental point of view, one can instruct some players to play the fix ESS strategy throughout the game. Regardless of when and how rich new players join the game, the latter will eventually loose. Loosing means that their fortune will be close enough to zero.

In setting up the game, one has to be careful not to allow strategies to change too fast. This can be overcome by restricting the change in force players apply to the levers to small values; that is, one needs to ensure that $|u_i(t) - u_i(t + \Delta t)| < \delta(u_i)$ where $\delta(u_i)$ is some small positive number. Note that for convenience a differential version of the game was used. It is more natural and convenient (from the perspective of an experimentalist) to set up a difference equations version of the game. This will allow time for computations, and restrict players' actions to 1 per period. The value of β_b can be regulated to allow different number of players to participate in the ESS coalition.

This game is ideal from the perspective of a gambling house owner. If you are allowed to set the parameters, solve for the game and you are guaranteed to never lose it (if you can make sure that players stay in the game long enough).

The game is set up such that the fortune may be defined as, say, the weight of an animal. Rewards can be energy, and penalties can be withholding of energy. In this case losers will become skinny (and thus may be removed from the game) and winners become fat.

Finally, a new twist of the game will be presented during the conference. This will answer the following questions: Suppose that a player knows nothing about the game (e.g., the underlying model), except that it has an ESS. What would be the strategy that will ensure that the player will never loose the game? How should the strategy dynamics be followed? An example with an algorithm that ensures winning will be discussed and demonstrated.

References

Darwin, C. (1859), *The Origin of Species*, J. Murray, London.

Maynard-Smith, J. (1982), *Evolution and the Theory of Games*, Cambridge University Press, Cambridge, United Kingdom.

Vincent, T., Van, M. and Goh, G. (1996), 'Ecological stability, evolutionary stability, and the ESS maximum principle', *Evolutionary Ecology* **10**, 567–591.

FORAGING GAMES BETWEEN FIERCE PREDATORS AND THEIR PREY

Joel S. Brown

Department of Biological Sciences, University of Illinois at Chicago
845 W. Taylor St., Chicago, IL 60607, USA

Abstract: Carnivores and their prey often represent behaviorally sophisticated games of
stealth and fear. However, traditional mass-action models of predator-prey dynamics treat
individuals as behaviorally unresponsive "molecules" in Brownian motion. Here, the goal is
to extend foraging theory to consider a predator-prey game of stealth and fear and then
consider the consequences of this game for predator-prey population dynamics. At the ESS
of the game, the prey select an optimal level of apprehension and vigilance while the predator
selects the optimal amount of time to remain in a hunting patch before moving onto another
patch of prey. *Copyright © 1998 IFAC*

Keywords: Behaviour, Dynamics, Ecology, Game theory, Knowledge acquisition, Pay-off
functions, Risk, Tradeoffs.

1. INTRODUCTION

In ecology, there have been two complementary yet
somewhat divergent approaches to studying the
interactions between fierce predators and their prey.
The first (Taylor, 1984) focuses on the fact that
predators kill prey: predators are lethal. The second
(Lima and Dill, 1990) focuses on the observation that
fierce predators scare their prey: predators have non-
lethal effects on prey behaviour. In fact, fierceness is
not a property of the predator but rather a property of
the prey. If prey exhibit conspicuous fear responses
to their predators then the predator is deemed fierce.

Interest in the lethal effects of predators have
produced models of the dynamics of predator and
prey populations (Rosenzweig and MacArthur, 1963;
Murdoch and Oaten, 1975). In these models,
predators influence the population size and stability
properties of the prey population. Interest in the non-
lethal effects of predators have drawn on models of

feeding behaviour (MacArthur and Pianka, 1966;
Emlen, 1966). Several models predict how feeding
animals should use habitat selection and vigilance
behaviours to balance conflicting demands for food
and safety (Gilliam and Fraser, 1987; Brown, 1988;
Abrams, 1991; Houston , *et al.*, 1993).

Actual systems of behaviourally sophisticated
predators and prey should simultaneously exhibit both
the lethal and non-lethal effects of fierce carnivores.
Such systems of predator and prey should provide
particularly rich foraging games (van Balaan and
Sabelis, 1993; Hugie and Dill, 1994). The predators
select their optimal allocation of effort among areas
with prey and the prey select their optimal level of
apprehension and vigilance in response to the
strategies of other prey and predators. My goal is: 1)
to develop a model for the foraging behaviour of
fierce carnivores and their prey, and 2) to consider the
consequences of this foraging game for population
dynamics. In this game, let the predators select their

residence times within areas with prey (prey patches) and the prey select their baseline level of apprehension and their level of vigilance. Furthermore, assume that there is an informational asymmetry in that the predators have close to perfect information on the number and whereabouts of prey while the prey have only imperfect information on the predators' whereabouts.

2. MODEL OF FIERCE PREDATORS AND THEIR PREY

2.1 The Catch-22 of Mass-action Predator-prey Models

Rosenzweig and MacArthur (1963) showed how predator-prey systems tend to show oscillatory dynamics that may dampen towards a stable equilibrium, expand towards a limit cycle, lead to the predator's extinction, or result in the extinction of both prey and predator. Such a model can have the following form for predator and prey respectively:

$$N_{t+1} = N_t\{EXP[r(K - N_t)/K - P_tH/N_t]\} \qquad (1)$$

$$P_{t+1} = P_t\{EXP(r'H-d)\} \qquad (2)$$

where N_t and P_t are the population sizes (at time t) of the prey and predators, respectively, r (intrinsic growth rate) and K (carrying capacity) describe logistic growth of the prey in the absence of predators, H is the predator's per capita harvest rate of prey, r' describes the conversion of prey consumed into the per capita growth of predators, and d is the predator's density-independent death rate.

Let an individual predator's harvest rate be given by Holling's (1965) disc equation:

$$H = aN_t/(1+ahN_t) \qquad (3)$$

where a and h are the predator's encounter probability and handling time on prey, respectively. The form of H means that the prey experience safety in numbers. As N increases the probability of any given prey being captured by a predator , H/N, declines.

In the above predator-prey model the intraspecific competition among prey (via logistic growth) stabilizes the predator-prey dynamics while the safety in numbers experienced by the prey as a result of the predator's harvest rate function (3) destabilizes the equilibrium point (N^*,P^*). Which effect outweighs the other determines stability and gives the prey isocline (combinations of N and P such that $N_{t+1} = N_t$). Safety in numbers prevails at low N while intraspecific competition prevails at high N. The model places the predators in an ecological Catch-22.

If the predator is very efficient at capturing prey then the equilibrium point is to the left of the peak in the prey's isocline and the system is intrinsically unstable (the equilibrium point is unstable). If the predator is inefficient then the system has a stable equilibrium to the right of the hump, but now the predator is susceptible to extinction from stochastic or catastrophic declines in the prey's environmental quality. Does behavioral flexibility and the non-lethal effects of predators on their prey via a foraging game of fear and stealth rescue the predator from its Catch-22?

2.2 Sample System of Mountain Lions and Mule Deer

Mountain lions (Puma concolor) and mule deer (Odocoileus hemionus) in the mountains of southern Idaho, USA approximate closely a single-prey single-predator system (Hornocker, 1970; Altendrof, 1997). Mountain lions capture deer on the boundaries and interiors of wooded patches, and the deer move from these wooded patches to open shrublands as bedding and feeding areas, respectively. Mountain lions move frequently among wood patches. However, mountain lions rarely harvest more than one deer from a patch of woods and frequently they capture no deer before they are obliged to move in search of deer in another woodlot. A mountain lion does not deplete the value of a patch through its removal of deer. Rather, a woodlot becomes unprofitable to a lion as the deer either become too wary to catch or as the deer escape from the woods.

For the following model, consider an environment in which the prey occur as isolated individuals within patches of suitable habitat. The value of a food patch includes both the presence and catchability of a prey. The longer the predator remains in the patch the more wary the prey becomes. The predator must decide how long to remain in the patch before giving-up on trying to capture the increasingly wary prey. The prey must select its optimal level of vigilance in response to their perception of whether there is a predator in its neighborhood. In particular, the prey must select a baseline level of apprehension that determines their level of vigilance in the absence of any tangible evidence of a predator's presence. Set the level too high and the prey misses valuable feeding opportunities, set the level too low and the prey is slower to respond appropriately to the actual presence of a predator within its patch.

2.3 Prey Vigilance

Let the prey's fitness function be written as:

$$G = pF \qquad (4)$$

where p is the probability that the prey survives predation and F is the prey's fitness given it survives predation (survivor's fitness). In the context of equation (1) for prey population dynamics and in the context of vigilance these equations can be written as:

$$p = EXP(-\mu) \qquad (5)$$

$$F = EXP[r(1-u)(K-N)/K] \qquad (6)$$

where u is an individual's level of vigilance and μ is predation risk. Vigilance, $u \in [0,1]$, can be thought of as the proportion of time an individual spends scanning for predators. As such, increasing vigilance reduces survivor's fitness, F, and reduces predation risk, μ (see below).

The finite growth rate of the prey as been written as a fitness generating function, G, (Vincent and Brown, 1988) because ultimately the fitness of an individual prey will be influenced by its own vigilance strategy, the vigilance strategies of other prey and the patch residence times of the predators. This represents the foraging game between and among prey and predators.

Let the instantaneous risk of predation be influenced by the prey's encounter rate with predators, m, the predator's lethality in the absence of vigilance, $1/k$. the effectiveness of vigilance in reducing predator lethality, b, and the prey's level of vigilance, u:

$$\mu = m/(k+bu) \qquad (7)$$

This model of vigilance follows that of Brown (1999). In the present model the prey's optimal level of vigilance is given by:

$$u^* = SQR\{mK/(K-N)br\} - k/b \qquad (8)$$

where m is the prey's estimate of its encounter rate with predators.

2.4 Prey Apprehension Under Imperfect Information

Imagine prey that are uncertain as to the actual whereabouts of the predators, but which are able to make an estimate of its encounter rate with predators based on cues emitted by the predator when it occupies a prey's patch. Let M be the actual prey's encounter rate with a predator when a predator currently occupies a prey's patch. If a prey has perfect information then it should let $m=0$ or let $m=M$ depending upon whether a predator is present in the patch or not. However, a prey with imperfect information may maintain some background level of apprehension, m', even in the absence of any nearby predators. This level of apprehension is determined by its baseline expectation of encountering a predator

in the absence of any cues of predation risk. When a predator actually enters a patch the forager acquires information regarding its possible presence and, on average, adjusts its apprehension higher the longer the predator remains in the patch. For example, following the arrival of a predator, the prey's expected level of apprehension may be approximated as a "learning curve":

$$m(t) = m' + (M-m')[1-EXP(-\alpha t)] \qquad (9)$$

where α is the rate at which a prey catches onto the presence of a predator and t is the time since a predator has actually been in the patch (the prey never actually knows this). As t goes from zero to infinite the prey's expectation of encountering a predator rises from m' to M. Under this model of imperfect information, the prey's optimal level of vigilance can be approximated from equation (8) as:

$$u^*_{ab} = u(m') \text{ in the absence of a predator} \qquad (10)$$

$$u^*_{pr} = u(m(t)) \text{ in the presence of a predator} \qquad (11)$$

The challenge for the prey is to select the optimal level of background apprehension, m'. Increasing apprehension reduces feeding rates and reduces and reduces predation risk. Set the level of apprehension too high and the forager misses out on valuable feeding during periods when there are no predators in the patch. Set the level too low and the prey experiences unacceptably high predation risk in the presence of a predator. The background level of apprehension must strike a balance between feeding rate in the absence of predators and safety in the presence of a predator. Using equation (11) as vigilance in the presence of a predator, substituting this into equation (7) for actual predation risk and integrating this risk over the stay time of a predator in the prey' patch yields the average risk of predation to the prey:

$$\underline{\mu} = p\int[M/(k+bu^*_{pr})]dt \qquad (12)$$

where p is the probability that a predator is actually in the prey's patch, and the risk is integrated from 0 to t (the giving up time of the predator).

The forager's average level of vigilance over time is given by:

$$\underline{u} = (1-p)u^*_{ab} + p\int u^*_{pr}(t)dt \qquad (13)$$

The values of $\underline{\mu}$ and $\underline{u}$ calculated from equations (12) and (13) can be substituted into the prey's fitness generating function (4). Given the probability that a predator is actually within the prey's patch, p, and given the predator's giving up time, t, it is possible to find the value of m' (background level of apprehension) that maximizes the prey's finite rate of growth. The model possesses the two extremes of prey with perfect information ($\alpha \to \infty$) and completely ignorant prey ($\alpha \to 0$). When $\alpha \to \infty$, then the optimal value for $m' \to 0$. As $\alpha \to 0$, the optimal level for apprehension, $m' \to pM$.

2.5. Likelihood of a Predator in a Prey's Patch

There are two aspects to the predators' whereabouts and the likelihood that a prey currently shares its patch with a predator. The first describes the proportion of predators currently in patches of prey, p, where at most one predator per patch. The second, describes the probability that a given predator is within a prey patch, q, or is searching for a prey patch. The former matters to the prey and the latter matters to the predator. If t is the average giving-up time of a predator and T the predator's average search time for a prey patch, then:

$$T = 1/aN(1-p) \qquad (14)$$

$$q = t/(t+T) \qquad (15)$$

$$p = qP/N \qquad (16)$$

Equations (14)-(16) can be solved to give an explicit expression for p (the expression is quadratic and yields a positive root for $0<p<1$). To find the prey's optimal level of vigilance, it needs to know the prey and predator population sizes, the probability of there being a predator within its patch, and the predator's giving-up time.

2.6 The Predator's Giving-up Time

Let the predator be aware of the numbers of prey and the average baseline level of apprehension found among the prey. This means that the predator can sense and respond to the general level of apprehension among the prey but it is unable to adjust its giving-up time to the m' of a specific prey that deviates from the prey's norm. Furthermore, let the predator's giving-up time, t, be the time it remains within a patch before seeking a new one. This means that it leaves a patch after this amount of time should it not capture a prey, and it also represents (somewhat unrealistically) the time it remains in a patch even when it has captured a prey in the patch which will occur on average at times less than $t/2$. This simplifying assumption is probably excusable for situations where the predator visits many patches prior to actually capturing a prey (average t is little influenced by the quick kill of a prey within a patch), or situations (like the mountain lion) where prey capture entails a fairly lengthy period of consuming the deer -- in which case all of the above parameters need to be modified by the numbers of predators consuming prey rather than hunting for prey.

When the prey have imperfect information the predator's expected harvest rate, $\mu(t)$, declines with time spent in the patch. The longer the predator pursues the prey, the less catchable it becomes. There may come a point, t^*, at which the prey is no longer worth pursuing and the predator is better off abandoning the prey and seeking another. The predator's optimal stay time within the patch before abandoning the hunt satisfied a marginal value theorem (Charnov 1976). The predator should abandon the hunt on a given prey at the point where its expected capture rate, $\mu(t)$, drops to its average capture rate from seeking and pursuing another prey. The optimal giving-up time, t^*, satisfies:

$$\mu(t^*) = H(t^*) \qquad (17)$$

where the predator's average harvest rate is given by the probability of making a kill within a given patch divided by the time required to find the patch and dispatch the prey:

$$H(t) = [1-EXP(-\int \mu(t)dt)]/[T+t^*] \qquad (18)$$

where the integral is evaluated from $t=0$ to $t=t^*$.

For any given N, P, t, and prey background level of apprehension, m', it is possible to find a given predator's optimal value for t^*. All else equal, increasing N increases a predator's harvest rate without altering the likelihood of capturing a prey within its patch. Hence, t^* should decline. Increasing P has the opposite effect by increasing the predator's travel time among patches, reducing its H and, hence, increasing t^*. Increasing the t of all the other predators decreases a predator's T, increases its H, and , hence, reduces t^*. Increasing m' has two opposing effects. The first is to decrease H and to decrease the overall quality of a patch. The net effect of increasing m' is to increase t^*.

2.7 The Predator-prey ESS

At the evolutionarily stable strategy (ESS, sensu Maynard Smith and Price, 1973) the system must be ecologically stable in the sense that the prey and predator population dynamics do not lead to the extinction of either prey or predator. And the system must be evolutionarily stable in the sense that no rare alternative strategy can invade either the prey or the predator population. At the ESS of the present model, the prey's fitness generating function must take on a maximum at m'^* when all other prey are using m'^*, when all predators are using t^* and when population sizes are at equilibrium (N^*, P^*; assuming the dynamics arrive at a stable equilibrium). Similarly, at the predator's ESS, t^* must maximize the harvest rate of a given predator in a population where the prey are using strategy m'^*, the predators are using t^*, and the population sizes are at equilibrium (N^*, P^*).

At the ESS four conditions must be satisfied and values must be found for $m'*$, $t*$, $N*$, and $P*$. For the ecological stability of prey and predator, respectively:

$$r(1-\underline{u})(K-N)/K - HP/N = 0 \qquad (19)$$

$$H = d/r' \qquad (20)$$

where equations (19) and (20) are found by setting the prey's (1) and predator's (2) finite growth rates equal to one.

For evolutionary stability the prey's fitness generating function (4) and the predator's average harvest rate of prey must take on maxima at $m'*$ and $t*$ when the population is using strategies $m'*$ and $t*$, and the population sizes are at $N*$ and $P*$.

The ESS values for the prey's baseline level of apprehension and the predator's giving-up time can be found iteratively by using the following steps. First, set the prey's background level of vigilance to zero, $m=0$ for all prey and set the predator's stay time to $t=1/\mu$ where $u=0$. For these values of m' and t, solve numerically for $N*$ and $P*$. If these values are not positive it may be necessary to adjust m' or t to generate a positive $N*,P*$ combination. Second, find $t*$ for the current values of m', $N*$, and $P*$ through successive iterations. Third, recalculate $N*$ and $P*$. Fourth, For the current values of $t*$, $N*$, and $P*$ use successive iterations to find the $m'*$ that maximizes individual fitness when all prey are using the same strategy. Repeat steps 1-4 until the values of $m'*$, $t*$, $N*$ and $P*$ converge on the ESS values. In summary, equations (19) and (20) can be used to find $N*$ and $P*$ (where average values of $\underline{u}$, μ and H must generally be found by numerically solving the integrals over t), then numerical iterations must be used to find the values of $m'*$ or $t*$ that satisfies the ESS Maximum principal -- a strategy of the individual must maximize its fitness when everyone else also uses that strategy (similar to a Nash solution).

2.8 An Example ESS with Population Dynamics

Another way to find the ESS for this predator-prey game is to construct the predator's and prey's isoclines. The prey's isocline considers all combinations of N and P such that equation (19) is satisfied and the prey's population size is at equilibrium. Furthermore, one can solve numerically for the optimal levels of $t*$ and $m'*$ given N and P are fixed (this is not the ESS for the full system, but it represents a transitional optimum for a given combination of populations sizes). Similarly, the predator's isocline considers all combinations of N and P such that equation (20) is satisfied and the predator's population size is at equilibrium. The intersection of the prey's and predator's isoclines not only provide insights into the stability of the $N*,P*$ equilibrium, but it also gives the ESS values for $m'*$ and $t*$. It is at once indicative of the ecological and evolutionary equilibrium. Such a solution assumes that the behavioral dynamics influencing μ, occur much more rapidly than the ecological dynamics influencing N and P.

For example, let, $r=.2$, $r'=.1$, $K=100$, $d=0.009$, $k=1$, $b=100$, $M=10$, $\alpha=.1$, and $a=.01$. The ESS occurs at $m'* = 1.4$, $t*=3.15$, $N*=52$, $P*=5.15$. At this equilibrium the prey have a level of baseline apprehension that is 14% of the maximum encounter rate with predators, m'/M. The equilibrium point is stable and the system is highly resilient. The prey's isocline is humped shaped but it has a very flat peak (y-intercept of $P=1.3$, x-intercept of $N=K=100$, and a peak value of $N=51$, $P=5.17$). As one moves along the isocline from low to high values for N, the prey's baseline level of apprehension declines, the predator's giving-up time at first declines sharply and then increase again as N approaches K, and the initial vigilance level of the prey at first increases and then declines with N. The predator's isocline has a positive slope over most of its range (in this example it has a shallow and positive slope for values of $N>>N*$). It has an x-intercept of $N=10.5$ and gently curves to $N=62$, $P=5.5$ before leveling off and then gradually declining as N approaches 100. Along the predator's isocline (from low to high N), prey apprehension at first increases and then declines, the predator's giving-up time also is humped shaped, and the prey's baseline level of vigilance increases monotonically.

3. CONCLUSIONS

The adaptive behaviours of the prey and predator, and the imperfect information of the prey break the Catch-22 of the earlier mass-action predator-prey models. The positive slope of the predator's isocline means that the predators are extremely efficient predators when predators are rare (the predator's isocline intercepts the prey's axis at $N = 10.5$. However, at high values of P the predators become very inefficient as the prey become increasingly wary and harder to catch. The ease of catching prey when the predators are rare buffers the system against extrinsic stochasticities and catastrophes. The inefficiency of the predators at catching prey when predators are abundant promotes a stable equilibrium and promotes intrinsic stability. In many predator-prey systems the dynamics of μ may be more important than dynamics on N and P. For this reason, management of fierce predator systems may want to focus more on prey vigilance behaviours and predator giving-up times and a little less on measurement of N and P.

REFERENCES

Abrams, P.A. (1991). Life history and the relationship between food availability and foraging effort. *Ecology* **72**,1242-1252.

Altendorf, K.B. (1997). Assessing the impact of predation risk by mountain lions (*Puma concolor*) on the foraging behavior of mule deer (*Odocoileus hemionus*). *Unpubl. M.S. Thesis*, Idaho State Univ. Pocatello, ID, USA.

Brown, J.S. (1988). Patch use as an indicator of habitat preference, predation risk and competition. *Behav. Ecol. Sociobiol.* **22**:37-47.

Charnov, E.L. (1976). Optimal foraging and the marginal value theorem. *Theor. Popul. Biol.* **9**:129-136.

Emlen, J.M. (1966). The role of time and energy in food preference. *Amer. Natur.* **100**,611-617.

Gilliam, J.F. and D.F. Fraser (1987). Habitat selection under predation hazard: a test of a model with foraging minnows. *Ecology* **68**,1856-1862.

Holling, C.S. (1965). The functional response of predators to prey density and its role in mimicry and population regulation. *Mem. Ent. Soc. Can.* **45**,1-60.

Hornocker, M.G. (1970). An analysis of mountain lion predation upon mule deer and elk in the Idaho Primitive Area. *Wildlife Monogr.* **21**,1-39.

Houston, A.I., J.M. McNamara and J.M.C. Hutchinson (1993). General results concerning the trade-off between gaining energy and avoiding predation. *Phil. Trans. Roy. Soc. London, Series B* **341**,375-397.

Hugie, D.M. and L.M. Dill (1994). Fish and game: A game theoretic approach to habitat selection by predators and prey. *J. Fish Biology* (Supplement A) **45**:151-169.

Lima, S.L. and L.M. Dill. (1990). Behavioral decisions made under the risk of predation: A review and prospectus. *Can. J. Zool.* **68**,619-640.

MacArthur, R.H. and E.L. Pianka. (1966). On optimal use of a patchy environment. *Amer. Natur.*, **100**,603-609.

Maynard Smith, J. and G.R. Price (1973). The logic of animal conflict. *Nature* **246**,15-18.

Murdoch, W.W. and A. Oaten (1975). Predation and population stability. *Adv. Ecol. Res.* **9**,1-131.

Rosenzweig, M.L. and R.H. MacArthur (1963). Graphical representation and stability of predator-prey interaction. *Amer. Natur.* **97**,209-223.

Taylor, R.J. (1984). *Predation.* Chapman and Hall, New York.

van Balaan, M. and M.W. Sabelis (1993). Coevolution of patch strategies of predator and prey and the consequences for ecological stability. *Amer. Natur.* **142**,646-670.

Vincent, T.L. and J.S. Brown (1988). The evolution of ESS theory. *Ann. Rev. Ecol. Syst.* **19**,423-443.

AN IMPLEMENTATION OF INTELLIGENT NODE BASED ON LONWORKS TECHNOLOGY

Wang Junjie , Zhang Wei , Xie Chunyan

Department of Automation
Tshinghua University
Beijing 100084 P.R. China

Abstract: In order to realize the automation of management of Beijing District Heating System, a thorough study on Fieldbus and one of its form—LonWorks technology have been conducted. And according to the heating power network's requirement of monitored control systems, a test model of distributed monitoring network has been designed. The system consists of seven intelligent nodes. Each node, adopting the LonWorks Neuron Chip as its model,is equipped with auxiliary circuit and application software. Especially, the FCU node, having used ETSK model and fuzzy control algorithm, is much more excellent in controlling accuracy than normal PID and fuzzy PID.*Copyright©1998 IFAC*

Keywords: Fieldbus, Monitored control systems, Intelligent instrumentation, Controller modulators, Fuzzy control.

1. PREFACE

LonWorks technology is a fieldbus technology put forward by the American Echelon Company at the beginning of the 1990's. It is a complete technological platform for developing controlling and monitoring network system. Its network system consists of several intelligent nodes. Communication can be made among nodes through various transmission media, such as twisted pair, power line or radio frequency. The OSI (open system interconnection) reference model laid down by the ISO (International Standardization Organization) is followed. LonWorks technology mainly includes the following four parts.

- Specialized device 3150 and 3120 Neuron Chip
- LonTalk protocol
- Specialized transceiver, controller, adapter, router and other function templates
- LonBuilder and NodeBuilder developing platform

LonWorks technology is especially fit for making a distributed, multi-node monitored control systems. Therefore, it's decided to apply the technology to the Beijing District Heating System. The Beijing District Tube Network is a typical distributed, multi-node system, with hundreds of heating power stations distributed all around the city. Each heating power station has temperature, pressure, flow quantity and many other parameters to measure and data to process. Meanwhile, it is also necessary to control some pumps and valves. The operating condition of all the heating power stations in the city requires to be monitored and managed in a central control room. In order to accumulate technology and experience, a test model of distributed controlling and monitoring network has be built in a laboratory which is based on LonWorks technology with heating power station as its background. The whole system consists of seven intelligent nodes. They are analog input (AI) node, analog output (AO) node, discrete input/output (DI/DO) node, frequency input (FI) node, fuzzy control unit (FCU) node, LED display node(LED) and network monitoring computer (PC) node.

2. THE FUNCTION OF THE TEST MODEL

The monitoring network, which consists of seven intelligent nodes, has adopted free topological structure. Every node is equal in the network, with no major-minor distinction. The kernel of the intelligent node is the TP/FT-10 free topological twisted-pair controlling module made by Echelon Company. It is actually a minimal system, with a 3150 Neuron Chip, a 32k bytes PLCC packaged Eprom Socket (in which the chip with the application program is plugged), FTT-10 transceiver (by which the free topological structure of the network is carried out). 5MHZ crystal oscillator, a 18-pin plug-in unit P1 connected with I/O port and a 6-pin plug-in unit P2 connected with UTP . Signal transmission uses non-polarized Manchester encoding with a transmission speed of 78Kbps. The maximal transmission distance of the free topological structure is 500M. The network system structure of the test model is shown in Figure 1.

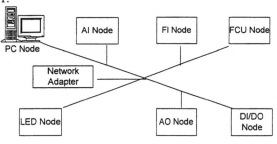

Fig 1: Free topological network system structure

The monitoring object of the whole test model is a hot water heating power system. The parameters of the hot water are the temperature of the inlet high-temperature water (T_1), the temperature of the outlet low-temperature water (T_2), flow frequency of the outlet water (F). They are sent respectively to AI and FI nodes. The frequency received by the FI node is the signal of the vortex flowmeter. It must be calculated with the coefficient of flow. Then the volume flow quantity Q can be got.

$$Q=F/\xi \qquad (1)$$

Then the value of this Q is sent to the AI node. In the AI node, complex calculations are to be made. First, the density ρ of the low-temperature water is evaluated through T_2. Then the mass flow quantity M can be worked out.

$$M=Q\cdot\rho \qquad (2)$$

Then it works out the enthalpy of the high-temperature water i_1, low-temperature water i_2 respectively according to T_1 and T_2, and the heat flow quantity H. Finally, the sum of heat quantity ΣH can be worked out

$$H=M(i_1-i_2) \qquad (3)$$

$$\Sigma H= \int_{t1}^{t2} M(i_1 - i_2) \qquad (4)$$

The calculating results of AI node are sent to LED display node and PC node respectively. LED node displays the value of T_1, T_2, F, M, H, ΣH, the values to be displayed are decided by the switching condition of DI node. When PC node has received data from AI node, it stores them in the data base, and at the same time, the bound value for alarm is sent out to compare with data in AI node. When the data go beyond the bound, signal will be sent to DO node to make alarm. The data of AI node will also be sent to FCU node. After comparing the data with the set value ,the FCU node ,according to the result of fuzzy control algorithm, controls the valve and regulates the flow quantity of backwater. Serving as the monitoring station of the controlling room, PC node collects all the operating parameters of the whole system, makes compilation, comparison and judgement, and changes the set parameter of FCU with the help of expert system, so that the whole system is running under the optimum state. The relation of network variables among nodes is shown in Figure 2.

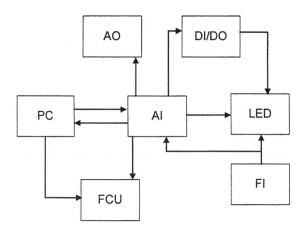

Fig 2: The relation of network variables

3. PC NODE

A 100MHZ IBM PC is chosen to serve as the network-monitoring computer. The monitoring software of the node, based on the operating system of Windows 3.1, is developed by Visual Basic. By using DDE Server, which is in Echelon Company's LonManager Series, the monitoring software communicates with PCNSS, which is plugged in ISA bus socket in the PC. In this way, the real-time communication between the monitoring software and the LonWorks network is carried out. Its software architecture is shown in Figure 3. The main function of the PC node is to monitor system parameters which need real-time display and to set alarm bound, the relevant parameters of FCU node, etc.

The application software of the PC node consists of initializing module, network parameter setting and monitoring module, FCU value setting module and historical data recording module.

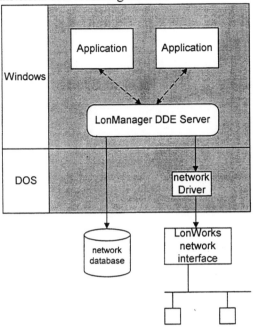

Fig 3: The software architecture of the PC node

4.   FCU NODE

As a fuzzy control unit with LonWorks intelligent node function, FCU node consists of two parts, as shown in Figure 4. An 80C196KC microprocessor made by Intel Company is used in the single loop control model. It has the basic structure and function of a regulator, including analog input/output, discrete input/output, keyboard processing and parameter display. It realizes fuzzy control algorithm through software and realizes both-way communication with LonWorks bus through serial interface connecting with TP/FT-10.

The fuzzy control algorithm of the FCU node adopts an extended TSK model—ETSK model, based on LR type fuzzy numbers and their operations. By replacing the consequent of its rule with a fuzzy linear function, ETSK model synthesizes the merits of LM model, TSK model and Fuzzy Regression Model and thus enhances its ability in describing the uncertainty.

First, It needs to set up and identify the ETSK model of the controlled object. Suppose the ETSK model of the controlled object is

IF   y(k-1) is Small THEN y(k)=$a_{11} \otimes$ y (k-1)$\oplus b_{10} \otimes$

u(k-d) $\oplus b_{11} \otimes$ u(k-d-1)   ALSO
IF   y(k-2) is Big THEN y(k)=$a_{21} \otimes$ y (k-1) $\oplus b_{20} \otimes$
u(k-d) $\oplus b_{21} \otimes$ u(k-d-1)   (5)

Here d is the delay time, a_{11}, a_{21}, b_{10}, b_{11}, b_{20} and b_{21} are LR type fuzzy numbers which are necessary to be given in advance. The separation of universe of discourse about the input signals is shown in Figure 5. SV in the figure is the set value of the circuit.

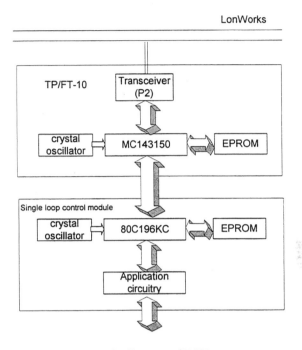

Fig 4: The schematic diagram of FCU

A further identification on the center-of-gravity parameter and area coefficient of the above coefficients can be made. Because of the complexity of the above presumptive ETSK model, the expression (5) can be changed into the weighted TSK model .

IF   y(k-1) is Small THEN y(k)=g_{a11}y(k-1) + g_{b10}
u(k-d) + g_{b11} u(k-d-1) WITH ω_1 (y(k-1))   ALSO
IF   y(k-1) is Big THEN y(k)=g_{a21}y(k-1) + g_{b20} u(k-d)
+ g_{b21} u(k-d-1) WITH ω_2 (y(k-1))   (6)

Then the first-order discrete model in consequent of its rule in expression (6)is changed into continuous model. The gain of the corresponding model (K), time constant (T) and delay time (τ) are evaluated. According to empirical formula, two pairs of regulator parameters (K_{c1}, K_{11}) and (K_{c2}, K_{12}) are worked out. Finally, the fuzzy control algorithm can be got:

IF   y(k-1) is Small THEN Δu(k)=K_{c1}[e(k)-e(k-1)]
+K_{11} e(k)   WITH ω_1 (y(k-1)) ALSO
IF   y(k-1) is Big THEN Δu(k) = K_{c2}[e(k)-e(k-1)]
+K_{12} e(k)   WITH ω_2 (y(k-1))   (7)

The contrastive experiment shows that FUC outdoes

E5AX-VAA regulator of the OMROM Company in speed and control accuracy.

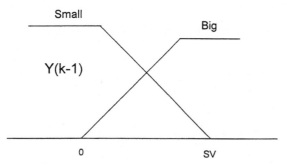

Fig 5: The separation of the universe of discourse Y (k-1)

5. CONCLUSION

Fieldbus and LonWorks technology represent the development trend of the industrial control technology. It will be the main technological method in industrial control in the coming century. It, above all, needs an open, standardized network communication protocol, to guarantee the reliability, compatibility and the high speed of the signal transmission. Furthermore, various intelligent node instruments need to be developed. These node instruments should not only have the functions of signal collection, amplification, transformation, but also of automatic compensation, linear adjustment, fault diagnosis, fuzzy identification, advanced control and other artificial intelligent technology, so that the technology level of the system can be raised. Our research work is just a first step. Much more tasks are waiting for us to fulfill.

REFERENCES

ECHELON COMPANY(1994). Neuron Chip Distributed Communication and Control Processors.

ECHELON COMPANY, (1995). Neuron C Programmer's Guide.

Pinto,J.,(1995). Fieldbus--an independent view, In:*control & instrument,*May.

Takagi T.,Sugeno M.,(1985). Fuzzy identification of systems and its application to modeling and control .In: *IEEE Transactions on Systems,Man and Cybernetics,* **15, 1985,** 116-132

Misir D.,Malki H. A.,Chen G.(1996). Design and analysis of a fuzzy proportional-integral-derivative controller, In: *Fuzzy sets and systems,* **1996,** 79(3):297-314

Mizumoto M.(1995). Realization of PID controls by fuzzy control methods,In: *Fuzzy sets and systems,* **1995, 70**(2):171-182

ROBUST FUZZY LOGIC-BASED CONTROL OF A HYDRAULIC FORGE PRESS

W. Garth Frazier

Materials and Manufacturing Directorate
Air Force Research Laboratory
Wright-Patterson AFB, Ohio, USA
fraziewg@ml.wpafb.af.mil

Abstract: The design of a robust fuzzy logic-based velocity versus stroke control law for a multiple-valve hydraulic forge press is presented. Simulation results illustrating the robustness of the control law are also presented. *Copyright © 1998 IFAC*

Keywords: Fuzzy control, manufacturing systems, hydraulics

1. INTRODUCTION

High performance robust control of hydraulic forging equipment is becoming increasingly important in achieving repeatable product quality in an environment where the same piece of equipment is often required to produce several different parts subject to different operating conditions in a single day. These variations in load versus stroke and speed versus stroke conditions combined with the relatively primitive control strategies traditionally used in these systems inhibit "set it and forget it" tuning of the control law. In research environments, this problem is further accentuated because the same process is rarely repeated more than a few times. In fact, oftentimes several equipment-tuning experiments are performed before one "true" forging experiment is performed.

Current industrial practice has partially alleviated this problem by keeping a database of control computer parameters that have worked well in the past for each particular forging process and then entering the correct parameters each time a process is performed. This strategy works well until an input error occurs, a new forging process is required or the inevitable wear and tear on the equipment begins to take its toll.

As an alternative to this experience-base/trial-and-error approach, it is desirable to be able to perform almost any forging process and have the forge press perform according to the desired velocity versus stroke profile. To achieve this, a feedback control strategy that is specifically designed to accommodate

non-constant ram speeds as well as changing load versus stroke conditions should be formulated.

In this paper a fuzzy logic-based control law is designed for a multi-valve hydraulic forge press. This control law is shown to be robust with respect to time-varying load versus stroke conditions, time-varying velocity versus stroke commands, and equipment parameter variations. The basis of the simulated analysis of the fuzzy logic-based control law is a validated, high-fidelity computer simulation of the forge press (Frazier, *et al.*, 1997).

2. DESCRIPTION OF THE FORGE PRESS

2.1 Overview

A simplified drawing of the press is shown in Fig. 1. The forge press is a 1000-ton vertical, double-acting type with a power plant consisting of 200 hp electric motor operating at 1200 rpm driving a radial-piston pump with a maximum capacity of 80 gpm. This corresponds to a maximum sustained ram speed of 0.58 ips. To help stabilize the nominal head pressure at 3800 psi and to support brief ram speed demands exceeding 0.58 ips, a 16 gal hydraulic accumulator is attached to a 120 gal nitrogen bottle charged to 3800 psi.

The ram is actuated by three hydraulic valves that allow fluid to flow from the high pressure side of the

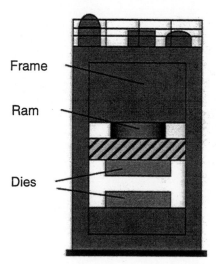

Fig. 1. Drawing of vertical hydraulic forge press

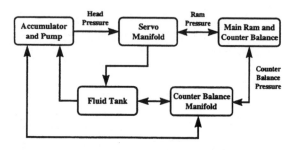

Fig. 2. Stylized hydraulic circuit for the forge press

manifold (head) to the low pressure side (ram). The two smallest valves are three-way servovalves that also allow fluid from the ram side of the manifold to be diverted to the tank (reservoir). This provides for rapid slow down capability as well as a means of returning fluid to the tank after a forging operation is completed. Since each of the valves has a different dynamic range, flow curve, and bandwidth, and since they are connected in parallel, the velocity of the ram has the potential of being controlled with high accuracy over a wide dynamic range. A stylized hydraulic circuit for the system is shown in Fig. 2.

The ram system weighs approximately 24 tons, but this varies based upon the weight of the particular tooling set that is installed. The ram is supported by two counter-balance pistons that yield when loaded with more than 26 tons. Maximum stroke is 15 in.

Sensors for feedback control include a linear ram position transducer, and head and ram pressure transducers. The control computer is a 80386-based industrial computer operated at a sampling rate of 200 Hz. A programmable-logic-controller (PLC) is used for supervisory control functions.

2.2 *Computer Model*

A computer model of the press was built using a commercially available software package. The from first principles of hydraulics and mechanical dynamics. Parameters were obtained from OEM data sheets and experimental tests. The control computer was modeled assuming 12-bit DAC's and sample period computational delay. The simulation model of the current control law was built from the original control computer source code. Further details and validation of the model are described elsewhere (Frazier, *et al.*, 1997). A proposed controller redesign using a modified PID-type control is described in a separate publication (Frazier and Medina, 1998).

3. ROBUST CONTROL LAW DESIGN

3.1 *General Strategy*

The general design strategy adopted for the new control law was to constantly adjust the "sensitivity" of the valves using feedback from the differential manifold pressure and the magnitude of the velocity error and then calculate the change in the valve commands based upon this sensitivity factor and the velocity error. A diagram illustrating this strategy is given in Fig. 3.

To achieve a good balance among the valves based on the total flow demand a cascading strategy was employed. If the demand on the high bandwidth small valve is outside a specified range, then the command to the large valve is change so that the requirements from the small valve are brought into line. In a similar fashion if the demand from the large valve is outside of its specified range, the command to the throttle valve adjusted. This strategy provides for a means of keeping the small valve out of saturation, thereby maintaining the effect of high bandwidth, high loop gain, and accuracy over the entire range of operating speeds and loading conditions.

3.2 *Fuzzy Logic-Based Control Law Design*

Using the strategy described above a fuzzy logic-based control law was designed. The universes of discourse for the inputs, differential manifold pressure (head pressure minus ram pressure) and velocity error, are shown in Figs. 4 and 5. They are defined as

$$V_{err} = V_{des} - V_{est} \qquad (1)$$

and

$$\Delta P = P_{head} - P_{ram}. \qquad (2)$$

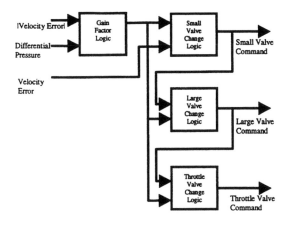

Fig. 3. General control strategy

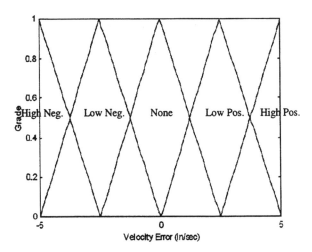

Fig. 4. Velocity error universe of discourse

Table 1. Gain factor consequence matrix

|Velocity Error|	ΔP		
	Low	Medium	High
None	3	2	1
Low Pos.	4	3	2
High Pos.	5	4	3

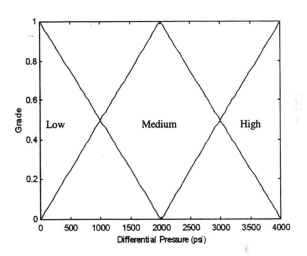

Fig. 5. Differential pressure universe of discourse

The universe of discourse for the intermediate variable, the sensitivity factor, is shown in Fig. 6. The consequence matrix relating the absolute value of the velocity error and the differential pressure is given in Table 1 where 1 represents the lowest gain factor and 5 the highest.

The universe of discourse for the small valve change is shown in Fig. 7. The universes of discourse for the large valve change and the throttle valve change are similar, except for the voltage ranges and are not given here.

As described previously, the large valve change and the throttle valve change are driven by deviations from a nominal setting of the small valve command and large valve command, respectively. The universe of discourse for the so-called valve command errors is the same in both cases, but is not given here. These "errors" are defined as

$$SV_{err} = 5.0 - SV_{cmd} \qquad (3)$$

and

$$LV_{err} = 5.0 - LV_{err}. \qquad (4)$$

Consequence matrices for the valve command changes are given in Tables 2-4. These values were

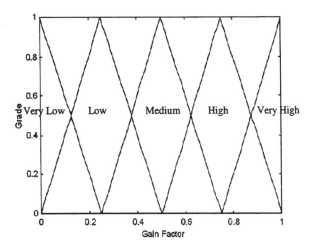

Fig. 6. Universe of discourse, gain factor

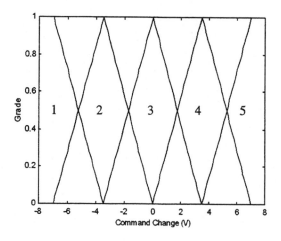

Fig. 7. Small valve command change universe of discourse

Table 2. Consequence matrix, small valve change

Velocity Error	Gain Factor				
	Very Low	Low	Med	High	Very High
High Neg.	2	1	1	1	1
Low Neg.	2	2	2	2	2
None	3	3	3	3	3
Low Pos.	4	4	4	4	4
High Pos.	4	5	5	5	5

Table 3. Consequence matrix , large valve change

Small Valve Error	Gain Factor				
	Very Low	Low	Med	High	Very High
High Neg.	3	4	4	5	5
Low Neg.	3	3	3	4	5
None	3	3	3	3	3
Low Pos.	3	3	3	2	1
High Pos.	3	2	2	1	1

Table 4. Consequence matrix, throttle valve change

Large Valve Error	Gain Factor				
	Very Low	Low	Med	High	Very High
High Neg.	4	4	4	5	5
Low Neg.	3	3	4	4	5
None	3	3	3	3	3
Low Pos.	3	3	2	2	1
High Pos.	2	2	2	1	1

manually determined through a systematic trial-and-error process.

4. SIMULATION RESULTS

In order to test the robustness of the new control law, several simulations corresponding to different velocity versus stroke and load versus stroke conditions were performed. Some of the simulation results illustrating the robustness of the control law are described below. The fuzzy logic functions were implemented using existing software tools (Beale and Demuth, 1994).

4.1 Constant Strain Rate, No Load

The press was given an exponentially decaying velocity versus stroke command without a forming load. The simulation results are shown in Figs. 8 and 9. Because of the initially large velocity error, the small valve immediately opens to its maximum value, causing the large valve to begin to open. Once the small valve detects the overshoot, it begins to decrease, while the large valve holds steady. When the small valve falls below its "desired" value, the large valve decreases appropriately. After the valves find their nominally correct range, the tracking is accurate out to the stopping point.

4.2 Constant Strain Rate, Disk Forging

In this second example, an exponentially decaying velocity command was given, but the load profile

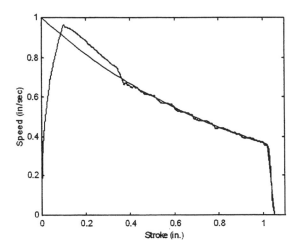

Fig. 8. Speed vs. stroke, no load case

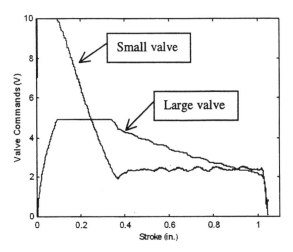

Fig. 9. Servo valve commands, no load case

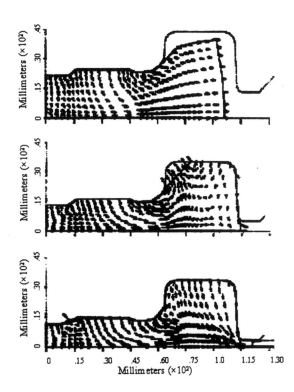

Fig. 10. Finite-element simulation

snapshots from a finite-element-based simulation of the forming sequence are shown in Fig. 10. The arrows correspond to the velocity field in the workpiece. The forge press simulation results for this case are shown in Figs. 11-13. In this case, the overshoot is much greater than in the unloaded case. However, this is primarily due to the coincidence that just as the press gets up to speed, the load on the press quickly levels out. This is because the flow stress of the material was reached at just the right (wrong?) moment. The plot of valve commands reveals a response similar to previous example. Figure 11 clearly reveals the velocity deviations that occur when the forging load increases rapidly. The rapid increase in load is due to the workpiece impacting the side of the die and the formation of flash as excess material is forced out of the die cavity.

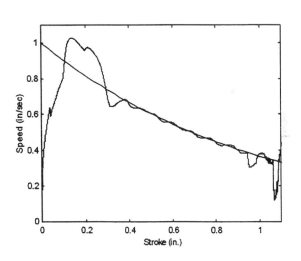

Fig. 11. Speed vs. stroke, disk forging case

4.3 Other Simulations

The control law was tested in a variety of other scenarios and was found to be very robust with respect to variations in loading profile, velocity command profile, and system parameter variations.

87

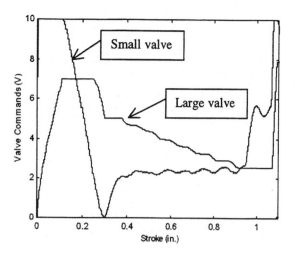

Fig. 12. Servo valve commands, disk forging case

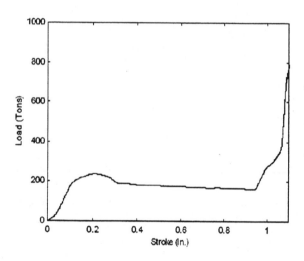

Fig. 13. Load vs. stroke, disk forging case

5. SUMMARY

The development of a robust fuzzy logic-based control law for a multi-valve hydraulic forge press has been presented. The resulting control law has been shown effective in handling a variety of forging scenarios without requiring adjustments that are typical in industrial practice. Application of modern control design techniques, such as fuzzy logic-based design, has much to offer to the control of heavy industrial equipment such as forge presses.

REFERENCES

Beale, M. and H. Demuth (1994). *Fuzzy Systems Toolbox*. PWS Publishing Company, Boston, Mass.

Frazier, W. G., E. A. Medina, J. C. Malas, and R. D. Irwin (1997). Modeling and simulation of metal forming equipment. *Journal of Materials Engineering and Performance*, **6**, p. 153-160.

Frazier, W. G. and E. A. Medina (1998), Modeling and control of metal forming equipment. *Transactions of the North American Manufacturing Research Institution of SME*, **26**, p. 55-60.

WAVELET PACKET ANALYSIS IN THE CONDITION MONITORING OF ROTATING MACHINERY

Kevin Bossley* R. Mckendrick* C.J. Harris** C.Mercer ***

* Parallel Applications Centre, 2 Venture Road, Chilworth, Southampton, SO16 7NP, UK.
**Image, Speech and Intelligent Systems Research Group, Department of Electronics and Computer Science, University of Southampton, S017 1BJ, UK
***Prosig Ltd, Link House, High Street, Fareham, PO16 7BQ, UK.

Abstract: In this paper, novel methods for performing condition monitoring for power station turbine shafts are presented. The objective of this work is to investigate methods for producing accurate turbine vibration fault alarms during turbine shaft rundowns. Wavelet packet analysis is employed to extract spectral features from healthy vibration signals and the probability density functions of these features are estimated. Both Gaussian models, using Bayesian inferencing, and mixture models are employed. Preliminary results show that the more computationally expensive mixture models produce more accurate density estimates and hence more reliable fault alarms. Copyright© 1998 IFAC

Keywords: Condition monitoring, conditional probability estimation and neural networks.

1   INTRODUCTION

Unscheduled shutdowns or abrupt failures of power station turbine shafts create significant economical losses. The work presented in this paper is directed at developing additional techniques for online monitoring of these systems, to minimise system downtime and maximise system and operator safety. The described methods are generic and are thus appropriate for monitoring many different types of rotating machinery.

A primary condition monitoring technique is to analyse the complex non-stationary vibrations signals, which are measured by accelerometers or velometers positioned on shaft bearing housing and by probes measuring the displacement of the shaft relative to the bearing housing. During operation the turbine shafts rotate at a constant speed (typically 3000 RPM). However, periodically the turbine shafts are rundown e.g. when power demands drop, for scheduled maintenance and so on. As a shaft decelerates normally unobservable vibration modes of the system, both rotational and structural are excited. This can exaggerate the effects of system

faults and, as a result, previously unobservable faults can be identified. This makes rundowns very important to the condition monitoring process. Similarly, as the unit is returned to service these phenomena may be observed as the unit runs up. Additionally, there may be other temporary faults occurring due to non-uniform temperature distributions as the unit starts up.

The initial aim of this work is to produce alarms that are automatically triggered by a faulty shaft. As illustrated in Fig. 1, the proposed system consists of two components, the data acquisition module, and the data analysis PC. The data acquisition module collects raw data, and ascertains the condition of the shaft. Under normal conditions, that is, when no faults are present the data acquisition module need only transmit a limited amount of information to the analysis PC. The details of this can be defined by the operator. However, if an alarm is triggered the time-series data is compressed using wavelet-packet compression and transmitted to the analysis PC. Alarms are then displayed on the analysis PC and the received data can be analysed in real-time and/or logged for off-line analysis. The use of wavelet

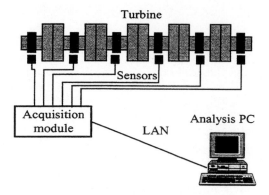

Fig. 1. The condition monitoring system.

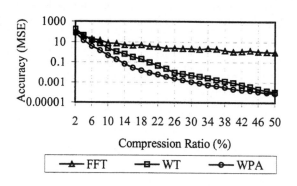

Fig. 2. Demonstration of the accuracy of wavelet packet analysis compression; FFT and WT are the Fast Fourier Transform and conventional wavelet transform, respectively.

based compression significantly reduces data volumes, thus saving valuable network bandwidth. Network traffic may well increase dramatically, during alarm conditions.

Many approaches can be employed to produce these alarms during shaft rundown. Generally, an aspect of the healthy system is modelled and significant discrepancies between expected and observed behaviour are used to trigger alarms. The emitted vibration signals may be modelled directly using data and knowledge driven non-linear speed-dependent models. Alternatively, a small number of representative features may be extracted from the vibration signals and data-driven statistical alarms may be constructed. The topics covered here are the use of wavelet-packet analysis to extract features, and the investigation of alarms based on density estimates.

Details of the extracted wavelet packet features are described in the following section. Alarms are constructed from estimated speed-dependent conditional probability densities for the extracted features. Two methods for performing probability density estimation are considered. Lastly, the method for generating the alarms from the probability density estimates is described. Examples are given as the various aspects of the system are introduced. For the purpose of this paper only one vibration signal (channel) is considered.

2 WAVELET PACKET ANALYSIS

Wavelet analysis decomposes a signal into a set of orthogonal basis functions, know as wavelets. These are local in both time and frequency, making them an ideal tool for analysing non-stationary signals. Wavelet packet analysis (Coifman, et al., 1992a; Saito, 1994) (WPA) is an extension to conventional wavelet analysis and decomposes a signal into sets of orthonormal subspaces that can each perfectly represent the signal. Furthermore, these subspaces are constructed by the recursive application of high and low pass FIR filters making them relatively rapid (cheap) to compute.

2.1 Wavelet Compression

Using energy estimates described by Newland (1993) and an appropriate selection criteria, a data driven algorithm, based on the work of Coifman and Wickerhauser (1992b), is used to optimally select a wavelet packet representation. As an example this algorithm is used to select the wavelet packet representation appropriate for compressing the vibration signals. The performance of the resulting compression is indicated in Fig. 2, where the accuracy of compressing vibration data produced from a typical rundown is demonstrated.

2.2 Wavelet Packet Features For Condition Monitoring

The algorithm used to select the optimal wavelet packet decomposition could be used to select features ideal for alarm generation as well as the discrimination between different faults and normal operation. However, alarms (and even fault diagnosis) based on individual wavelets is inappropriate for turbine shaft condition monitoring. Power station turbine shafts decelerate relatively slowly producing pseudo-stationary signals. Also, due to the large inertia of the rotating shaft transient behaviour such as that produced from faulty gearboxes is rarely observed.

These observations initially suggest that wavelet analysis may not be an appropriate method for extracting features from the vibration signals. However, WPA also performs efficient multi-resolution frequency decomposition. The WPA decomposition is produced by the recursive application of high and low pass filters; each followed by down-sampling operations. These filter and down-sampling operations can be augmented into convolution-subsampling operators H and G, which represent the high and low pass operations, respectively. The recursive application of these filters results in a binary tree, which can be represented by a rectangle of coefficients (Coifman, et al., 1992a), see Fig 3a.

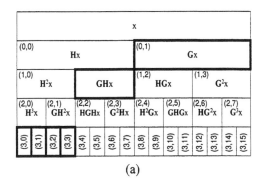

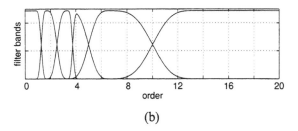

(a)

(b)

Fig. 3. Example wavelet packet decomposition, (a) shows the coefficient rectangle and (b) is an example spectral decomposition.

By the recursive application of the operators H and G, orthonormal basis are produced. Each basis represents the correlation of the original signal with a set of wavelets, and contains the energy in the original signal. There are 2^N possible orthonormal bases, the highlighted boxes in Fig. 3a represent an example.

Synchronously sampling (i.e. sampled at equal angles) is used when analysing signals from rotating machinery. This allows orders to be extracted from the signals, which represent the number of cycles per revolution and are thus ideal for representing speed-dependent vibrations.

During normal operating conditions nearly all the energy of the vibration signals emitted from the turbine shaft is contained in the first four orders. For this reason, conventional condition monitoring systems are based on observing these four orders. However, faults often manifest themselves elsewhere in the spectrum, e.g. oil film instability on the shaft produces sub-synchronous components around 0.46 order. Therefore, in this work the monitored features are chosen to be bands of the complete energy spectrum. For a gas turbine other "pulsations" derived from the burners may be significant. Other obvious examples of different frequency bands would be the blade passing frequency of a fan.

The chosen bands are those defined by the highlighted nodes of the WPA decomposition shown in Fig. 3a. To produce order bands, which are approximately centred at the first four orders, a sample rate of 40 samples per revolution is chosen. Using Beylkin wavelet filters (Beylkin, *et al.*, 1991) with 18 coefficients the order bands shown in Fig. 3b are produced. The Beylkin wavelet is chosen as it has a relatively fast cut-off and flat pass band, and a local frequency spectrum, when compared to many of the other popular choices of wavelet.

Example data for features 0 and 1 (which are approximately centred on the first two orders), produced from simulated turbine rundowns, are given in Fig. 4. As the shaft runs down, the vibrations excite two structural resonances, one centred at 30 Hz and another at 10 Hz. Therefore, as the various order components of the vibration signal run through these resonances the energy in the filter bands change accordingly. This is clearly shown in Fig. 4a, where the first order component excites the two resonances at 1500 and 500 RPM, respectively. As the filters are not exact (i.e. compact and centred on the orders), all four order components can be observed in the different features. For example in the given speed range the second order only excites the 30Hz resonance, at 750 RPM, but the first order exciting the resonances can also be observed.

More accurate order banks could be produced by using a FFT, and simply summing energies in the different order bands. However, cross-talk between order banks is still inevitable due to the resonances bandwidth and spectral smearing. The wavelet approach is computationally cheaper, and also leads onto a method for compressing the data. An FFT requires $O(N\log_2 N)$ operations, while the wavelet packet decomposition is linear with respect to the size of the signal.

3   PROBABILITY DENSITY ESTIMATION

For generating alarms each of the features is considered a deterministic random variable whose probability density function is estimated from healthy data. Due to the speed-dependent nature of the vibration signals these density functions are considered conditional on speed, that is the functions $p(f_i|s)$ are modelled, where $i=0,...,5$ and s is the speed. The identification of these functions is an off-line process, the results of which are used to produce on-line alarms.

There are many methods for modelling conditional distributions. This is a task of statistical inference. Two methods are considered in this paper, both of which draw on techniques from the statistical and neural-network literature.

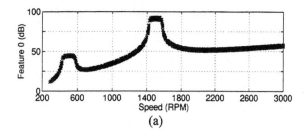

(a)

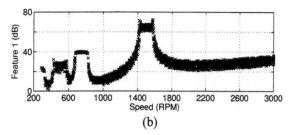

(b)

Fig. 4. Example feature data produced from healthy rundowns.

3.1 Gaussian Model

Here, it is assumed that each of the features is described by a Gaussian probability function given by:

$$p(f_i|s) = \frac{1}{\sqrt{2\pi\hat{\sigma}(s)}} \exp\left(-\frac{(f_i - \hat{\mu}(s))^2}{2\hat{\sigma}(s)^2}\right),$$

where the speed-dependent mean and variance are estimated from healthy rundown data. Here, we have chosen to estimate the mean by a B-spline generalised linear model (Bossley, 1997) given by:

$$\hat{\mu}(s) = \sum_{j=1}^{p} b_j(s) w_j$$

where $b_j(s)$ are the B-spline basis functions and w_j are the parameters or weights identified from the data. This model is particularly attractive, as its output is a linear function of the weights making weight identification a relatively trivial task. Also, B-splines are compact basis functions, producing a computationally cheap sparse representation, with little weight interference.

Conventionally, weights are identified using maximum likelihood estimation, which result in the least mean squares solution. However, this can produce estimates that generalise poorly. Bayesian inference, resulting in maximum a posterior (MAP) estimation, produces more robust solutions, and also a speed-dependent variance, $\hat{\sigma}(s)$.

A prior p.d.f. is placed on the weights which constrains the mean to be smooth. This successfully controls erratic function estimates, which typically characterise poor generalisation. The optimal weight estimate is identified by maximising the posterior p.d.f. From Bayes theorem this is given by:

$$p(w|D) = \frac{p(D|w,\beta)p(w,\alpha)}{p(D|\alpha,\beta)}$$

where α are β are hyper-parameters representing the reciprocals of the variances of the Gaussian prior and likelihood function, respectively. Finding a

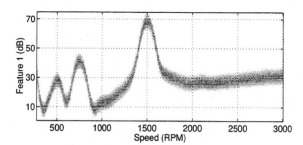

Fig. 5. Gaussian p.d.f. estimate for feature 1.

suitable balance between these hyper-parameters, and the resulting MAP weight estimate, is a non-linear optimisation problem. Fortunately, a set of relatively computationally inexpensive re-estimation formulae have been derived, which converge to consistent solutions (Mackay, 1991, Bossley, 1997).

As a result a Gaussian probability function for the feature is identified with a univariate B-spline model as the mean and with the variance given by:

$$\hat{\sigma}(s) = \beta^{-1} + \mathbf{b}(s)^{\mathrm{T}} \mathbf{H}^{-1} \mathbf{b}(s)$$

where $\mathbf{b}(s)$ is the p-dimensional vector of basis function outputs and $\mathbf{H}$ is the Hessian of the MAP cost function produced from the log of the posterior p.d.f. Bayesian estimation has been used by many authors; one of the pioneers was Mackay (1991) who successfully applied it to neural networks. The reader is referred to chapter 6 in Bossley (1997) for details of the priors and re-estimation formulae employed in this paper.

Using relatively flexible generalised linear models, consisting of 30 quadratic B-splines evenly spaced across the speed range Gaussian models for the 6 features were produced. This results in a piecewise quadratic estimate of the mean. The probability density function estimate for Feature 1 is shown in Fig. 5.

3.2 Gaussian Mixture Models

If the distribution of the features differs from a Gaussian, the Gaussian model described above will give a poor representation of the data and unreliable alarms will be produced. To model non-Gaussian densities more flexible models are required. A popular method is to use mixture models where the conditional distribution is defined by the weighted sum of Gaussian kernels (Bishop, 1995):

$$p(f_i|s) = \sum_{k=1}^{M} \alpha_k(s) \phi_k(f_i|s)$$

where

$$\alpha_k(s) = \frac{\exp(\hat{\alpha}_k(s))}{\sum_{j=1}^{M} \exp(\hat{\alpha}_j(s))}$$

are the mixing coefficients and the Gaussian kernels are given by:

$$\phi_k(f_i|s) = \frac{1}{\sqrt{2\pi\sigma_k(s)}} \exp\left(-\frac{(f_i - \hat{\mu}_k(s))^2}{2\sigma_k(s)^2}\right)$$

where

Table 1 Performance of the probability density estimates in terms of $J_t(w)$.

	Feature		
	0	1	2
Gaussian	1.92545	2.68126	3.08679
Mixture	1.78772	2.63506	2.70901

	Feature		
	3	4	5
Gaussian	3.17461	4.003	3.77392
Mixture	2.91649	3.97866	3.7157

$$\sigma_k(s) = \exp(\hat{\sigma}_k(s)).$$

As with the Gaussian model described in section 3.1, the various speed-dependent parameters ($\hat{\alpha}_k(s)$, $\hat{\mu}_k(s)$ and $\hat{\sigma}_k(s)$) of this model can be modelled by B-spline generalised linear models.

Maximum likelihood is the conventional approach to identifying these B-spline generalised linear models from the data. The likelihood is used to construct the following cost function:

$$J_i(\mathbf{w}) = -\sum_{t=1}^{T} \ln\{p(f_i(t)|s(t))\}, \qquad (1)$$

where t is the index to the T data pairs. The mixture model is found by minimising this function with respect to the parameters of the 3M B-spline models, which is a difficult non-linear optimisation problem.

An iterative algorithm similar to the EM-algorithm has been developed to identify these weights. Initially, the parameters of the mixture model are set to produce a Gaussian density estimation model, as described in the previous section. This is achieved by setting the mean and variance of the Gaussian kernels to be equal to the mean and variance produced for the Gaussian model. The mixing coefficients are set to be 1/M and the resulting mixture model represents a Gaussian model. The weights producing this Gaussian estimate are perturbed and an ad-hoc iterative training cycle is performed. Each iteration the parameters for the set of models (i.e. $\hat{\alpha}_k(s)$, $\hat{\mu}_k(s)$ and $\hat{\sigma}_k(s)$) are identified in turn. The Truncated Newton non-linear optimisation algorithm is used to identify these parameters. While one set of models is identified the others are held at their current optimal values, and the iterative cycle converges to a solution.

This iterative approach is employed, as the construction of the required Hessian for a complete mixture model is prohibitively computationally expensive. However, for reasonably sized models the iterative method is still relatively slow to converge, and is hence significantly computationally more expensive than identifying a Gaussian model. Also, maximum likelihood estimation is performed and as mentioned in the previous section this can produce ill-defined models. This could be improved by the introduction of Bayesian inference, but inevitably this would increase computational costs. Methods for improving the training speed of these mixture models and identifying these within a

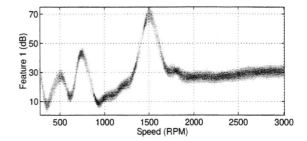

Fig 6. Mixture model p.d.f. estimate for feature 1.

Bayesian framework remain key areas for future work.

Mixture models consisting with 3 kernels and B-spline models with 30 evenly spaced quadratic basis functions to model the speed-dependent parameters, have been identified to estimate the feature p.d.fs. The resulting density estimate for feature 1 is shown in Fig 6, which can be compared with Fig. 5.

The probability density functions produced can be compared with the Gaussian models, by using the negative log likelihood cost function, equation (1). For 20 healthy rundowns, all independent of the training data, the results shown in Table 1 were produced. It can be seen that the mixture model produces lower cost functions for all the features, suggesting that these p.d.fs are a better match to the true densities.

4   ALARMS

Probability density estimates for the features during healthy rundowns can be used to construct alarms. At a given speed the p.d.f. for the feature is univariate, and the alarm should be triggered if the likelihood of the feature value falls below a given threshold. These thresholds are set by considering the probability of data falling in the resulting acceptance and rejection regions.

For example healthy features could be defined by a 99% probability threshold, and data falling in the most unlikely 1% of the density estimate are deemed abnormal (outliers). To reduce false alarms, a feature alarm will only be triggered if a number, say 10, consecutive features have fallen into the rejection regions.

Unfortunately, there is no closed form solution to finding these regions from the probability density estimates. A relatively cheap approximation for a Gaussian model is available, based on the inverse of the error function, but for mixture models this task is non-trivial. For example the Gaussian mixture models may be multi-modal. Initially, an iterative algorithm has been used to set these acceptance regions but a more efficient solution is required to guarantee truly real-time computation.

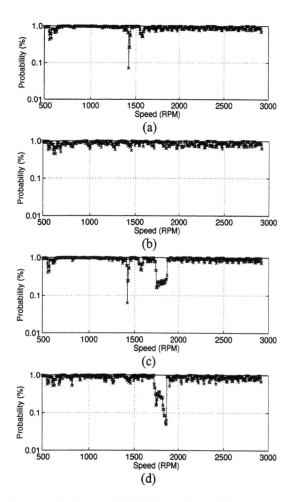

Fig. 7. Rejection probabilities for Gaussian and mixture density estimates, (a) and (b) are from the Gaussian and mixture alarms, for a healthy rundowns, while (c) and (d) for the same alarms for for a faulty rundown.

These alarms can be accessed by considering the rejection probability[1] required, to ensure an alarm is not triggered during a normal shaft rundown. The rejection probability for a rundown, produced from the Gaussian and mixture density estimates for feature 3 are shown in fig. 7a and fig. 7b. For the Gaussian model this rejection region threshold should be set considerably lower, making the resulting alarm less sensitive to faults.

This can be demonstrated by simulating a loose sensor mounting which resonants, with a small gain and narrow bandwidth, at 120Hz. This resonance is excited by the fourth order component at 1800 RPM. In fig. 7b and 7c this is clearly shown by the introduction of outliers around 1800 RPM. However, for the Gaussian model these outliers are greater than the threshold prescribed by the healthy rundown.

---

[1] Here, the rejection probability is defined as one minus the max probability of the last ten consecutive features. Feature probability is defined as the probability in the regions where the density is greater or equal to the density at the feature value.

5   CONCLUSIONS

In this paper preliminary results of employing WPA to extract spectral features from vibration signals suitable for condition monitoring are presented.

Irrespective of the type of features extracted from the vibration signals data-driven conditional probability density function estimation is required to ascertain the condition of the shaft. These features may not be Gaussian, and the more flexible mixture models, as demonstrated in this paper, can produce more robust probability density estimation. However, these improvements come at a price; mixture models are considerably computationally expensive, both for off-line identification and producing on-line alarms.

This work is part of a larger study into general condition monitoring, and alarm generation techniques. Using non-linear data-driven models of the vibration signals is one of the alternative methods. Recent results suggest that this method produces reliable alarms and is more appropriate for real-time operation. Naturally a comparative study based on real-time operational data is required.

ACKNOWLEGEMENTS

The authors are grateful to EPSRC for their financial support during the preparation of this work.

REFERENCES

Bossley, K.M. (1997). *Neurofuzzy modelling approaches in system identification*. PhD thesis, Faculty of Engineering and Applied Science, Southampton University, UK.

Bishop , C.M. (1995). *Neural Networks for Pattern Recognition*. Clarendon Press - Oxford

Coifman R.R., Y. Meyer, and M.V. Wickerhauser (1992a). Wavelet analysis and signal processing. In: *Wavelets and Their Applications* (M.B. Ruskai, G. Beylkin, R. Coifman, I. Daubechies, S. Mallat, Y. Meyer, and L. Raphael (Ed)), pages 153-178. Jones & Bartlett, Boston.

Coifman R.R. and M.V. Wickerhauser (1992b). Entropy-based algorithms for best-basis selections. *IEEE Trans. Info. Theory.* 38(2), pages 713-719.

Mackay, D.J.C. (1991). *Bayesian Methods for Adaptive Models*. PhD thesis, California Institute of Technology, Pasadena, California.

Newland, D.E. (1993). *An introduction to random vibrations, spectral and wavelet analysis*. Third Edition. Longman Group Limited, Essex, England.

Saito, N. (1994). *Local feature extraction and its applications using a library of Bases*. PhD thesis, Faculty of the Graduate School of Yale University.

Monitoring and Control of Rugate Filter Fabrication Using the Orthogonal Functional Basis Neural Network

Steven B. Fairchild, Y. Cao[1],
C. L. Philip Chen[2], Senior Member, IEEE
Steven R. LeClair

Materials Directorate, Wright Laboratory
Wright-Patterson Air Force Base, Ohio 45433

ABSTRACT

The Orthogonal Functional Basis Neural Network (OFBNN) is applied to the problem of monitoring and control of the rugate filter fabrication process. Rugate filters are thin film optical filters that are fabricated using Physical Vapor Deposition (PVD). During the deposition process, optical monitoring is used to generate spectral patterns which correspond to increasing thickness values of the film. The OFBNN is trained to predict the film thickness value associated with each spectral pattern. The results of this theoretical study demonstrate that the OFBNN can achieve significantly higher prediction accuracies than the simple curve fitting technique currently in use. *Copyright © 1998 IFAC*

1. INTRODUCTION

1.1 Fundamental Rugate Filter Concepts

A rugate filter consists of an optical coating having a sinusoidal refractive index profile deposited on a substrate [21]. This refractive index variation is continuous and periodic throughout the coating. Numerous identical cycles will yield one highly reflective band (Fig. 1). There are five independent parameters describing the sine wave refractive index profile (Fig. 2). These parameters are:

n_a = average refractive index
n_{pv} = peak-to-valley refractive index excursion
$n_a P$ = period of sine wave in optical thickness
θ = phase at the substrate in radians
N = number of sine wave cycles

The equation for sine wave refractive index profile, $n(x)$, is given by:

$$n(x) = n_a + n_{pv}/2 \; sin\{2\pi x/n_a P + \theta \} \quad (1)$$

where x is the distance in optical thickness, which is physical thickness multiplied the by

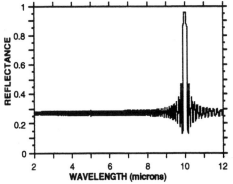

refractive index. The reflection band can be tailored in amplitude, bandwidth and wavelength

Figure 1. Reflectance spectrum generated by a rugate filter designed to reflect incident energy at a wavelength of 10 microns. The peak height is determined by the number of cycles in the refractive index profile, and the peak width is determined by the peak-to-valley refractive index excursion.

[1] Ph.D. Student, Dept. of Computer Science& Engineering, Wright State Univ., Dayton, Ohio 45435
[2] NRC Fellow, Department of Computer Science& Engineering, Wright State Univ., Dayton, Ohio 45435

location, with higher order reflection bands being absent. The wavelength that spectrally locates this reflection band is determined by:

$$\lambda = 2 \, n_a \, P \qquad (2)$$

Therefore, the location of the reflection band is determined by the average value of the refractive index and the length of the period in the refractive index profile.

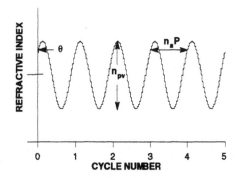

Figure 2. Fundamental rugate filter parameters. θ is the initial phase angle, n_{pv} is the peak-to-valley refractive index excursion, n_aP is the period. The reflectance peak produced by this refractive index profile is determined by the equation $\lambda = 2n_aP$.

1.2 Physical Vapor Deposition of Rugate Filters

Physical vapor deposition (PVD) is a thin film deposition process in which a target material vaporized from a source is transported through vacuum to a substrate where it condenses and film growth takes place. Target vaporization can be induced by either a thermal source or an electron beam source. A thermal source utilizes resistive heating to accomplish vaporization, where an electron beam source uses a high flux electron gun. Shutter control is used to control the amount of vapor released and thus the amount of the material deposited. Optical monitoring can be used to determine the thickness of the film as it is being grown.

Making a rugate filter requires the ability to produce refractive index values that nature does not readily provide. PVD allows this to be accomplished by varying the ratio of two different materials with different refractive indices being simultaneously deposited. The maximum amplitude excursion of the sine wave is determined by the difference in refractive

index between the two materials. Continuous values between the maximum and minimum peak values are obtained by the monitoring and control of the deposition rates which determine the percent composition of the two materials. Limitations in monitoring and control capabilities present obstacles to the fabrication of narrow bandwidth, high reflectance filters.

2. PROCESS CONTROL OF RUGATE FILTER FABRICATION

A successful rugate deposition is determined by the system's ability to match the theoretical refractive index profile by continuously varying the deposition rate of each material. Thermal and electron beam sources are used simultaneously, one for each of the materials being depostied. In-situ sensors monitor the two deposition rates. The rate of the thermally driven material is held relatively constant, while the rate of the e-beam driven material is varied sinusoidally. The critical parameter that must be monitored for accurate control of the e-beam source is the optical thickness of the filter as it is being deposited. This thickness value, along with the detected deposition rate from the thermal source, is used to determine the required deposition rate from the e-beam source. These values must be correlated to produce the desired refractive index value (Figure 3).

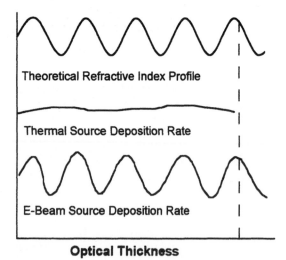

Optical Thickness

Figure 3. The theoretical refractive index profile of the rugate filter, and the deposition rates of the thermal and electron beam sources. Knowing the instantaneous values of the thermal and e-beam sources allows for the determination of the refractive index value of the material being deposited. Once the film thickness is determined, the ideal refractive index value from the theoretical profile can be compared with the instantaneous value and necessary adjustments can be made.

Monitoring the optical thickness of the filter as it is being deposited is difficult. A broad band spectral monitor is used to generate the real time spectral response of the filter as it is grown. A small region of the spectrum away from the main reflectance band is used for monitoring. At specific time intervals, corresponding to increasing values of film thickness, a spectral profile of the filter's optical response is obtained (Figure 4). Each succeeding film thickness value has a unique spectral profile based on the number of sinusoidal periods that have been generated in the film. A curve fitting routine is then used to predict the optical thickness from a section of the spectral response curve. However, this prediction technique is not very accurate. The standard deviation in the RMS error in the optical thickness prediction may be as high as 100 angstroms. This high prediction error is a limiting factor in fabricating rugate filters whose experimental performance match their theoretically predicted response.

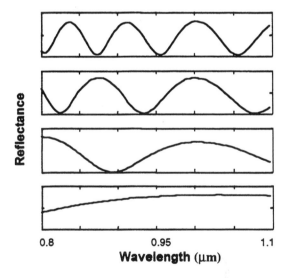

Figure 4. The theoretical spectral profiles of a rugate filter for different thicknesses as it is being grown. The spectrum changes considerably as the filter increases in thickness. Starting at the bottom, the spectra correlate to optical thicknesses of 0.5, 2.0, 3.5, and 5.0 microns

3. ORTHOGONAL FUNCTIONAL BASIS NEURAL NETWORK (OFBNN)

The orthogonal functional basis neural network (OFBNN) has been applied to the optical thickness prediction problem to see if a better theoretical fit can be achieved. The OFBNN utilizes an orthogonal transformation, subset selection technique and regularization to improve the generalization performance and computational tractability of a neural network.

Traditional methods of function approximation or regression involve a linear combination of the product of single variable or fixed basis functions (e.g., polynomial, spline, and/or trigonometric expansions). From Barron [17], the problem with traditional methods is that there are exponentially many orthonormal functions but unless all of these orthonormal functions are used in the fixed basis, there will remain functions that are not well approximated, i.e., the order of the squared approximation error is $1/n^{(2/d)}$, where n is the number of basis functions and d is the number of input variables. This problem is avoided by tuning or adapting the parameters of multi-variable basis functions to fit the target function as in the case of neural networks, wherein the order of the squared approximation error is $1/n$.

The biological origins of neural networks [18] as chronicled by Pao [19] established the multi-variable sigmoid as 'the' basis function for neural networks. Today the suite of multi-variable basis functions employed in neural networks is without bound, but the most commonly used are the sigmoid and radial basis functions. Radial basis function neural networks typically employ subset selection to identify a set of Gaussian basis functions. Broomhead and Lowe [4] have tried to choose such a subset randomly from the entire given set. In lieu of random selection, Rawlings [1] has proposed a systematic approach that employs forward selection to choose the subset that best explains the variation in the dependent variable incrementally. Based on this concept, Chen et al. presented an efficient implementation of forward selection using the orthogonal least square method (OLS) [8]. Subset selection can also be used to avoid overfitting by limiting the complexity of the network. From the literature, overfitting may be avoided when combining subset selection with other methods such as regularization [2,3], and as contributed by Mark Orr, combining OLS and regularization [9]. For the OFBNN, the subset is selected from a combination of functional basis evolved from a set of heterogeneous basis functions. Using this approach, all the information provided in the given data set is utilized, resulting in a more efficient neural network architecture. For a detailed explanation of the OFBNN, see reference [22].

4. SIMULATION STUDY ON RUGATE FILTER DEPOSITION DATA

The goal of the neural network is to reduce the variance in the optical thickness prediction error. Simulations were performed with a training set made from theoretical rugate filter deposition data. The refractive index of the rugate filter varies between 1.6 and 1.9, and the reflectance peak is centered at 700 nano-meters. The training set consists of 678 spectral patterns between 0.5 and 5.25 microns of optical thickness. Each pattern is a reflectance spectrum between 0.8 and 1.1 microns, and has an associated optical thickness value (Figure 4.). The spectra are sampled at 100 points. These spectral patterns are the input patterns to the neural network, and it learns to predict the optical thickness that corresponds to each spectrum.

The curve fit routine currently used for the optical thickness prediction during actual rugate filter fabrication is proprietary and not available for an exact performance comparison with the OFBNN. Therefore, a back propagation neural network was used for comparison with the OFBNN. Each network was trained on the 678 patterns consisting of 100 inputs (sampled spectra) and one output (optical thickness). For testing, gaussian noise, with a maximum amplitude of plus or minus 10% was added to the data to simulate actual conditions. Three testing sets were used. The first was the 678 training patterns with the added noise. The second and third had 977 and 1289 patterns with added noise. The second and third testing sets had more patterns over the same range as the training set, and provided insight into the interpolation capability of the trained neural network.

The back propagation network had sinusoidal functional enhancement nodes on the input layer to help capture the sinusoidal nature of the data. For each input, two more inputs were created consisting of the sine and cosine of the input value, increasing the number of inputs from 100 to 300. The following tables show the training and testing results for both networks.

Table 1. Training and testing errors for back propagation (BP) network with sinusoidal functional enhancements.

Training MSE Error = 1.9778e-07

Testing Error:

Number of Test Patterns	Max Error	Min Error	Standard Deviation
678	0.2794	-0.3103	0.0662
977	0.6814	-0.5245	0.0744
1289	0.3919	-0.3054	0.0706

Table 2. Training and testing errors for OFBNN network.

Training MSE Error = 1.1516e-07

Testing Error:

Number of Test Pattens	Max Error	Min Error	Std. Dev.
678	0.0031	-0.0068	0.0019
977	0.0058	-0.0497	0.0041
1289	0.0027	-0.0376	0.00014

The OFBNN produces a much lower prediction error than the multi-layer network running BP algorithm for each test case. Since the target value for the prediction is the optical thickness of the filter, the units for the error values can be microns or angstroms (0.0031 microns = 31 angstroms). For 678 test patterns, the standard deviation of the prediction error is 19 angstroms for the OFBNN net, and 662 angstroms for the BP net. For 977 test patterns, the standard deviation is 41 angstroms for the OFBNN net and 744 angstroms for the BP net. For 1289 test patterns, the standard deviation is 1.4 angstroms for the OFBNN net and 706 angstroms for the BP net.

Based on these results, the OFBNN network shows considerable promise for improving the optical thickness prediction performance that is needed for rugate filter fabrication. The standard deviation of the prediction error is considerably lower than values achieved with the current curve fit prediction method.

5. Conclusion

The fabrication of rugate filters with PVD presents challenging monitoring and control problems. The ability to accurately predict the optical thickness of the film as it is being grown is crucial to ability to produce very high reflectance, very narrow bandwidth filters. Currently, a curve fitting technique is used for predicting the optical thickness during fabrication. This technique produces a mean squared prediction error of approximately 100 angstroms, which limits the ability to produce filters whose reflectance spectrum closely matches that predicted by theory. With simulated data, the OFBNN has demonstrated a superior ability to predict the optical thickness of the film from the spectral patterns that are generated during film growth. A mean squared prediction error of less than 20 angstroms was achieved in two out of the three trials reported.

Due to proprietary issues with the manufacturer, making an exact performance comparison of the OFBNN with the curve fit routine currently being used during filter fabrication was not possible. Therefore, we used a multi-layer neural network with the back propagation algorithm for comparison with the OFBNN. The OFBNN utilizes an orthogonal transformation, subset selection technique and regularization to improve the generalization performance of a neural network. The OFBNN is not only more computationally tractable but also gives better generalization performance. The OFBNN demonstrates a superior performance over the back propagation network on the simulated data. The OFBNN offers the potential for more accurate control of the rugate filter fabrication process by improving the optical thickness prediction during film growth.

6. Acknowledgments

The Authors would like to acknowledge the pioneering work in OLS, ROLS, and RRBF methods by S. Chen and M. Orr and the Functional-Link Net by Y. H. Pao. We would like to thank W. H. Southwell of Rockwell Science Center for providing data as well as suggesting that neural networks be applied to the improvement of monitoring and control of the rugate filter fabrication process.

References

[1] Rawlings, J. O., 1988, *Applied Regression Analysis.*, Wadsworth & Brooks/Cole, Pacific Grove, CA.

[2] Barron, A. R., and Xiao, X., 1991, Discussion of "Multivariable adaptive regression splines" by J. H. Friedman. *Ann. Stat.* 19, pp. 67-82

[3] Breiman, L., 1992, *Stacked Regression.*, Tech. Rep. TR-367, Department of Statistics, University of California, Berkeley

[4] Broomhead, D. S., and Lowe, D., 1988, *Multivariable functional interpolation and adaptive methods.* Complex Syst. 2, pp. 321-355

[5] Park, J., and Sandberg, I. W., 1991, "Universal approximation using and radial-basis-function networks," *Neural Computation*, Vol. 3, No.2, pp. 246-257.

[6] Igelnik, Boris and Pao, Yoh-Han, 1995, "Stochastic Choices of Basis Functions in Adaptive Approximation and the Functional-Link Net", *IEEE Trans. on Neural Networks*, Vol. 6, No.6, pp. 1320-1328.

[7] Funahashi, K., 1989, "On the approximate realization of continuous mappings by neural networks," *Neural Networks*, vol.2, pp.183-192.

[8] Chen, S., Cowan, C. F. N. and Grant, P. M., 1991, "Orthogonal least squares learning algorithm for radial function networks," *IEEE Trans. on Neural Networks*, Vol.2, No.2, pp.302-309.

[9] Orr, M. J. L., 1995, "Regularization in the Selection of Radial Basis Function Centers", *Neural Computation.*, 7, pp. 606-623.

[10] Tikhonov, A. N. and Arsenin, V. Y., 1977, *Solutions of Ill-Posed Problems.*, Winston, Washington.

[11] Press, W. H., Teukolsky, S. A., Vetterling, W. T.,and Flannery, B. P., 1992, *Numerical Recipes in C*, 2nd ed. Cambridge University Press, Cambridge, UK.

[12] Golub, G.H., Heath, M., and Wahba, G., 1979, "Generalized cross-validation as a method for choosing a good ridge parameter," *Technometrics* 21 (2), pp. 215-223.

[13] Mackay, D. J. C., 1992. Bayesian interpolation. *Neural Computation* 4 (3), pp. 415-447.

[14] AllebmD, N., 1974, "The relationship between variable selection and data augmentation and a method for prediction," *Technometrics* 16 (1), pp. 125-127

[15] Berliner, L. M., 1987, "Bayesian control in Mixture Models", *Technometrics*, November 1987,Vol.29, No.,4 pp 455-460

[16] Chen, S., Chng, E.S., and Alkadhimi, K., 1995. "Regularized orthogonal least squares algorithm for constructing radial basis function networks," *International Journal of Control*, submitted.

[17] Barron,A.R. 1993. "Universal Approximation bounds for superpositions of a sigmodal function," ," *IEEE Trans. on Information Theory*, 39(3), pp.930-945.

[18] McCulloch,W.S. and Pitts,W., 1943. "A logical calculus of the ideas immanent in nervous activity," *Bulletin of Mathematical Biophisics,*. pp.115-133

[19] Pao.Y.H., 1996. "Memory based computational intelligence for material processing and design,"Wright Laboratory Technical Report WL-TR-96-4062. Wright Patterson AFB, OH, pp.1-14

[20] Allen,D.M, 1974. "The Relationship between variable selection and data augmentation and a method for prediction," Technometrics, Vol.16, No.1, pp.125-127

[21] Johnson, W. E., and Crane, R. L., 1993, *Introduction to Rugate Filter Technology.*, SPIE vol. 2046, pp. 88-108.

[22] C. L. P. Chen, Y. Cao, and S. R. LeClair, *Orthogonal Functional Basis Neural Networks*, submitted to *IEEE Trans. on Neural Networks*, 1998

BLACK-BOX MODELING OF A RAPID SAND FILTER

H.L.H. van Ginneken[*]   R. Babuška[*]   J.Th. Groennou[#]
J.W.N.M. Kappelhof[#]   H.B. Verbruggen[*]

[*]*Delft University of Technology, Faculty of Information Technology and Systems*
Control Engineering Laboratory, Mekelweg 4, P.O. Box 5031, 2600 GA Delft
The Netherlands, fax: +31 15 2786679, e-mail: r.babuska@et.tudelft.nl
[#]*KIWA Research and Consultancy, Groningenhaven 7, 3430 BB, Nieuwegein*
The Netherlands, e-mail: groennou@kiwaoa.nl

Abstract: Black-box modeling of a rapid sand filter is addressed. Two model structures that are suitable for black-box modeling are used: global nonlinear models of a polynomial type and local linear Takagi–Sugeno fuzzy models. The global models are built using the statistical significance of the regressors. The Takagi-Sugeno fuzzy models are built by fuzzy clustering. The two methods are compared in terms of accuracy and interpretability of the obtained models. *Copyright © 1998 IFAC.*

Keywords: Black-box modeling, fuzzy modeling, nonlinear regression, fuzzy clustering, filtration, water purification.

1. INTRODUCTION

This paper is devoted to black-box identification of a rapid sand-filtration process. A regression model has been built from data in order to predict the iron concentration in the effluent as a dynamic function of the filtration rate. The aim is to use the obtained model in a model-based predictive controller. By proper control, the iron concentration can be reduced which improves the water quality and reduces the maintance costs of the water distribution system.

The article is organized as follows: Section 2 describes the different types of nonlinear black-box models used in this application. They are divided in global nonlinear models of the polynomial and exponential type, and local linear fuzzy models of the Takagi-Sugeno type. The main steps of the model-building procedure, i.e., the selection of regressors and the estimation of parameters are addressed. Section 3 presents the application to the rapid sand filter. The choice of the relevant inputs, and the regressors selection are described. The obtained models are evaluated in terms

of their prediction accuracy and interpretability. Section 4 concludes the paper.

2. MODEL REPRESENTATION

Many multiple-input single-output (MISO) nonlinear dynamic processes can be represented by (Ljung and Glad, 1994) a regression model

$$y(k) = f(\mathbf{x}(k)), \qquad (1)$$

where the regression vector is given by

$$\mathbf{x}(k) = [u_1(k-1)\ldots u_1(k-nu_1)\ldots u_m(k-1)\ldots$$
$$u_m(k-nu_m)\ldots y(k-1)\ldots y(k-ny)] \quad (2)$$

The dynamical order of the inputs and output is given by nu_i and ny. The function $f(\cdot)$ is usually unknown and in black-box modeling is approximated by some flexible structure. In this paper, the following two model structures are used:

1. *Global models.* These models are typically linear in the parameters, but the regressors are nonlinear

functions of the inputs, such as products, powers or exponents. The parameters can be estimated by least-square methods.

2. *Local linear models.* These models approximate the process as a collection of several local linear models. In this paper, a fuzzy mechanism decides, depending on the regression vector, which models to combine to generate the output.

2.1. Global models

Polynomial and exponential models are considered. In polynomial models the output y depends on the products and powers of the elements of x, such as in the following example:

$$y = a_1 x_1 + a_2 x_1^2 x_2 + a_3 x_2 + a_4 x_1^2 + a_0. \quad (3)$$

Here, x_1 and x_2 are the elements of the input vector x, and a_j are parameters. An exponential model may look like:

$$y = a_1 e^{x_1} + a_2 x_2 + a_3 e^{-x_3} + a_0. \quad (4)$$

With these structures, the identification is simplified to the selection of regressors that give the most information about the process. Four standard selection methods are (Draper and Smith, 1966):

1. Evaluation of all possible models. This procedure involves fitting all possible model equations to the measurement data, followed by the selection of the model that meets some selection criteria best. It is clear that the computational load for this exhaustive search is very high, as the number of models to be evaluated is 2^n, with n being the number of regressors. Note that the number of regressors is much larger than the number of inputs.

2. Backward elimination. In backward elimination, all possible regressors are initially incorporated in the model. Then, based on the statistical significance, regressors are sequentially deleted from the model. The significance is assessed by the partial F-test:

$$F_i = \left(\frac{a_i}{s(a_i)} \right)^2, \quad (5)$$

where a_i is the least squares estimate for the i-th parameter (see equation 24) and $s(a_i)$ is the corresponding standard deviation. The lowest F-value is compared to the F-value corresponding to a given significance level α, which is usually between 0.01 and 0.1. If the regressor is not significant, it is deleted. If the lowest F-value is still significant the model is complete. This method is faster than the evaluation of all possible models, but it requires the solution of large regression problems.

3. Forward selection. Forward selection starts with the smallest possible model (a constant), and adds sequentially extra statistically significant regressors to the model. The regressor which, in combination with the regressors already incorporated in the model, minimizes the error of the model output is selected. If the partial F-test for this regressor is above the significance level α, the regressor is added to the model. If not, the procedure terminates as no new regressors can be added. This procedure is faster than the previously introduced procedures, and it avoids the solving of regression problems that are larger than necessary. However, the effect of a newly introduced regressor on the ones added earlier ones is not taken into account, which can introduce some redundancy.

4. Stepwise regression. To overcome the drawback of introducing redundancy in forward selection, stepwise selection re-examines the regressors introduced earlier. Like forward selection, it starts with a model containing only a constant term. Then, regressors whose significance is above the level α_i are added, but also the significance of the regressors inserted earlier is examined. If the significance of a parameter drops below level α_r, the corresponding regressor is removed from the model. The significance level α_i for the insertion of regressors is not necessarily equal to the level α_r for rejection, but mostly this is the case.

This selection procedure has the simplicity and speed of forward selection as well as the ability to reject regressors which have become insignificant, like the backward elimination procedure does. One drawback of the stepwise selection is that it selects all significant regressors, which can lead to a large number of regressors. For some objectives, restrictions may exist on the maximum number of regressors. For these situations, a slight modification of the procedure is introduced. The procedure is like the normal stepwise regression, except that if a regressor is added and the maximum number of regressors is exceeded, the regressor with the lowest partial F-value is rejected, even though its significance may be above α_r. After this, a new regressor is selected and the significance levels are calculated. All regressors that are not significant are rejected and if all regressors are significant, the regressor with the lowest significance is deleted from the model. This procedure continues until the newest regressor does not provide a significant improvement of the model or until there are no regressors left to be examined.

The rejection of a significant regressor can be advocated as follows. The F-value of parameter a_i is a measure for the augmentation of the model error if regressor x_i is rejected. For parameter a_i, the Sum of Squares due to a_i is defined as follows

$$SS_{a_i} = SS_a - SS_{a \backslash a_i} \quad (6)$$

where

$$SS_a = \sum (y - \bar{y})^2 \qquad (7)$$

and

$$SS_{a \setminus a_i} = \sum (y^* - \bar{y})^2 \qquad (8)$$

In these last two equations y represents the model output for the model containing all parameters in a, and y^* represents the model output for the model without a_i, but with the other parameters in $\mathbf{a}$. The mean square error due to a_i is given by

$$MS_{\beta_i} = \frac{SS_{\beta_i}}{df} \qquad (9)$$

where df is the number of degrees of freedom, which is equal to 1 as only the i-th parameter is considered. A similar definition with respect to the error of the model output is given in the following equation

$$SS_\varepsilon = \sum (y_m - y)^2 \quad , \quad MS_\varepsilon = \frac{SS_\varepsilon}{n - p} \qquad (10)$$

where y_m is the measurement data, n the number of data samples and p the number of parameters in $\mathbf{a}$. By definition,

$$F_{a_i} = \frac{MS_{a_i}}{MS_\varepsilon}, \qquad (11)$$

which makes it possible to conclude that by deleting the regressor corresponding to the lowest F-value, the model error will increase the least.

2.2. Local linear models

Another method for building nonlinear black-box models is the piecewise linear approximation of the function by local linear submodels. An advantage of this model is the simple structure, which makes it easy to understand and interpret the model. This paper deals the Takagi-Sugeno fuzzy models (Takagi and Sugeno, 1985) which consist of fuzzy *if-then* rules in the following form

$$R_i : \textbf{If } \mathbf{x} \text{ is } A_i \textbf{ then } y_i = \mathbf{a}_i^T \mathbf{x} + b_i, \quad i = 1, 2, \ldots, K \qquad (12)$$

Here, R_i is the i-th rule of the total set of K rules, $\mathbf{a}_i$ and b_i are the consequent parameters and A_i is the fuzzy set for the i-th rule. A_i is defined by the membership function

$$\mu_{A_i}(\mathbf{x}) : \mathbb{R}^p \to [0, 1] \qquad (13)$$

with p being the number of inputs of the model. For a TS model according to (12) and (13), the output of the model is given by:

$$y = \frac{\sum_{i=1}^{K} \mu_{A_i}(\mathbf{x}) \, y_i}{\sum_{i=1}^{K} \mu_{A_i}(\mathbf{x})}. \qquad (14)$$

Building a TS model from data is the approximation of a nonlinear problem by a collection of local linear models. The identification of the local models is basically divided in two steps:

1. *Constructing the antecedent membership functions $\mu_{A_i}(\mathbf{x})$.* This step consists of finding the location of the local models. In this paper, fuzzy clustering of the product space in hyperplanar clusters is applied (Babuška, 1998).

2. *Estimating the consequent parameters $\mathbf{a}_i$ and b_i.* After the membership functions have been constructed, the consequent parameters of the local models can be estimated from the data by least-squares methods.

Because of the lack of space, the above method cannot be detailed here. For details, the reader is referred to (Babuška, 1998).

3. MODELING A RAPID SAND-FILTRATION PROCESS

Filtration of groundwater is the process of purifying water, whereby it slowly flows through a porous substance. During this passage, the water quality improves as impurities are removed.

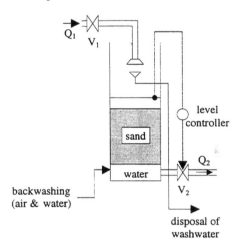

Fig. 1. Schematic diagram of a rapid sand filter.

A schematic diagram of a rapid sand-filter is given in Figure 1. A filter generally consists of a large tank with a double bottom. The part above the upper bottom is filled with a layer of sand of about 1.5 meters, which is called the filter bed. Using valve V_1 the amount of water that is sprayed onto the filter is set. The level of water above the filter bed is kept constant by the level controller. In the filter bed, oxidation of iron takes place as the water flows slowly, typically 3 to 8 meters per hour, from the top to the bottom of the filter. The oxidized flakes accumulate on the grains and in the openings between the grains. The purified water flows through fine orifices in strainers in the upper bottom to the space in between the two bottoms. After some time, typically 24 to 72 hours, the filter bed gets clogged, the resistance increases and the filter needs to be cleaned. This is done by *backwashing* the filter, which means that water and air are pumped through the filter from bottom to top at high speeds, carrying the impurities out of the filter bed. The drain is placed below the sprayer and is closed during filtration.

A measure of the quality of the effluent is the iron concentration which is influenced by

- the filtration rate, i.e., the flow rate of the water through the filter;

- the runtime, i.e. the time that has past since the last backwashing procedure;

- the backwashing procedure;

- the iron concentration in the influent.

As a high iron concentration in the effluent causes colored water and possibly blockage of the water distribution system, it is recommended to keep the iron concentration as low as possible. If the iron concentration between two backwashing procedures needs to be minimized, the filtration rate is the only control variable. If one can predict the performance of the filter as a function of the filtration rate and the runtime, it is possible to decide on a strategy for the filtration rate during the run, based on the demand prognosis.

3.1. Developing the models

Experiments were designed in order to obtain measurements of the filtration rate and the turbidity of the effluent. To be able to describe the filtration process according to (1), the elements of the regression vector (2) need to be determined. In Figure 2, a small set of data is plotted. It can be seen that the filtration rate and the turbidity are to some extent correlated. Every change in the filtration rate is followed by a turbidity change about 10 to 15 minutes later. Also the size and the direction of the filtration rate change play an important role. The large positive change at t=50 results in a steep change in turbidity (with an overshoot), whereas the turbidity slowly reacts to the negative change at t=95.

The filtration process is a dynamical process. These dynamics are manifested by the difference in effect filtration rate changes from t= 80 to t=125 have on the turbidity. Two similar successive rate changes show a difference in the height of the turbidity peak. For the modeling purposes, the state of the filter is assumed to be given by the amount of accumulated iron flakes in the filter bed. This means that it depends on the amount of iron supplied in the influent and the amount of iron that has passed the filter and entered the effluent. Hence, the state of the filter can be described as follows

$$State \sim \left(\alpha \int v \cdot dt - \beta \int T \cdot v \cdot dt \right) \quad (15)$$

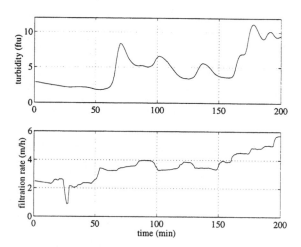

Fig. 2. Sample of data to visualize the correlation between the filtration rate and the turbidity.

where T is the turbidity of the effluent, v the filtration rate, α the iron concentration in the effluent and β the linear relation between the turbidity and the iron concentration in the effluent. As α and β are unknown, only the integrals are used as model-inputs. By using the past values of the turbidity, a regressive model is obtained with the regression vector (2) given in Table 1.

Table 1 Elements of the regression vector.

Input variable	Symbol	
Turbidity of the effluent	T	
Filtration rate	v	
Filtration rate change	dv	
Amount supplied iron	Fe_{in}	$= \int v \, dt$
Amount discharged iron	Fe_{out}	$= \int T \cdot v \, dt$

As the cross-correlation between the filtration rate and the turbidity has its maximum value of 0.25 when the turbidity lags 11 minutes behind the filtration rate, the process can be written as follows:

$$T_t = f\left(T_{t-11}, v_{t-11}, dv_{t-11}, (Fe_{in})_{t-11}, (Fe_{out})_{t-11}\right) \quad (16)$$

where T_t is the turbidity for time t. Note the large lag for the turbidity. Even though a smaller delay would provide a better auto-regressive component, the expected error resulting from more iterations for a simulation justifies the above choice.

Having identified the input variables, the methods described in Section 2. can be applied to build a model from measurement data. To train the model, the data set shown in Figure 3 is used. As the experiments, during which these measurements were taken, were also meant to examine other aspects of the filtration process, the data cannot be used for validation. Because of extreme changes in the backwashing procedure, the process parameters of two successive filter runs are not similar, which makes the validation of the model difficult.

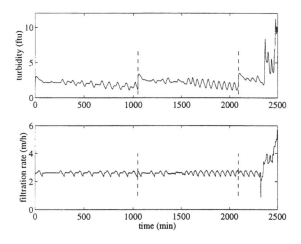

Fig. 3. Plot of the training data.

In Figure 3, the turbidity in the effluent and the corresponding filtration rate during a period of about two days are shown. The backwashing moments are removed from the regressors and the regressand. These moments are given by the vertical dashed lines.

3.1.1 Global model

As discussed in Section 2.1., building a model is reduced to the selection of the best regressors, once the type of regressors is known. Using little knowledge of the process, one can decide what type of regressors are more likely to be able to describe the process. Once this has been done, the regressors are formed and the best are selected. In this article, only the polynomial model is discussed. Comparable models, such as exponential models can be derived in a similar manner.

Table 2 Results of the modeling by polynomials.

RMS	used regressors				
27.0	offset	x_2^2			
20.5	offset	x_2^2	$x_2 x_4$		
19.7	offset	x_2^2	$x_2 x_4$	$x_3 x_5$	
19.1	offset	x_2^2	$x_3 x_4$	x_1	x_4
19.0	offset	x_2^2	$x_3 x_4$	x_1	x_4	x_3^2
where $\mathbf{x}=\begin{bmatrix} T & v & dv & (Fe_{in}) & (Fe_{out}) \end{bmatrix}$					

The results of the selection of polynomials by the stepwise regression procedure are given in Table 2. The Root Mean Square value (RMS) is used as a measure for the error of the model with respect to the measurements. The model was built by selecting the most significant regressors from the superset of regressors, which consists of the input variables (as given in equation 16) and their powers and products. Models containing 1 to 5 regressors plus a constant factor were

developed. In Table 3, the values and variances of the parameters of the most elaborate model are given. In Figure 4, the one-step head prediction of the model as well as the measurement data are displayed for two specific time-intervals. The upper plot is an enlargement of an interval during which the filtration rate was more or less constant. The lower plot displays the interval with large filtration rate fluctuations. Note the difference in scale, both in time and magnitude, in the two plots.

Table 3 Values and variances of the parameters of the model containing 5 regressors and an offset.

regressor	value	variance
offset	-4.76e-01	8.99e-04
x_2^2	1.65e-00	6.11e-04
$x_3 x_4$	-1.01e-02	5.22e-07
x_1	1.47e-01	1.82e-04
x_4	-7.66e-04	4.93e-10
x_3^2	6.81e-00	9.10e-01

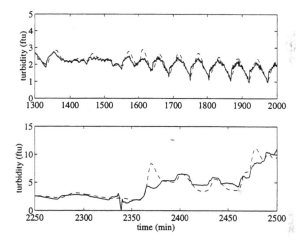

Fig. 4. Output of the polynomial model with 5 regressors for the training data (solid line: model output, dash-dotted line: training data).

As one can see in Figure 4, the model is capable of accurately predicting the turbidity for the interval during which the variation of the filtration rate is relatively small. With only 5 regressors and an offset, the model cannot give an accurate prediction for situations during which the filtration rate heavily changes. The parameter value for T_{t-11} is only approximately 0.15, which indicates that the delay may be too large for the auto-regressive input, which was predicted in Section 3.1.1.

3.1.2 Local linear model

Building a TS-model starts by locating the fuzzy clusters in the 6 dimensional product-space of the output variable T_t and the 5 input variables $T_{t-11}, v_{t-11}, dv_{t-11}, (Fe_{in})_{t-11}$ and $(Fe_{out})_{t-11}$. Knowing these locations, it is possible to construct the antecedent membership functions. Using these, the local linear functions for T_t depending on the inputs can be obtained. In Figure 5, the effect of the choice of the number of clusters on the RMS value of the model is shown.

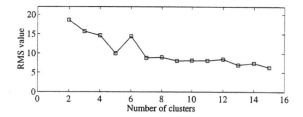

Fig. 5. RMS value as a function of the number of clusters.

This figure shows that, in general, a larger number of clusters results in a better model. However, by constantly increasing the number of clusters, the improvement of the error decreases. This justifies the choice of 5 clusters to build a model. For this model the RMS value is about 9.5, which is a half of the RMS for the polynomial model. In Figure 6, the results of the modeling are shown. The parameters of the TS-model are obtained by a global least-squares estimation, as this estimation compensates the interaction between the clusters and provides the most accurate overall prediction. The parameters of the local models and the centers of the clusters are shown in Tables 4 and 5, respectively.

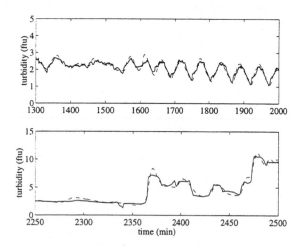

Fig. 6. Output of the TS model with 5 clusters for the training data (solid line: model output, dash-dotted line: training data).

Figure 6 shows that, like the polynomial model, the TS model can give a accurate prediction for the stable filtration rate, but it also shows that the model is capable of accurately predicting the turbidity as a result of large filtration rate changes. As this model consists of local submodels, it is capable of predicting the turbidity for different circumstances. From Table 5, it

can be concluded that cluster 1 is used to describe the filter for the time just after a backwashing procedure. Cluster 2 and 3 describe the filter during a relatively constant filtration rate and cluster 4 and 5 describe the filter during large filtration rate changes. Note the insensitivity to the filtration rate of cluster 1 in Table 4.

Table 4 Consequent parameters for the TS model.

	T	v	dv	Fe_{in}	Fe_{out}	offset
1	1.26e-1	0.60	3.80	2.32e-2	-9.30e-3	2.0
2	3.54e-1	3.47	-5.78	9.83e-3	-4.29e-3	-26.7
3	4.71e-1	5.45	-6.03	5.36e-4	-5.08e-4	-55.0
4	1.65e-1	7.13	-7.21	6.99e-2	-1.44e-2	-19.7
5	-2.67e-1	5.36	-7.90	4.80e-2	-8.34e-3	-10.3

Table 5 Centers of the 5 clusters of the TS model.

	T	v	dv	Fe_{in}	Fe_{out}
1	2.71	3.54	1.02e-3	1.58e2	0.44e3
2	2.36	2.60	1.59e-4	4.28e2	1.05e3
3	1.82	2.61	-7.13e-5	1.10e3	2.48e3
4	4.92	5.62	5.35e-3	4.62e2	1.49e3
5	6.96	4.24	6.78e-3	5.07e2	1.74e3

4. CONCLUSIONS

From the result of the modeling as given in Figures 4 and 6, one can conclude that the TS-model gives better predictions than the global model, especially when the filtration rate changes heavily. This is because this model has different linear submodels for specific situations. Also, the parameters of the local models can be interpreted more easily than those of a global nonlinear model. The available data, however, does not contain enough variation to be able to draw final conclusions. The filtration rate was more or less constant during most of the experiments, which makes it impossible to tell how the models perform if the filtration rate changes within the domain of interest. To obtain measurements during which the filtration rate changes more, new experiments need to be done.

REFERENCES

Babuška, R. (1998). *Fuzzy Modeling for Control.* Kluwer Academic Publishers, Boston.

Draper, N.R. and H. Smith (1966). *Applied Regression Analysis.* New York: John Wiley & Sons.

Ljung, L. and T. Glad (1994). *Modeling of Dynamic Systems.* Englewood Cliffs, New Jersey: Prentice-Hall.

Takagi, T. and M. Sugeno (1985). Fuzzy identification of systems and its application to modeling and control. *IEEE Trans. Systems, Man and Cybernetics 15*(1), 116–132.

DESIGN BY REIFICATION:
PROCESS-CENTRIC AND SYSTEM-CENTRIC MODELS IN PROCESS CONTROL

Robert G. Wilhelm, Jr.

Objective Control, 9637 Sunset Drive, Powell OH 43065 USA

Abstract: The structure of most process control software is only very tenuously related to the process it is designed to control. This has subtle negative consequences, both for process operation and for the evolution and maintenance of the software itself. A similar argument can be made for the tools used to configure and manage the hardware and software of the control system software and hardware. This paper discusses a framework for organizing control software around the orthogonal concepts of an "intelligent process" and an "intelligent control system." *Copyright © 1998 IFAC*

Keywords: Control-oriented models, Object modeling techniques, Model-based control, Architectures, Software engineering.

1. INTRODUCTION

The vast majority of process control systems extant today are conceived in terms of a "controller" which is attached to the process via sensors and actuators (Fig. 1). This model is based on the assumption that the process possesses the resources needed to produce its product except for one critical element: the intelligence and knowledge required to carry out the task. In this traditional model, the controller (often in conjunction with human operators) provides the missing intelligence. In effect, the controller serves as the brain of the process.

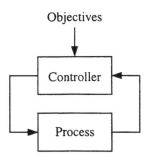

Fig. 1. Conventional model of control.

Every computer-based control system has a hardware architecture and a software architecture which largely determine how process control applications are implemented on that system. While the *functionality* of a process control application is (presumably) driven by the process it is designed to control, the *architecture* of traditional control software is decidedly not. Quite naturally, the architecture of software for process control is influenced primarily by three driving forces:

- the building blocks of control algorithms and logic from which it is to be constructed;
- the artifacts of the software technology with which it is implemented (procedures or objects, tasks or processes, data structures, etc.);
- and the hardware architecture of the system in which the software is executed.

The programming environments for modern process control systems manage to hide most of the software and hardware artifacts and provide high-level abstractions for the construction of control algorithms and logic. However, most control software bears only a very indirect relationship to that of the process it is designed for. In fact, except for the choice of names associated with I/O points, it is generally quite difficult to discern anything about the process from an examination of the control system software. Such an architecture can be called "control-centric."

Having made the analogy that the controller serves as the brain of the process, we note in passing that the architecture of the human brain also bears only a very indirect relationship to that of the body and organs which it controls.

The control-centric approach to process control has a subtle but significant influence on process operation. The concern of process operation is, by definition, to affect the way the process is performing – to obtain the desired results in terms of throughput, quality, and efficiency. Operators and process engineers achieve these goals largely through the control system by manipulating the setpoints and tuning parameters of the controllers and by logical switches which turn control loops on or off, or change their mode of operation. That is to say, process operation is achieved largely through indirect means – at arms length, so to speak, through the control system.

Ideally, if the control system is, in effect, the brain of the process, it should create the illusion that the operator is dealing with an "intelligent process." That is, the operator should be able tell the process what to do and how to behave directly. This illusion can be (and usually is) largely achieved by effective design of graphical user interfaces. The user interface, however, is merely a veneer on the underlying system. As a result, we have a user-interface (which looks like the process) and the real process itself sandwiching the control software which has an entirely different structure.

Reification is the act of treating something abstract as a material object. While the process itself is already physical, the "intelligent process" is not. Instead of imposing the control intelligence on the process from without, an alternative model for process control is to embed the control intelligence and process interface in a model of the process itself – most likely an object model. In such a model, each element of the process encapsulates the requisite intelligence to control its own functions. Such an architecture can be called "process-centric."

AI applications in process control often employ an object model of the process because such a model forms the basis for reasoning about the process elements and their relationships to each other. There is no reason, however, to limit this practice to AI applications. Even when the control logic is in the more traditional form of hard-coded algorithms, many benefits accrue from the intelligent process model. This has recently been recognized by the batch control community to the extent that it is being

made part of international standards for batch automation (ISA, 1995).

A control system designer must also decide on the configuration of hardware and software of the system. What types and how many processors are needed? What network topography is appropriate? Which processors will be responsible for which control functions? Such decisions take on even more weight when the system is not supplied by a single vendor. Support for the design, deployment, and management of these aspects of the system argues for a "system-centric" model.

Each of these models, control-centric, process-centric, and system centric has a role to play in control system design and operation. Because process technology, control technology, and computer technology evolve along completely different paths and timelines, it is desirable that the process-centric model, control-centric model, and system-centric model be orthogonal to each other. That is, to the extent possible, changes in one should not impact the others. This maximizes reusability of the elements of each model and minimizes the effort required when any of them is modified.

The remainder of this paper discusses a framework for effective integration of the process-centric, system-centric, and control-centric views to make process control systems easier to design, use, maintain, and evolve.

2. CONTROLLER AS INVERSE PROCESS

It is well known in control theory that the ideal controller is, *in some sense*, an inverse of the process. See, for example, (Morari and Zafiriou, 1989). We emphasize the qualifier "in some sense" because strict mathematical inversion is feasible for only a very limited class of systems. In other cases it may lead to instability, extreme sensitivity to noise, or may simply be unrealizable as in the inverse of a pure time delay. Nonetheless, every good controller represents some form of approximation to the inverse of the static and/or dynamic behavior of the process and its disturbances.

When we take into account the sensors, actuators, and I/O devices, the control loop of Fig. 1 can then be represented as in Fig. 2. We put "inverse" in quotes to emphasize that considerable flexibility is allowed in its definition.

2.1 Object Modeling

A well-established practice in object-oriented design is to construct a system from software components (objects) which represent elements of the real-world problem domain. This technique is widely accepted in both business and technical applications. There is a subtle point in object modeling, however, which is

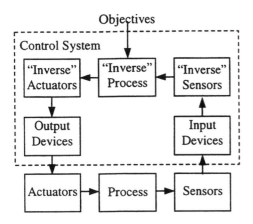

Fig. 2. Control as an inverse of the process.

seldom pointed out. If an object's purpose is to *simulate* or stand in for the real-world component it represents, its behavior must be designed to mimic that of its real-world counterpart. If, on the other hand, the object's purpose is to *control* the operation of its real-world counterpart, then, like any controller, the object's behavior must, in some sense, invert the behavior of the real-world entity.

We can easily conceive of the entire control system as an object model of the process it controls, where each element of the model represents some part of the process, and has responsibility for controlling it. The behavior of each of these elements is, in some sense, an inverse of the behavior of its real-world counterpart.

Starting with the sensors, a real-world sensor produces an electrical signal proportional (in the broad sense) to the value of a process variable such as pressure or temperature. The sensor object in the control system is the only component which should logically have the responsibility for knowing the characteristics of the sensor, and inverting them to produce an estimate of the process variable in engineering units, given the value of the electrical signal in, say, volts. An example of an industrial system extensively employing the object model in this way is discussed in (Gilbert and Wilhelm, 1993).

If, in the real process, a measurable product quality is affected by changes in certain process variables, the process object, acting as controller, determines appropriate control actions to regulate that product quality by using some form of inverse of the process characteristic.

2.2 Coordination and Levels of Abstraction.

The major advantage of the object model has little, if anything, to do with control, *per se*. Rather, it is the organization it imposes upon the software which implements the control strategy. As we do with most complex systems, we usually conceive of processes as being organized hierarchically into major systems containing subsystems, eventually containing the

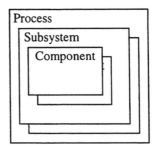

Fig. 3. Hierarchical organization of the process.

lowest level components of interest (Fig. 3). This works very well for organizing most control strategies.

In all but the simplest of systems some control actions require the coordination of actions by several elements. Some coordination may be achieved by direct communication between peer elements. More complex forms of coordination, however, must be performed at a different level of abstraction in the process model. The concept of a "system" or "subsystem" is, arguably, somewhat more abstract than that of a leaf-level physical component. In a boiler house, for example, a valve or a pump is quite concrete, but the feedwater system, which is a collection of many valves, pumps, tubes, and so forth, exists as an entity only because we define it to be so. Nonetheless, the actions required to increase the feedwater flow to a boiler may require coordination of actions on several valves, pumps, and other elements, which is easiest to associate logically with a feedwater system object.

Another attribute of the more abstract layers in a model is that a lower level component may belong to more than one higher level subsystem simultaneously. For example, the riser tubes in a boiler may be considered to be simultaneously part of the feedwater system and part of the steam generator system. A well-constructed object model provides a convenient way to model this "multiple containment," as each object has its own unique existence, yet it can be referenced by many other objects either as a peer or as a contained component.

2.3 Role of Control-Centric Models

The object model which reifies the intelligent process is neutral with respect to the control technology used within it. Each component merely acts as a container for a set of control capabilities associated with the corresponding part of the physical process. Its interfaces define its capabilities in terms of the information it can provide and the operations it can perform, but encapsulates and hides the implementation of those operations.

The object model, therefore, does not do away with the need for the constructs and building blocks used in control-centric models. It simply provides a process-centric structural framework within which to place the control. Inside of each of the objects in the process-centric model we expect to find control algorithms and logic constructed of building blocks associated with control-centric models. These may be simple, traditional algorithms such as PID, model-based control, rule-based systems, or whatever the designer deems appropriate.

Relationship to model-based control. The traditional control-centric model is often extended to a "model-based" structure, meaning that some kind of mathematical model is used to enhance the capability of the controller. There are a number of possible reasons for explicitly including a model of the process behavior, including:

- estimation and filtering, whereby the model can estimate state information which cannot be directly measured, or can estimate the true value of variables corrupted by noisy measurements;
- time-delay compensation, whereby the model can predict the process response to control actions before it can be measured;
- model reference control, whereby the "reference model" determines a feasible trajectory for a "controller" which is an inverse of the process dynamics.

It is important to realize that the model used in a model-based controller is not the controller itself, and is not an inverse of the process behavior, but a simulation or predictor of that behavior in the usual forward sense. In our process-centric structure, such a model will be contained within one of our inverse process components. In other words, it will be a model within the model.

A generalized description of a model-based controller is given in Fig. 4. From the traditional control-centric viewpoint, the composite of the model and control algorithm (everything within the dashed line in Fig. 4) is known as a "model-based controller." The controller output affects both the real process and its model. Process output (measurements) are fed into the model to update it,

and the controller acts on output from the model, rather than from the process measurements directly.

From the process-centric point of view, the entire traditional model-based controller (everything within the dashed lines in Fig. 4) would be encapsulated in the inverted process object model component which controls the corresponding process.

3. A PROCESS-CENTRIC CONTROL SYSTEM

A process-centric control system framework is currently under development for the control of submerged-arc furnaces used in the production of various alloys and metals. A highly simplified diagram of the furnace system is shown in Fig. 5. The furnace system may be conceived as being composed of two major interacting subsystems. One of these is the metallurgical subsystem, in which ore and other additives are converted into molten metal with the desired properties and, of course, some waste byproducts. The other major subsystem is the electrical system, consisting primarily of three electrodes which deliver energy to the furnace via plasma arcs to fuel the metallurgical reactions, and transformers which deliver the electrical power to the electrodes (Wasbø, 1996).

The electrodes are equipped with hydraulic positioning "heads" which can raise and lower them individually to control the arc, and with "slipping" systems which reposition the electrodes within the head to compensate for electrode consumption. The

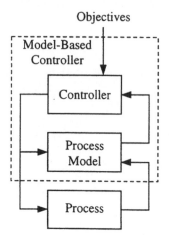

Fig. 4. Model-based controller.

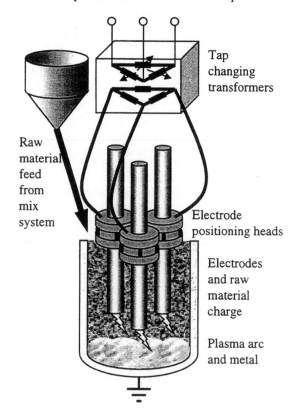

Fig. 5. Submerged-arc furnace process.

transformers are equipped with on-load tap changers to control the voltage delivered to the electrodes.

The control objectives vary according to operating conditions, but under normal operation one of the objectives is to maximize the production of metal by operating at the highest possible load subject to the electrical and thermal constraints of the furnace. Our purpose here is not to discuss the control strategies and their implementation, but to focus on the influence of the process-centric model.

Each important element of the process has a corresponding object in the model which is responsible for its control. For example, an electrode head object is responsible for controlling the positioning mechanisms which raise and lower the electrode.

As we indicated earlier, some of the coordination required among actions taken on the elements may be achieved by direct communication between peer objects. For example, "slipping" the electrode to account for consumption has the same effect on the process as lowering the electrode head. In order to keep the penetration of the electrode constant, the slipping mechanism informs the head of any moves it makes, so that the head can compensate by raising an equivalent amount.

More complex forms of coordination, however, must be performed at a different level of abstraction in the process model. For example, furnace load control involves coordinating all three phases of the transformers and electrodes. This takes place at the level of the "electrical system" object, which is the container of the electrodes and transformers.

The electrical system is a good example of reification. Although an obvious abstraction to any engineer, it is not a single physical entity, but a collection of component parts and their interconnections. Treating such a system as a concrete object, and endowing it with its own intelligence and behavior to augment those of its components, provides the most appropriate domain for higher levels of control which coordinate the functions of the lower level elements. In this particular case, the electrical system represents the level of the process at which the overall circuit topology is known, thereby allowing it to compensate for coupling between the phases and to coordinate the actions required of transformers and electrodes during load changes.

4. SYSTEM-CENTRIC MODELS

A completely different aspect of control system design is the configuration of the control system hardware and software itself. All but the simplest systems today involve multiple processors and a variety of I/O devices distributed over multiple networks. Control software and process I/O must be distributed so as to satisfy performance and reliability goals.

The hardware/software configuration is, of course, a complex system in its own right, and the equipment requires a certain amount of management and control itself. In a distributed system, modules may need to be brought in and out of service independently, and may support on-line reconfiguration.

Although all control systems provide tools for configuration, we assert that most of them suffer from the same malady that conventional control software does. Just as most control software lacks an explicit model of the process it controls, most system configuration and management tools lack an explicit model of the system they manage. In other words, the on-line management and control of the hardware/software configuration can benefit from the same approach that we have taken with the process-centric model. That is, the control of the software and hardware elements can best be handled by an object model of those elements themselves.

4.1 Relationship between system-centric and process-centric models.

The relationships among the process, the system, and their corresponding models can be depicted as in Fig. 6. We call this the "four quadrant pattern."

The real physical process interfaces with the control system via "transducers" – sensors and actuators (not shown in Fig. 6). Each such transducer is a two-ported device which is connected to both the process and the control system. On the process side, sensors respond to process variables and actuators affect process variables. On the system side, sensors produce signals which can be read by input hardware, and actuators respond to signals from the system generated by output hardware.

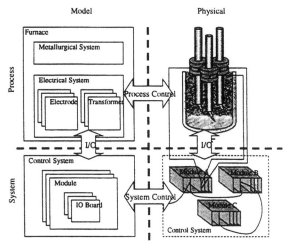

Fig. 6. The four quadrant pattern.

The elements of the system model are responsible for the control of their physical counterparts. In the I/O system this includes sampling inputs and causing the hardware to act upon outputs. The actual communication path between the process and its control model passes through the control system and its control model, but at a higher level of abstraction we may ignore this fact and pretend that the interaction between the process and its process-centric control model is direct.

4.2 Maintaining Orthogonality

The secret to maintaining orthogonality among various models and abstractions lies in defining the interfaces between them in an architecturally neutral way. Fundamentally, this means that one model should not contain assumptions about the structure of another model. The process model and system model need to exchange information, but they can do so without any dependencies upon each others' structure.

The dual-ported nature of transducers is conceptually critical to maintaining the orthogonality of the process-centric view and the system-centric view. The transducers are represented as objects on both sides of the Process/System boundary, but the interfaces of the objects on the process side are concerned only with process variables, whereas the interfaces of the objects on the system side are concerned only with the signals understood by the I/O system. These objects, representing the dual aspects of the transducers are associated with each other at run-time by virtue of a shared "tag name" – a unique identifier for each I/O point in the system.

The I/O system configuration associates a tag name with each physical I/O channel, which is a direct analog to the wiring of transducers to I/O devices in the physical system. The process model configuration associates a tag name with each sensor or actuator, which is a direct analog to the identification of a transducer in the physical process.

In the furnace control system presently under development, the basic design will be adapted to many furnace variations with many different types of I/O hardware and software. In some of these systems, process I/O is managed by SCADA software. While different SCADA systems have different application programming interfaces (APIs), it makes no difference to the control model. The I/O framework includes a "SCADA object" class which provides the standard neutral interface to the process model, and is easily subclassed to handle the API of any selected SCADA system.

5. CONCLUSIONS

Process engineers and process operators are interested in running a process, not running a control system. Ideally, the artifacts of both the control technology and the computer technology should be completely invisible to the process engineers and operators (Wilhelm, 1996). At the same time, control systems engineers can only serve the needs of these process people by making effective use of control technology and computer systems technology.

By encapsulating the control software within an explicit object model of the process itself, the modularity and objectives of the process are better preserved. Control strategies are more easily adapted to changes to the process, or to similar but slightly different processes. If interfaces between the process objects are well designed, different control technologies can readily be mixed and matched within the same system, as the object model is completely neutral to the control technology encapsulated by any given object.

Similarly, by encapsulating system configuration and management tools in an explicit object model of the computer system itself, systems become more understandable, and easier to maintain and evolve. Ironically, constructing an explicit model of the computer system makes it easier to rid the control application software of any artifacts of the computer system architecture.

ACKNOWLEDGEMENT

The author would like to thank Dr. Magne Fjeld, Adjunct Professor with NTNU in Trondheim, Norway for his contributions during discussions of this subject.

REFERENCES

Gilbert, J.W., and R.G. Wilhelm, Jr. (1993). "A concurrent object model for an industrial process control application." *Journal of Object-Oriented Programming* **7**, pp. 35-44.

ISA (1995). *Batch Control. Part I: Models and Terminology*. Standard ISA-S88.01-1995. The International Society of Measurement and Control (ISA). Research Triangle Park, North Carolina.

Morari, M., and E. Zafiriou (1989). *Robust Process Control*, ch. 4. Prentice-Hall, Englewood Cliffs, New Jersey.

Wasbø, Stein O. (1996). *Ferromanganese Furnace Modelling Using Object-Oriented Principles*. Doctoral Thesis, Division of Engineering Cybernetics, NTNU, Trondheim, Norway.

Wilhelm, R.G, Jr. (1996), "In search of the Cheshire cat: the invisible automation system paradox." *ISA Transactions* **35**, pp. 321-327.

VISUAL SURVEILLANCE AND TRACKING OF HUMANS BY FACE AND GAIT RECOGNITION

Ping S. Huang, Chris J. Harris and Mark S. Nixon

Image, Speech and Intelligent Systems research group
Department of Electronics and Computer Science
University of Southampton
Southampton SO17 1BJ, UK

Abstract: Increased emphasis on automated real time intelligent surveillance system has led to the need to identify and track people in complex environments. Independent features such as face, gait provide valuable clues as to identity, which coupled with data fusion and tracking algorithms offer a potential solution to this problem. In this paper we address the first problem of recognizing humans in real time, data fusion and tracking will be performed by neurofuzzy state estimators. A new approach which combines eigenspace transformation with canonical space transformation is proposed here. This method can be used to reduce data dimensionality and to optimize the class separability of different classes simultaneously. *Copyright © 1998 IFAC*

Keywords: Image matching, Pattern recognition, Statistical analysis, Eigenvectors, Data fusion, Tracking application

1. INTRODUCTION

Increased emphasis on automated real time intelligent surveillance system has led to the need to identify and track people in complex environments. Independent features such as face, gait provide valuable clues as to identity, which coupled with data fusion and tracking algorithms offer a potential solution to this problem. In this paper we address the first problem of recognizing humans in real time, data fusion and tracking will be performed by neurofuzzy state estimators (Harris and Wu, 1997).

Recently, (Niyogi and Adelson, 1994) distinguish different walkers by extracting their spatiotemporal gait patterns obtained from the curvefitting "snakes". (Little and Boyd, 1997) use frequency and phase features from optical flow to recognize different people by their gait. However, these feature-based methods, which use boundaries, lines, edges or optical flow are dependent on the reliability of the feature extraction process. Using the human shapes and their temporal changes in walking, (Murase and Sakai, 1996) propose a template-matching method which uses the parametric eigenspace representation applied in face recognition (Turk and Pentland, 1991) to recognize different human gait. For recognizing certain persons by their gait, this is a more feasible feature compared to those approaches above. Based on Principal Component Analysis (PCA), eigenspace transformation (EST) has actually been demonstrated to be a potent metric in automatic face recognition and gait analysis, but without using data analysis to increase classification capability.

Face image representations based on PCA have been used successfully for face recognition described in (Turk and Pentland, 1991). However, PCA based on the global covariance matrix of the full set of image data is not sensitive to class structure in the data. In order to increase the discriminatory power of various facial features, (Etemad and Chellappa, 1997) use Linear Discriminant Analysis (LDA) - also called Canonical Analysis (CA), which can be used to optimize the class separability of different face classes and improve the classification performance. Unfortunately, this approach has high computation cost when using large images. Hence, it was only tested with small images. In this paper, we call this approach *canonical space transformation*(CST).

Combining EST with CST, our approach reduces the data dimensionality and optimizes the class separability of different classes simultaneously. Experimental results in gait and face recognition show that this method is superior to using EST or CST alone.

2. METHOD

In a nutshell, we combine two transformations - EST based on PCA and CST based on CA. Images in high-dimensional image space are converted to low-dimensional eigenspace using EST. The obtained vector thus is further projected to a smaller canonical space using CST. Recognition is accomplished in canonical space. Patently, the reduced dimensionality results in concomitant decrease in computation cost.

Assume that there are c classes for training and each class represents a gait sequence or a face class of a single person. $x_{i,j}$ is the j-th image in class i, and N_i is the number of images in i-th class. The total number of training images is $N_T = N_1 + N_2 + \cdots + N_c$. This training set is represented by

$$[\mathbf{x}_{1,1}, \cdots, \mathbf{x}_{1,N_1}, \mathbf{x}_{2,1}, \cdots, \mathbf{x}_{c,N_c}] \quad (1)$$

where each sample $\mathbf{x}_{i,j}$ is an image with n pixels. The mean vector for the full image set is given by

$$\mathbf{m_x} = \frac{1}{N_T} \sum_{i=1}^{c} \sum_{j=1}^{N_i} \mathbf{x}_{i,j}. \quad (2)$$

By subtracting the mean from each image, the image set can be described by a $n \times N_T$ matrix $\mathbf{X}$, with each image $\mathbf{x'}_{i,j} = \mathbf{x}_{i,j} - \mathbf{m_x}$ forming one column of $\mathbf{X}$, that is

$$\mathbf{X} = [\mathbf{x'}_{1,1}, \cdots, \mathbf{x'}_{1,N_1}, \cdots, \mathbf{x'}_{c,N_c}]. \quad (3)$$

2.1 Eigenspace Transformation

Let $\mathbf{R}$ be a $n \times n$ matrix and represented by

$$\mathbf{R} = \mathbf{X}\mathbf{X}^{\mathbf{T}}. \quad (4)$$

EST uses the eigenvalues and eigenvectors generated by data covariance matrix to rotate the

original data coordinates along the direction of maximum variance. If the rank of $\mathbf{R}$ is K, then the K nonzero eigenvalues of $\mathbf{R}$, $\lambda_1, \ldots, \lambda_K$, and their associated eigenvectors $\mathbf{e}_1, \ldots, \mathbf{e}_K$ satisfy the fundamental eigenvalue relationship

$$\mathbf{R}\mathbf{e}_i = \lambda_i \mathbf{e}_i, \qquad \text{i=1,\ldots,K.} \quad (5)$$

In order to solve equation (5), we need to calculate the eigenvalues and eigenvectors of the $\mathbf{R}$ which is computationally intractable for typical image sizes. Based on *singular value decomposition* theory(Murakami and Kumar, 1982), we can compute another matrix $\tilde{\mathbf{R}}$ instead, that is

$$\tilde{\mathbf{R}} = \mathbf{X}^{\mathbf{T}}\mathbf{X}, \quad (6)$$

in which the matrix size is $N_T \times N_T$ and is much smaller than $n \times n$ in practical problems. Suppose the matrix $\tilde{\mathbf{R}}$ has nonzero eigenvalues $\tilde{\lambda}_1, \ldots, \tilde{\lambda}_K$ and associated eigenvectors $\tilde{\mathbf{e}}_i, \ldots, \tilde{\mathbf{e}}_K$ which are related to those in $\mathbf{R}$ by

$$\begin{cases} \lambda_i = \tilde{\lambda}_i \\ \mathbf{e}_i = \tilde{\lambda}_i^{-\frac{1}{2}} \mathbf{X}\tilde{\mathbf{e}}_i \end{cases}, \quad (7)$$

where $i = 1, \ldots, K$. The K eigenvectors are used as an orthogonal basis to span a new vector space. Each image can be projected to a single point in this K-dimensional space. According to the theory of PCA, each image can be approximated by taking only the $k \leq K$ largest eigenvalues $|\lambda_1| \geq |\lambda_2| \geq \cdots \geq |\lambda_k|$ and their associated eigenvectors $\mathbf{e}_1, \ldots, \mathbf{e}_k$. This partial set of k eigenvectors spans an eigenspace in which $\mathbf{y}_{i,j}$ are the points that are the projections of the original images $\mathbf{x}_{i,j}$ by the equation

$$\mathbf{y}_{i,j} = [\mathbf{e}_1, \cdots, \mathbf{e}_k]^{\mathbf{T}}\mathbf{x}_{i,j}, \quad (8)$$

where $i = 1, \ldots, c$ and $j = 1, \ldots, N_c$. We called this matrix $[\mathbf{e}_1, \cdots, \mathbf{e}_k]^{\mathbf{T}}$ the *eigenspace transformation matrix*. We call these eigenvectors *eigengaits* in gait analysis and *eigenfaces* in face recognition.

2.2 Canonical Space Transformation

Based on canonical analysis (Fukunaga, 1990), CST is presented as follows. We suppose that $\{\Phi_1, \Phi_2, \cdots, \Phi_c\}$ represents the classes of transformed vectors by eigenspace transformation and $\mathbf{y}_{i,j}$ is the j-th vector in class i. The mean vector of the entire set is given by

$$\mathbf{m_y} = \frac{1}{N_T} \sum_{i=1}^{c} \sum_{j=1}^{N_i} \mathbf{y}_{i,j}, \quad (9)$$

and the mean vector of the i-th class is represented by

$$\mathbf{m}_i = \frac{1}{N_i} \sum_{\mathbf{y}_{i,j} \in \Phi_i} \mathbf{y}_{i,j}. \qquad (10)$$

Let $\mathbf{S_w}$ denote *within-class matrix* and $\mathbf{S_b}$ denote *between-class matrix*, then

$$\mathbf{S_w} = \frac{1}{N_T} \sum_{i=1}^{c} \sum_{\mathbf{y}_{i,j} \in \Phi_i} (\mathbf{y}_{i,j} - \mathbf{m}_i)(\mathbf{y}_{i,j} - \mathbf{m}_i)^{\mathbf{T}}$$

$$\mathbf{S_b} = \frac{1}{N_T} \sum_{i=1}^{c} N_i(\mathbf{m}_i - \mathbf{m_y})(\mathbf{m}_i - \mathbf{m_y})^{\mathbf{T}}$$

The objective is to minimize $\mathbf{S_w}$ and maximize $\mathbf{S_b}$ simultaneously, that is to maximize the criterion function known as the *generalized Fisher linear discriminant function* and given by

$$\mathbf{J(W)} = \frac{\mathbf{W^T S_b W}}{\mathbf{W^T S_w W}}. \qquad (11)$$

Suppose $\mathbf{W}^*$ be the optimal solution and $\mathbf{w}_i^*$ be its column vector which is a generalized eigenvector and corresponds to the i-th largest eigenvalue λ_i. According to (Fukunaga, 1990), equation (11) can be solved and represented as

$$\mathbf{S_b w}_i^* = \lambda_i \mathbf{S_w w}_i^*. \qquad (12)$$

After the *generalized eigenvalue equation* is solved, we will obtain $(c - 1)$ nonzero eigenvalues and their corresponding eigenvectors $[\mathbf{v}_1, \ldots, \mathbf{v}_{c-1}]$ that create another orthogonal basis and span a $(c - 1)$-dimensional canonical space. By using this basis, each point in eigenspace can be further projected to another point in this canonical space by

$$\mathbf{z}_{i,j} = [\mathbf{v}_1, \cdots, \mathbf{v}_{c-1}]^{\mathbf{T}} \mathbf{y}_{i,j}, \qquad (13)$$

where $\mathbf{z}_{i,j}, j = 1, \ldots, N_i$ represent the new points and new trajectory in canonical space. We called this orthogonal basis $[\mathbf{v}_1, \cdots, \mathbf{v}_{c-1}]^{\mathbf{T}}$ the *canonical space transformation matrix*. By merging equation (8) and equation (13), each image can be projected into one point in the new $(c - 1)$-dimensional space by

$$\mathbf{z}_{i,j} = [\mathbf{v}_1, \ldots, \mathbf{v}_{c-1}]^{\mathbf{T}} [\mathbf{e}_1, \ldots, \mathbf{e}_k]^{\mathbf{T}} \mathbf{x}_{i,j}. \quad (14)$$

The *centroid* of each training sequence in canonical space is given by

$$\mathbf{C_i} = \frac{1}{N_i} \sum_{j=1}^{N_i} \mathbf{z}_{i,j} \qquad (15)$$

Fig. 1. Background Image

Fig. 2. Original Image

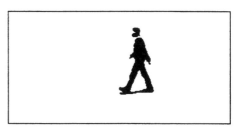

Fig. 3. Preprocessed Image

3. PREPROCESSING

For face recognition, no further preprocessing is needed after brightness normalization. However, for human gait analysis, it is essential to extract the required human outline by eliminating irrelevant background data from each image. We make two assumptions in the human walking sequences. The first is that individuals are walking laterally before a static camera, and secondly, that the body is not occluded. Naturally, to isolate the human silhouette, we can simply subtract the background from each image. The background image is generated by calculating the mean of all training images. Obviously, the difference image thus obtained is not binarized. To simplify the representation, a binary image is obtained by means of region growing technique (Dubuisson and Jain, 1995). Figures 1, 2 and 3 show the background image, an original human walking image in a sequence and the preprocessed image. In order to eliminate redundancies, the extracted human silhouette is fitted into a fixed 64×64 image template as described in (Murase and Sakai, 1996) which is illustrated in Figure 4.

| (a) | (b) | (c) | (d) | (e) |

Fig. 4. Image Templates of a Walking Person

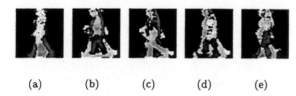

| (a) | (b) | (c) | (d) | (e) |

Fig. 5. The First Five Eigengaits

4. EXPERIMENTS

4.1 *Gait Recognition*

The sample human gait data came from the Visual Computing Group, University of California, San Diego. There are 5 people and 5 sequences of each. One walking sequence is selected from each person as the training sequence and remaining 20 sequences served as test sequences. We choose the first 100 eigenvalues which accumulate 80% of the total signal energy and their corresponding eigenvectors as the eigenspace transformation matrix. Figure 5 shows the first 5 eigenvectors which we called *eigengaits* used to approximate each walking template in the experiments.

Figure 6 shows the trajectories in the eigenspace that represent training gait sequences of five persons projected into the eigenspace. It is obvious that all the five trajectories overlapped and their centroids are close to each other. *Spatio-temporal correlation* (Murase and Sakai, 1996) is used here to recognize an input gait sequence in the eigenspace. Figure 7 shows the trajectories in the canonical space that represent the training sequences of five persons after projecting their transformed vectors to the canonical space. From this figure, all the five trajectories are widely separated into 5 different clusters. To recognize a human walking sequence from a database in the canonical space, the *accumulated distance to each centroid* is used. This will eliminate matching problems caused by velocity changes and phase shifts. The distance between the test vector sequence, $h(t)$, and training vector sequences, $\mathbf{z}_{i,j}$, is

$$d_i^2 = \sum_{t=1}^{T} \|h(t) - \mathbf{C}_i\|^2, \qquad (16)$$

where $\mathbf{C}_i$ is the centroid of class i. To match a test sequence $h(t)$ to a training sequence i

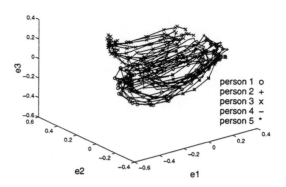

Fig. 6. Trajectories in The Eigenspace

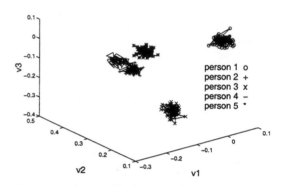

Fig. 7. Trajectories in The Canonical Space

Recognition results for 20 test sequences		
method	recognition rate	class separability
EST only	100%	overlapped
EST+CST	100%	widely separated

Table 1. Different Approaches in Gait

can be accomplished by choosing the *minimum* d_i^2. The comparison of 2 different approaches is shown in Table 1. Clearly, the feature vectors generated by the combined EST and CST yields the best classification result, since the classes do not overlap and a high recognition rate can be achieved.

4.2 *Face Recognition*

Our face data came from the Olivetti Research Laboratory in Cambridge, UK. There are 10 different images of 40 distinct subjects. Figure 8 shows some sample faces from the test data. Due to the enormous computation cost in applying CST alone, all the face images are reduced to 20×30. Figure 9 shows the first 5 eigenfaces used to span the eigenspace and to approximate each face image.

We have conducted one set of tests which chooses 2, 4, 6 and 8 faces separately from each subject and the rest as testing patterns. To recognize a

(a)   (b)   (c)   (d)   (e)

Fig. 8. Sample Face Images

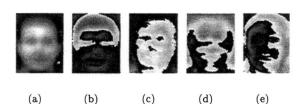

(a)   (b)   (c)   (d)   (e)

Fig. 9. The First Five Eigenfaces

method	no. of training faces			
	8 faces	6 faces	4 faces	2 faces
EST only	92.50%	88.75%	82.50%	75.94%
CST only	91.25%	88.75%	72.08%	67.81%
EST+CST	95.00%	93.75%	85.83%	72.81%

Table 2. Comparison of Different Tests in Faces

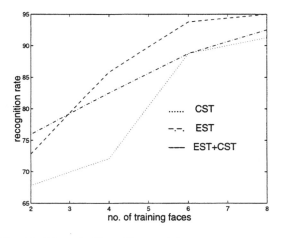

Fig. 10. Tests in Face Recognition

human face from a database with 40 different subjects, *nearest-neighbor rule* in eigenspace is used. The distance between a test vector, ω, and the trained vectors after eigenspace transformation, $\mathbf{y}_{i,j}$, is equated to

$$d^2_{i,j} = \min_{j} \|\omega - \mathbf{y}_{i,j}\|^2 \qquad i=1,\ldots,40. \quad (17)$$

The desired match for test face ω is derived by choosing i which minimizes $d^2_{i,j}$. An equivalent measure is used for recognition in canonical space, except that $\mathbf{y}_{i,j}$ is replaced by $\mathbf{z}_{i,j}$. The recognition results of this test for 3 different approaches are shown in Table 2 and in Figure 10. Test results show that the combined EST and CST approach produces best feature set for face recognition.

5. CONCLUSIONS

In this paper we propose a new approach which combined CST with EST which can be used to reduce data dimensionality and to optimize the class separability of different classes simultaneously. This will greatly improve the performance of eigenspace approach. Experimental results in this paper show that 5 training gait sequences is transformed into 5 widely separated clusters in the new space. In comparison with the results of EST and CST independently, our new approach appears to provide better results. This new approach has been also successfully applied to face recognition.

Given that individuals can be partially recognized by the independent features of face and gait, improved recognition can be achieved by sensor integration by developing state space models of the individual features and associated Kalman filters parameterized by a neurofuzzy feature based modeling algorithm (Harris and Wu, 1997). This approach has been successfully demonstrated for optical flow tracking of overtaking vehicles (Roberts *et al.*, 1995), extensions to face/gait tracking in the canonical sensor feature space is currently being pursued. Sensor integration is achieved by adding the weighted individual identity estimates from gait/face via the equation

$$\hat{x} = \frac{(\hat{x}_f P_f^{-1} + \hat{x}_g P_g^{-1})}{P_f^{-1} + P_g^{-1}}, \quad (18)$$

where $(\hat{x}, P)$ are the state estimates and covariances of the individual feature Kalman filters. A future area of research is active sensor control, whereby following recognition of an individual by the above gait/face algorithms, the camera is servoed to follow the identified "target".

6. REFERENCES

Dubuisson, M-P and A.K. Jain (1995). Contour extraction of moving objects in complex outdoor scenes. *International Journal of Computer Vision* **14**(6), 83–105.

Etemad, K. and R. Chellappa (1997). Discriminant analysis for recognition of human face images. *Journal of the Optical Society of America - Series A* **14**(8), 1724–1733.

Fukunaga, K. (1990). *Introduction to Statistical Pattern Recognition*. 2nd ed.. Academic Press.

Harris, C.J. and Z.Q. Wu (1997). Neurofuzzy state estimators and their applications. In: *Proc. of IFAC, AIRTC'97, Plenary Paper*. Kuala Lumpur, Malaysia. pp. 7–15.

Little, J. and J. Boyd (1997). Global versus structured interpretation of motion: Moving light displays. In: *Proc. IEEE Nonrigid and Articulated Motion Workshop*. San Juan, Puerto Rico. pp. 18–25.

Murakami, H. and V. Kumar (1982). Efficent calculation of primary images from a set of images. *IEEE Trans. on Pattern Anal. Machine Intell.* **4**(5), 511–515.

Murase, H. and R. Sakai (1996). Moving object recognition in eigenspace representation: gait analysis and lip reading. *Pattern Recognition Letters* **17**, 155–162.

Niyogi, S.A. and E.H. Adelson (1994). Analysis and recognizing walking figures in xyt. In: *Proc. IEEE Conf. on Computer Vision and Pattern Recognition*. Seattle, WA, USA. pp. 469–474.

Roberts, J.M., D.J. Mills, D. Charnley and C.J. Harris (1995). Improved kalman filter initialisation using neurofuzzy estimation. In: *Proc. of 4th Int. Conf., ANN, Proc. IEE*. Cambridge, UK. pp. 329–334.

Turk, M. and A. Pentland (1991). Eigenfaces for recognition. *ournal of Cognitive Neuroscience* **3**, 71–86.

PROCESS MODELS FOR INTERFACIAL MATTER

John F. Maguire

Southwest Research Institute, 6220 Culebra Road, San Antonio, Texas 78238-5166

Abstract: Materials transformation models are essential in the development of artificial intelligence for real-time control of materials processing. The quality of the machine-based decision making process depends critically on the degree to which such models capture the essential attributes of the physico-chemical transformation. In developing process models, the current state-of-the-art is to incorporate relevant source terms (e.g., chemical kinetic or thermal) into a diffusion-like equation while taking into account additional terms representing material or radiation transport. In this paper, it is shown that the above approach is inadequate in the processing of material systems which are not at the thermodynamic limit. A systematic approach to refinement of the models for small systems is presented. Such "small" systems include important applications such as superconducting thin films and materials processing in space-based microgravity environments. The latter will be of particular importance in that exploitation of the unique advantages of microgravity will require remote intelligent process controllers. *Copyright © 1998 IFAC*

Keywords: Intelligent, control, system, process, models, interfaces.

1. INTRODUCTION

It has been well established from the results of a number of space-based crystallization experiments that in space one tends to obtain larger more nearly perfect crystals than can be obtained on earth (Walter, 1987). This is especially true for biological macromolecules such as proteins (Bi *et al.*, 1994) but has also been found to be the case in the directional solidification of metals (Baltaile, *et al.*, 1994) and even semi-conductors display a range of similar phenomena (Gillies *et al.*, 1993). The observation that it is possible to produce better polymers, metals, and semiconductors in space is a major driving force to explore more thoroughly the underlying physical mechanisms for many of these processes. If these mechanisms can be understood in a qualitative and quantitative fashion, then the feasibility of developing space-based manufacturing technology could be more realistically assessed. Clearly, space-based manufacturing will be required to take advantage of microgravity conditions and such facilities will require artificial intelligence for real-time control of processing operations. While the theoretical basis for understanding such systems has not yet been fully developed, recent advances have laid a foundation which, although not entirely solid, is firm enough for many of our purposes. Also, it will be clear that the establishment of a well-founded physico-chemical basis for the treatment of small systems far from equilibrium will be an essential requirement for the application of artificial intelligence in machine-based control of manufacturing processes.

Given that the gravitational contribution to the interatomic potential is so minuscule, it might seem, at least at first sight, difficult to reconcile the above well established experimental observations with the current theories of crystal nucleation and growth (Haymet, 1992). While convection phenomena are of well known importance on earth, and while microgravity reduces convection, the detailed microscopic mechanisms by which such an effect occurs is not entirely obvious. For example, interatomic or intermolecular correlations near surfaces are of order of the range of the surface potential, i.e., say ten molecular diameters. The time scale of molecular "collisions" with the surface are of order picoseconds. Both the distance scale and the time scale are orders of magnitude below those at which a macroscopic or hydrodynamic treatment might be expected to hold. A surface area element will undergo millions of collisions in a <u>local</u> environment which will be quite unaware of any much slower and longer range macroscopic convective motions which might be taking place. While microgravity certainly reduces or eliminates convective terms the <u>local</u> flux may be more sensitive to the gradients in the local chemical potential and it is this quantity which should be of primary interest in developing a good model for the interaction of soft matter with hard surfaces. It will, of course, be quite impossible to unambiguously differentiate between these effects on earth so that both earth-based and space-based experiments will be needed. There remains a great deal to be discovered here and while the basic theory has yet to be developed in detail, there are a number of pointers which indicate that nucleation and growth may be more influenced by non-linear phenomena than had been thought originally (Cvitanovi• , 1989). This will of course have a major implication in the processing of thin films and interfacial materials such as superconductors. It will also help explain at the molecular level the effects of gravity on biological systems.

We shall see from the discussion in Section 2 that there are two essential ingredients if we are to observe the effects of small local fields on matter at the molecular level. First, the matter must exist in what is formally thermodynamically "small" system. This is a system in which the spatial correlation is long range (critical point) or in which the external field varies on the same distance scale as the correlation length. For a solid-fluid interface we have a small system because the surface potential term falls off approximately as r^{-6} and so varies on the same distance scale as the interatomic potential (see Figure 1). In such circumstances the free energy density has both a gravitational and a surface term. Small changes in the gradients of such terms may induce huge effect on structure. In some circumstances this surface field may induce a new form or structure of matter close the surface.

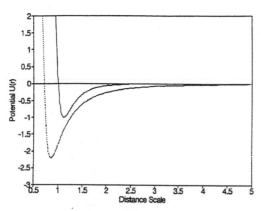

Fig. 1. Interatomic pair potential (solid line) and surface potential (dotted-line) for the reduced Lennard-Jones solid. Note the longer range nature of the surface potential.

In this paper we review a number of experiments and theoretical treatments which probe the fundamental nature of materials in fluids and at heterogeneous surfaces.

In order to discuss the formation of novel interfacial structures within the contemporary paradigm of thermodynamics of interfaces, it is useful to review briefly some modern ideas in two emerging, overlapping, and active areas of theory. These are the fields of non-equilibrium thermodynamics (including non-linear systems) and the thermodynamics of statistically small systems. It will be argued that it is in the overlap between these two fields that the effects of small perturbations such as microgravity and surface perturbations have their most profound impact on the structure and properties of interfacial matter. Let us point out at the outset that despite the fundamental nature of the theoretical ideas, such work is of decidedly practical significance and no proper discussion of such technologically important areas as thin film growth, polymer interfaces, nucleation, and crystal growth phenomena, phase transitions, or critical phenomena is possible without explicit recognition that these are all examples of non-linear systems and are therefore amenable to a common description. Moreover, both interfacial and critical point phenomena are examples of thermodynamically small systems so they may be described using similar theoretical approaches. It is in this interplay between "smallness" and nonlinearity that even very weak symmetry breaking fields may have enormous influence on the structure, thermodynamic, and transport properties of solids and interfaces. It is, for example, now becoming increasingly clear that fields of study such as polymer materials processing, the gas liquid critical point, and macromolecular interaction and adhesion on the cell wall, which might just a few years ago have been regarded as at opposite ends of a very wide spectrum, display a remarkable underlying universality.

2. NON-EQUILIBRIUM THERMODYNAMICS

There has been considerable progress in the understanding of both equilibrium (Rowlinson and Swinton, 1961) and non-equilibrium (deGroot and Mazur, 1984) thermodynamics. Attention has now focused on the extension of classical notions in thermodynamics to systems of such complexity that until recently, they would have been thought generally to be of such complexity as to defy any attempt at rigorous analysis. To be sure, the real enabling advance has been in the mathematical analysis of non-linear systems and these methodologies have opened the door for far more detailed "universal" analysis of complex systems. Of course, the notion of universality has been well established in classical thermodynamics for a very long time and arises from the fact that the equilibrium properties of the system are governed by the extremum of one of the thermodynamic potentials. Universality fails if the system is removed from equilibrium (e.g., crystal growth) for now some kinetic description must be attempted to take into account the return to equilibrium

$$\frac{\partial x}{\partial t} = -F(x, \lambda) \qquad (1)$$

In this vector equation, x represents the set of state variables $x = (x_1, x_2, x_3...)$, F is a suitable evolution operator, and λ is a control parameter which acts on the system from the surroundings. In the realm of linear irreversibility the notion of microscopic reversibility to link the flux of conserved variables to the generalized force is used,

$$\frac{\partial x}{\partial t} = -\Gamma \cdot \frac{\partial \Phi}{\partial x} \qquad (2)$$

in which Γ is a definite positive matrix and Φ is the thermodynamic potential. For example, if x_i were the number density in a small but statistically macroscopic volume element in a dense fluid, then if the volume element undergoes a density fluctuation, the rate at which equilibrium is restored is proportional to the generalized force. If the system is forced to be out of equilibrium by virtue of the boundary conditions or other externally imposed constraints, then so long as the non-equilibrium state is not too far removed from the equilibrium state, then Prigogine's minimum entropy production theorem can be used with the proviso that Onsager's reciprocal relations remain valid and that the generalized transport coefficients are independent of both the state variables and generalized forces.

This was the state-of-the-art up until the mid-seventies. At this time, the mathematical theory of non-linear systems advanced significantly (Cvitanovi• , 1989) and the role of bifurcation, complexity, and deterministic chaos began to be appreciated more widely within the thermodynamic context. For example, any thermodynamic system which is brought close to a "pitchfork" bifurcation may be described by a single order parameter, z

$$\frac{dz}{dt} = (\lambda - \lambda_c) z - uz^3 \qquad (3)$$

where λ is the control parameter and the value of the parameter u depends on the system under study. The set of state variables follow z via equations of state. In analogy with Equation (2)

$$\frac{dz}{dt} = -\frac{\partial u}{\partial z} \qquad (4)$$

where u is the so-called kinetic potential. While the potential u has nothing to do with a normal thermodynamic potential and is related to the kinetic evolution of the system, integration over z yields,

$$u = -\frac{1}{2}(\lambda - \lambda_c)z^2 + \frac{1}{4}uz^4 + C \qquad (5)$$

where C is a constant of integration. Equation (5) is remarkably similar to the well known Landau theory of phase transitions. In fact, even for very complex systems such as polymerization reactions, the observation of a pitchfork bifurcation (Maguire et al., 1994) allows an essentially thermodynamic discussion of instability and pattern formation. For our purposes here it is sufficient to recognize that the theory of non-equilibrium systems has undergone sufficient advance as to allow a proper treatment of pattern formation and "structure." This "structure" could be the evolution of the dynamic structure factor of a liquid mixture undergoing spinodal decomposition or it could be the pattern formed by material on the surface and interfacial region in a graphite reinforced composite structure. Both are examples of non-linear small systems and the science learned from one plays directly and surprisingly transparently into the other. In order, therefore, to appreciate the scientific basis and import of such phenomena in materials processing, as well as the central role of microgravity in space applications, it is appropriate that we review briefly, but reasonably thoroughly, the precise technical meaning of "small" in the thermodynamic sense.

2.1 The Thermodynamics of Small Systems.

In the usual elementary treatment of thermodynamics and statistical mechanics, the basic idea is that one is calculating averages for systems that are very large and which are not subject to an applied field. In reality, of course, no system is infinitely large and even very large systems are confined by vessels in which the interfacial region with the wall of the container must be considered small. For example, if p is the pressure, and T is the temperature, a schematic of the projections of the equation of state is usually depicted as shown in Figure 2(a). Also, only if the system is infinitely large will the liquid-gas and solid-liquid coexistence regions be perfectly horizontal and sharp. There are three lines of first order transitions which intersect in a triple point and the liquid-gas curve terminates in a bicritical point. Figure 2(b) shows a schematic of the chemical potential as a function of temperature (or any field variable) as the melting transition is traversed. If we take the solid liquid transition as an example, then if

$\mu > \mu_T$, the system will be completely solid and if $\mu < \mu_T$, it will be completely fluid. However, as pointed out by Rowlinson and Swinton (1961), if $\mu = \mu_T$, exactly, the system can be completely solid, completely liquid, or any arbitrary mixture of both. We have then the apparent paradox that a large system is homogeneous even if it contains more than one phase. Such considerations have, interestingly enough, been largely ignored in the past by computer simulators. The reasons for this are well-known results of fluctuation theory. In a large system in the macrocononical ensemble, the fluctuation in the number of particles, N, is proportional to $\sqrt{N}$ in each of the pure phases so as N gets large, the relative fluctuation becomes minuscule, $\sim 1/\sqrt{N}$. However, when $\mu = \mu_T$, the fluctuations in N become of order N and are thus never small. In such situations the fluctuations in the local thermodynamic properties can completely dominate the behavior of the system. At the solid-fluid transition, for example, the system could be all solid, all fluid, or any mixture in between. At this point, measurement of the size and range of the density fluctuations at and near a surface will provide very detailed information on the kinetics of the growth of the interface.

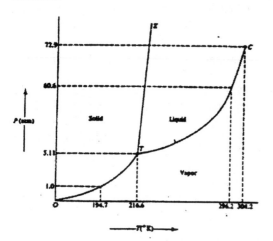

(a) PT projection of equation of state

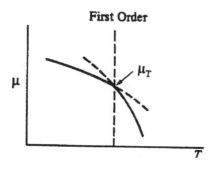

First Order

(b) Chemical potential as a function of temperature

Fig. 2

In order to produce the non-uniform system of our everyday experience, one must apply an external field of which the very weak gravitational and surface fields are the most ubiquitous, i.e.,

$$\mu = \mu(r) + w(r) + v(r) \qquad (6)$$

and

$$w(r) = mg(h - h_o) \qquad (7)$$

where m is the mass of a single molecule, g is the acceleration due to gravity, and h is the height above some reference height, h_o and $v(r)$ is the surface potential. For example, for the Lennard-Jones solid the surface potential is given by

$$v(r) = 8\pi \in \rho \left[\left(\frac{\sigma}{r}\right)^9 - \left(\frac{\sigma}{r}\right)^3 \right] \qquad (7a)$$

where r the distance above the surface (which usually defines the reference height), ρ is the density and the other parameters are constants. This is the expression used in Figure 1. The surface potential provides a perturbation which is longer range (r^{-3}) than the intermolecular interaction (r^{-6}). At the point of transition it is the gradient in this surface term plus that of the gravitational term which is responsible for the development of mass fluxes within the interfacial region. It is this gradient which is responsible for the structure and morphology of the interface.

2.2 The Point Approximation.

The surface term provides a sizable perturbation to the fluid, and the gravitational term, $w(r)$, is also important. Note that if $w(r)$ or $v(r)$ have spatial **gradients**, even minute gradients, then the nonlinear system will have two or more phases. We may, for example, have a bulk phase, a surface phase, and a fluid phase in equilibrium. There has been a good deal of recent interest (Maguire and Leung, 1991) and more than a little controversy (Strandberg, 1988) regarding the detailed nature of the melting transition in two dimensions and the existence, or otherwise, of an orientationally ordered hexatic mesophase in two dimensions. In this regard it is probably worth pointing out that in order to observe such a phase experimentally it will be essential that the measurements are carried out on the surface in-situ during the melting or growth process. As pointed out by Maguire and Leung (1991), if such a system is subjected to a shear force above a critical value it will flow like a cold plastic solid in which hexatic vortices may form. The presence of orientational *per se* will not be sufficient to establish the existence of a thermodynamic transition. It is this ability of the gradient of even minute applied fields to induce huge density difference that lies at the heart of the physics of small systems. Moreover, we might, even on the basis of these simple arguments, infer that the magnitude and form of this applied gradient may have an effect, possibly a substantial effect, on the morphology and structure of the materials which are formed. Notice that in the above case the system is, strictly speaking, a small system in that the length scale is determined by $dw(r)/dr$. For a liquid alkane at the triple point, this "gravitational" length kT/mg is about 10 km. For a liquid system, the pair radial distribution function g(r) decays to unity on a length scale of order 2.0 nm. Under these circumstances, it is essentially exact to assume that the chemical potential is a functional of the local density and write

$$\mu(r) = \mu\left[\rho(r), T\right] \qquad (8)$$

Equation (8) states that the chemical potential and therefore all of the physical properties of the system are equal to those in the field-free state in a fluid of the appropriate density and temperature. It can be used, therefore, to calculate the chemical potential in bulk phases.

The important point for our purposes is to recognize that there are two common situations in which Equation (8) is **not** valid. The first is near a critical gas-liquid point or consolute point in a fluid. Here, the correlation length is no longer determined by the range of the intermolecular potential, but diverges to an infinite length. The other is at a solid-fluid surface. In this case, the external field is a combination of both gravity and the surface potential. In such circumstances, the external field varies on the same distance scale as the oscillation in the radial distribution function so that here too, especially at the point of transition, minor variation in the field may have a huge effect on structure. It is for this reason that microgravity plays such an important role in crystal growth. In order to explain interfacial phenomena and growth, it is necessary then to extend the argument to take into account these gradient effects. There are two levels of approximation at which this may be attempted.

2.3 The Local Approximation.

The chemical potential, $\mu(\mathbf{r})$, and the free energy density, $\phi(\mathbf{r})$, have formally exact definitions that are not useful except for the purpose of conducting an evaluation using molecular dynamics or Monte Carlo calculations (Frenkel and Maguire, 1981 and Frenkel and Maguire, 1983). The crudest approximation is the point approximation of Equation (8), which will hold only when the inhomogeneity is weak on the scale of

the correlation length.

Following earlier attempts by Laplace and Rayleigh, van der Waals (Henderson, 1992) recognized that it is a free energy that is required to describe an interface, and developed the well known "square gradient" approximation for the free energy density,

$$\phi(r) = \phi\left[\rho(r), T\right] + \frac{1}{2}m\left|\nabla\rho(r)\right|^2 + \cdots \qquad (9)$$

where m is given by

$$m = -\frac{1}{6}\int r_{12}^2 \, U_{att}(r_{12}) dr_{12} \qquad (10)$$

and U_{att} is the attractive part of the intermolecular potential (i.e., the tail in Figure 1).

For many years, the significance of this result was not appreciated until it was independently derived in slightly different form by Cahn and Hilliard (1958). The Cahn-Hilliard equation has been widely used to describe small systems such as wetting phenomena (Rowlinson and Widom, (1982). The need to extend the treatment beyond the point approximation for crystal growth has been well illustrated in, for example, the recent theoretical investigations of Kupferman et al. (1995) on the coexistence of symmetric and parity-broken dendrites in a channel. These authors have solved the diffusion equation for crystal growth using the assumption that μ can be approximated by the Gibbs-Thompsom equilibrium relation. Interestingly enough, they find that for a large range of parameters stable symmetric and stable parity broken solutions co-exist. The branches of parity-broken solutions have their origin in symmetric solutions and arise from standard bifurcations. Figure 3(a) shows their solid-fluid interface for the numerical solution of the phase field model under isotropic conditions. Figure 3(b) shows an experimental picture of a dendritic phase. The agreement is qualitatively impressive and one can hardly doubt that the essential physics is captured by solution of a diffusion equation with an appropriate non-local term for the chemical potential. It is, of course, the non-linearity of the latter which leads to bifurcation and pattern formation in the presence of fluctuations.

(a) Model

(b) 2-D experimental streptavidin

Fig. 3

2.4 Non-Local Approximation.

The recent monograph edited by Henderson (1992) presents an excellent treatment of the non-local approximation within the context of "non-homogeneous fluids." The free energy density is regarded as a functional expansion of the formally exact expression of the free energy density.

$$\phi(\mathbf{r}_1) = \phi\left[\rho(\mathbf{r}_1)\right] -$$
$$\frac{1}{2}kT\rho(\mathbf{r}_1)\int c(\mathbf{r}_1,\mathbf{r}_2)\left[\rho(\mathbf{r}_2)-\rho(\mathbf{r}_1)\right]d\mathbf{r}_2 \quad (11)$$

where $c(\mathbf{r}_1, \mathbf{r}_2)$ is the direct correlation function between points 1 and 2, which is related to the total correlation function

$$h(r_1, r_2) = g(r_1, r_2) - 1 \quad (12)$$

where $g(\mathbf{r}_1, \mathbf{r}_2)$ is the two body distribution function, and $h(\mathbf{r}_1, \mathbf{r}_2)$ is given by the Ornstein-Zernike equation

$$h(r_1,r_2) = c(r_1,r_2) +$$
$$\int c(r_1,r_3)\rho(r_3)h(r_3,r_2)\,dr_3 \quad (13)$$

The non-local approximation contains correlation functions such as $c(\mathbf{r}_1, \mathbf{r}_2)$, which we do not know much about. Note that in Equation (13) the density, which now depends on position, is inside the integral (sometimes called the OZ2 equation) and that this identity requires the use of closure relations, such as the hypernetted chain approximation (HNCA), the mean spherical approximation (MSA), or the Percus-Yevick approximation (Kupferman *et al.*, 1995).

This discussion has provided an outline of the essential physics of small interfacial systems and has emphasized the necessity of taking into account both surface and gravitational effects with a suitable local or non-local approximation for the free energy density or chemical potential. It is the gradients in this local free energy density which drive diffusion, phase separation, crystal growth, and the growth of thin films in such systems. We may, therefore, write a generalized Smoluchowski-Langevin equation,

$$\frac{\partial}{\partial t}c(r,t) = D\nabla^2\mu + \delta \quad (14)$$

where δ represents a fluctuation term. In this equation, μ is given by Equation (6), which includes <u>both</u> a gravitational term and a suitable approximation for $\mu(r)$, as described in Section 2. Near the transition point, the gravitational term will have a huge effect on the gradient, and the fluxes will be completely dominated by gravitational effects. Clearly, in order to test the various theories for $\mu(r)$, it will be essential to have data in which the gravitational term $w(r)$ (Equation 6) can be turned off (i.e., space-based processing) and the consequences examined in terms of the structure and morphology of the system. The presence of both surface and gravitational terms in equations (6) and (14) implies that in a small system such as the interfacial zone the gravitational perturbation may greatly affect the surface and near surface spectrum of density fluctuations.

Alternatively, it may well be possible to apply external fields, possibly of variable symmetry, which enhance the development of a particular structure. Clearly, this raises a number of important issues in relation to the design of intelligent systems for process control. In particular the process model must take the non-equilibrium thermodynamic properties into account.

Notice that Equation (14) does not include a convective term. If such an effect were present, it will, of course, be quite difficult to sort out the relative importance of the above contributions.

3. CONCLUSION

It can be seen that the nature of surface and interfacial interactions, especially in the presence of small applied fields such as gravity, may have a very important influence on the structure and properties of matter. When it is recognized that all materials and process operations are conducted in systems which are not at equilibrium and that the thermodynamic criteria of "smallness" are met at the interfaces, it is perhaps less surprising that what may appear to be rather minor perturbations in process conditions may have what might be considered a disproportionally large influence on the structure and thereby the properties of a material.

For example, Figure 4 shows scanning tunnelling microscope images of n-alkane molecules. Here molecules exist as flexible coils in the usual liquid phase but they stiffen, straighten, and form a "rank and file" formation when brought close to the surface of highly ordered pyrolytic graphite (Lupkowski and Maguire, 1994). Figure 5 shows a Raman image of the surface of a superconducting thin film. In this image the different grey scales represent areas of differing chemical stoichiometry. In both cases the material can be thought of as having "aggregated" in a particular fashion. Predicting and controlling this "aggregation" on the basis of models based on Equation 14 will prove a challenging but necessary prerequisite for space-based intelligent processing.

It is clearly essential that intelligent control systems be aware that decisions and directives that may take the system into a region of phase space where there is likely to be great sensitivity to initial conditions should probably be avoided. On the other hand, it may be possible to use this very condition to apply a small external field of the appropriate magnitude and symmetry to prepare a structure of matter that had not previously been possible. In any case, it is clear that the application of these approaches will prove exceedingly interesting in the exploration and processing of new forms of matter both on earth and in space.

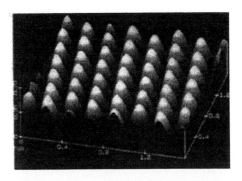

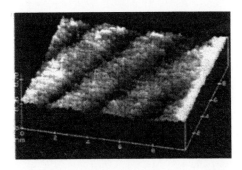

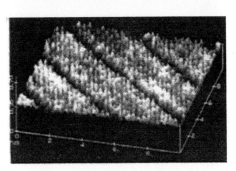

Fig.4.   Scanning tunnelling microscopy images of (a) structure of graphite surface; (b) Hexadecane adsorbed on graphite surface; and   (c) Heptadecane adsorbed on graphite surface.

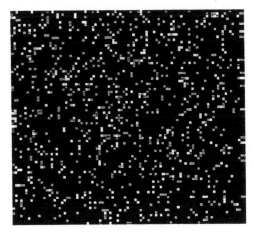

Fig. 5. Raman image of the surface of a superconducting thin film.

REFERENCES

C. C. Baltaile, R. N. Grugel, A. B. Hmelo, and T. G. Wang, *Mater. Trans. A*, **25A** (4), 86570, 1994.

R-C. Bi *et al.*, "Protein Crystallization in Space," *Microgravity Science and Technology*, **11** (2), 203-206, 1994.

J. W. Cahn and J. E. Hilliard (1958). Free Energy of a Non-Uniform System. I. Interfacial Free Energy, *J. Chem. Phys.*, 28:258. J. W. Cahn and J. E. Hilliard (1959). II. Thermodynamic Basis, *J. Chem. Phys.*, 30:1121. J. W. Cahn and J. E. Hilliard (1959). III. Nucleation in a Two-Component Incompressible Fluid, *J. Chem. Phys.*, 31:688.

P. Cvitanovi• , "Universitality in Chaos," Adam Hilger editor, 2nd edition, 1989.

D. Frenkel and J. F. Maguire, *Physical Review Letters*, **47** (15) 1981.

D. Frenkel and J. F. Maguire, *Molecular Physics*, **49** (3) 1983.

S. R. deGroot and P. Mazur, "Non-Equilibrium Thermodynamics", Dover Publications Inc., New York, 1984.

D. C. Gillies *et al.*, "Bulk Growth of II-VI Crystals in the Microgravity Environment of USML-1," Proceedings SPIE Symposium on Optical Instrumentation and Applied Science, San Diego, CA, July 1993.

A.D.J. Haymet, "Fundamentals of Inhomogenous Fluids," Chapter 9, Marcel Dekker, New York, 1992.

D. Henderson, editor, "Fundamentals of Inhomogenous Fluids," Marcel Dekker, New York, 1992.

R. Kupferman, D. A. Kessler, and E. B. Jacob, *Physica*, **A213**, 451-464, 1995.

M. Lupkowski and J. F. Maguire, *Composite Interfaces*, **2** (1), 1-14, 1994.

J. F. Maguire and C-P. Leung, *Physical Review B*, **43** (7B), 1991.

J. F. Maguire, P. L. Talley, and M .Lupkowski, *J. Adhesion*, **45**, 269-290, 1994.

J. S. Rowlinson and F. L. Swinton, "Liquids and Liquid Mixtures," 3rd ed., Butterworths, London, 1982a; I. Prigogine, "Introduction to Thermodynamics of Irreversible Processes," Wiley, New York, 1961b.

J. S. Rowlinson and B. Widom, "Molecular Theory of Capillarity," University Press, Oxford, 1982.

K. J. Strandberg, *Rev. Mod. Phys.*, **60** (16), 1988.

H. U. Walter, "Fluid Sciences and Materials Sciences in Space," Springer Verlag, Berlin, 1987.

An Image Blocking Neural System for Children on the Internet

Yukari Kamakura, Yoshiyasu Takefuji

Graduate School of Media & Governance
Keio university
5322 Endo, Fujisawa Kanagawa
2520816 Japan
kamachan@sfc.keio.ac.jp, takefuji@sfc.keio.ac.jp

Abstract: This paper proposes an image blocking neural system on the Internet for children. The proposed system is used for discriminating an image whether it is a pornography or not. The system is composed of two phases. In the first phase, image segments including naked human body are extracted from a target image using Munsell color order system. In the second phase, each of segments is classified into a pornography or others by using CombNet. In our simulation, 42 tested images are used and compressed to 12x24 pixel respectively are used. Our system can automatically discriminate the pornography images with 67% accuracy. *Copyright©1998 IFAC*

Keywords: Image recognition, Pattern identification, Backpropagation, Self-organizing system, Classification, Hybrid vehicles

1. INTRODUCTION

The explosion of the Internet causes serious problems on copyright, pornography, violence information and so on. The freedom of expression and the control is also one of the social problems in establishing security infrastructure for the network society. To solve the problems they come up to arguments for the rules and mechanism among many countries. World Wide Web Consortium (W3C) has developed PICS (Platform for Internet Content Selection) as a tool for solving the problem. However, it does not work well to recognize the content of the image because the recognition system depends only on the tags of HTML (http://www.w3c.org/Pics).

Although several database companies have been developing the systems to search, rate and make URL-database including specific context or images, it is a time consuming task for the conventional manual systems to classify a lot of images. Net Shepherd corporation is the first experimental rating community of 1500 people who review the content (http://www.netshepherd.com/Media/media.htm). However, they cannot catch up with the growth speed

of web pages. Though QBIC(Query by Image Content) system developed by IBM (http:// wwwqbic. almaden.ibm.com/) has functions to distinguish color, texture, and shapes of images, at final decision human discrimination is needed.

Through the present social condition, it is highly demanded to have image-content based recognition systems. Margaret M.Fleck and David A.Forsyth have developed the system to identify naked people by only image-content(Fleck, *et al.*, 1996; Forsyth, *et al.*,1996). The algorithm is to identify the naked people by grouping or connecting with the cylindrical parts filtered by skin color. Their system obtains a good results of 43% accuracy through all processing. However it is difficult to recognize the images without human parts, arms and legs which are exposed skin.

The goal of our system is to generate the good result of the clustered skin color by using CombNET (Iwata, *et al.*, 1990; Iwata, *et al.*, 1991) which is strong against the complicated pattern recognition.

2. OUR APPROACH

The proposed system consists of two phases as

follows;

In the first phase, image segments including the naked human body are extracted from a target image according to the Hue, Chroma and Value of the Munsell color order system.

In the second phase, each of segments is classified into a pornography or others by discriminating whether the area of skin color is clothed or not. In the proposed model, Kohonen's self-organization and backpropagation neural network are combined to recognize a complicated image.

2.1 Extraction of naked human body

The most of images on the Internet are colored and there are many noises in images so that it is important to specify the area of human body by skin color. And it is also needed to solve the problem to normalize the size and direction of the target image for image recognition. To make the problem simply this system deals with only the target images including one person which fronts and stands up.

Clustering skin color

Many of researches for human detection use Munsell color order system for extraction of skin areas. Munsell color order system is defined as a model closed to human vision and expressed by Hue, Chroma, and Value. And its numerical value is able to convert images from RGB expression. The clustering model by Munsell color order system is defined called Godlove Color Clustering Equation shown as a following ;

$$I = \sqrt{2C_a C_b \{1 - \cos(\Delta H)\} + (\Delta C)^2 + (4\Delta V)^2} \quad (1)$$

$$\Delta H = |H_a - H_b|$$
$$\Delta C = |C_a - C_b|$$
$$\Delta V = |V_a - V_b|$$

It is said that human skin color is composed of the combination with blood(red) and melanin(yellow and brown), and the hue of human skin color is restricted (Rossotti, 1983) and also the value of color is pretty strong. For the reasons of the above mentioned the proposed system uses the following equation changed Godlove Color Clustering Equation to make the hue more weighted and the value more lightened.

$$I' = \sqrt{10C_a C_b \{1 - \cos(\Delta H)\} + (\Delta C)^2 + (\Delta V)^2}$$
$$(2)$$

It is the point to decide the standard skin color, Ha, Ca,

Va. In this system it is used its color, RGB=(221, 161, 117), which is the average of skin color by 100 images selected at random.

Using the equation(2), the color images can change the grayscale images(Figure1) and make skin color clustering by making binary images at 230 of threshold shown as Figure2.

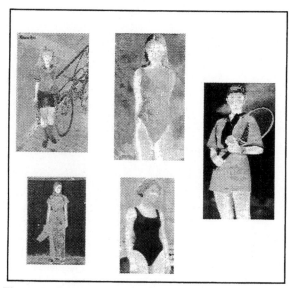

Figure 1. Grayscale images changed from RGB color data by using Godlove Color Clustering Equation based on Munsell color order system

Figure 2. Binary images clustering by skin color

Extraction of a part of human body

Standard human body size is able to shown as Figure 3. Under this size it is nearly possible to extract the part of human body using the principal axes of inertia and central moment.

The principal axes of inertia normalizes the

direction of the target image. A long axis shows vertical line of human body and short one shows horizontal line of human body. The central moment nearly points to the naval of the human body.

According to the measurement, each target image is able to extract pointed as nearly same place of human body as Figure 4.

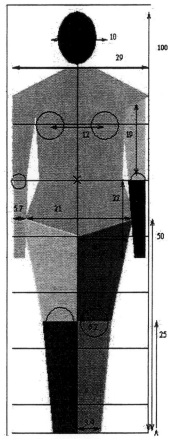

Figure 3. Standard human body

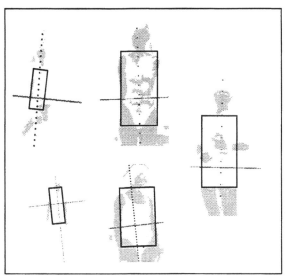

Figure 4. Extraction of a part of human body

2.2 Discrimination naked body using CombNet

CombNET is the hybrid neural model combined with Kohonen's self-organization and backpropagation, developed for the recognition of hand-written characters. It is also implicated for the human detection (Chashikawa, et al., 1998). First part of self-organization is called Stem Network, and second part of backpropagation is called Branch Network. The model is suitable for extracting features of patterns in big fluctuations.

Processing by CombNet

All input images, compressed 12x24 pixels, classified into several groups which contain similar pattern by Stem Network. And the result of the grouping data is discriminated by backpropagation for training of Branch Network. Branch Network is consists of three layered hierarchical Network. The output of the networks are between 0 and 1(Figure 5). For the evaluation, if the output is above 0.5, it is identified naked human body. And other is identified non-naked human body.

3. TESTED RESULT

14 images were prepared as naked body pattern and 28 images as non-naked human body pattern for training CombNET. Our result shows that with 67% we could correctly discriminated given images.

	Total	Correct Result	Accuracy%	
Naked Images	14	8	57	67
NonNaked Images	28	20	71	

4. DISCUSSION AND CONCLUSION

Our proposed system is effective to solve the problem by the result of 67% accuracy. However it still has some problems; sensitive our system is against the images' light and shade. To avoid the problem, it is needed to increase the numbers of data and more learning or to make clustering more strictly by skin color.

And other problems are; how to extract the features of 3 dimensional objects from static images, how to cope with enlargement, reduction and rotation of target images for pattern matching. A part of the 3-D problems might be solved by the model of M.Fleck (Fleck, et al., 1996). And for the orientation imaging problem (enlargement, reduction and rotation), it might be solved by non-linear neural oscillation for human detection(Oka, et al.,1998).

Our proposed system doesn't handle well with the shape of the image recognition problem. Taking into these ideas, we may be able to improve the system ability for the target problem.

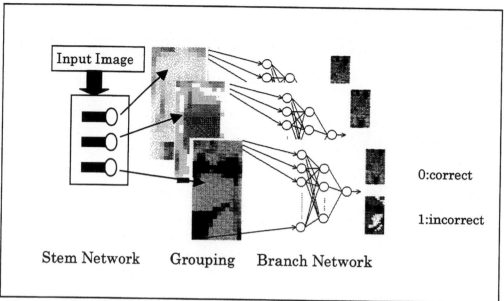

Figure 5. CombNET system

REFERENCES

Chashikawa, T, K.Fujii, Y.Ajioka, Y.Takefuji(1998).
   "Human Detection from Camera-Images by
   CombNET", *Proc. of EANN`98*.
Forsyth, D. A., Margaret M.Fleck(1996).
   "Identifying Nude Pictures", *Proc. Third IEEE*
   Workshop on Applications of Computer Vision.
   WACV`96.
Fleck, M.M., David A.Forsyth, Chris Bregler(1996).
   "Finding Naked People",*European Conference on
   Computer Vision*, **vol II**, pp.592-602.
Iwata, A., T.Tohma, H.Matsuo, N.Suzumura(1990).
   "A Large Scaled Neural Network"CombNet" and
   its application to Chinese Character Recognition",
   Proc.of INNS90.
Iwata, A., K.Hotta, H.Matsuo N.Suzumura,(1991).
   "A Large Scaled Neural Network"CombNet"",
   Proc. of IATSTED, July, 1991.
Oka, S, Y.Ajioka, Y.Takefuji(1998)."A Self-Organized
   Oscillation in Detecting Time-Varing Human
   Faces", *Proc. of EANN`98*.
Rossotti, Hazel(1983)."*Color: Why the World isn't
   Grey*", Princeton University Press, NJ.

HUMAN DETECTING SYSTEM
BY HYBRID SELF-ORGANIZATION CLASSIFICATION SCHEME
AND CombNET

Takakazu Chashikawa* **, Keizo Fujii*, Yoshiyasu Takefuji*

*Graduate School of Media and Governance, Keio University,
5322 Endo, Fujisawa, 252-0816,JAPAN
**Nittan Co.,Ltd.,
1-11-6 Hatagaya, Shibuya-ku, 151-8535, JAPAN

Abstract: This paper presents a system for detecting front view face in images using an artificial neural network. The system consists of three phases: extracting skin, making facial candidate patterns and recognizing face. We develop the hybrid self-organization classification scheme to extracts skin from input images and the CombNET to learn facial pattern. 100 images are used for training process and 840 images varying illumination, background and distance between people and camera are used for testing. The correct recognition rate is 78.2%. *Copyright©1998 IFAC*

Keywords: Pattern recognition, Hybrid vehicles, Self-organizing systems, Learning algorithms, Large-scale systems

1. INTRODUCTION

Recently, some human detecting systems have been reported by computer vision researchers. However, they remain the critical problems in terms of flexibility and computational time. In some of conventional human detecting system, the gray scale patterns are considered. These patterns include effect of noise: e.g. varying illumination. In our system, the demand for robust detection has been realized by multi-spectrum sensors. A visible camera and an infrared camera are used to obtain a hyper-spectrum images for human recognition.

Our system consists of three phases. In the first phase, human skin is detected from the hyper-spectrum images. In the second phase, a facial candidate patterns are generated. In the final phase, the CombNET proposed by A.Iwata et.al(1990) is used for learning facial patterns and recognizing them. In our architecture, as a result of second phase the face detecting problem is converted to a simple pattern recognition problem.

In the pattern recognition problem, it is a hard task for only a back-propagation network model to learn a lot or patterns, because the network becomes extremely large and easy converges to local minimum. The CombNET is well-known as a good model for large scale network. The CombNET consists of a vector quantizing network as a stem network and many backpropagation networks as a branch network . The stem network roughly classifies input patterns into several classes where each class contains similar features according to the self-organizing map algorithm. Each branch network precisely classifies an input pattern by the backpropagation. Since the stem network reduces a large number of categories, the computational time is much reduced in the branch network.

In this paper, we mainly describe the second phase and the third phase. The goal of our system is to detect human faces.

2.SYSTEM ARCHITECTURE

An overview of the proposed system is shown in Fig.1. The system consists of three phases: Skin detecting, Making patterns and Recognition.

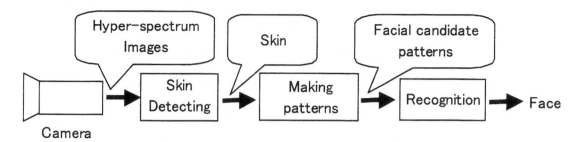

Fig. 1. Over view of the proposed system

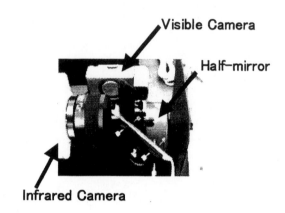

Fig. 2.Hyper-spectrum Camera

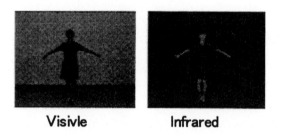

Fig. 3.Input images

2.1. Skin detectng phase

The skin detecting phase has been developed by T.Chashikawa, et al. (1998). We have developed a hyper-spectrum camera by combining visible camera and infrared camera (Fig.2). This camera can obtain a hyper-spectrum image with the same area simultaneously (Fig.3). The algorithm of detecting skin is based on hybrid self-organization classification algorithm (S.Oka, et al. 1998). We use the maximum neuron model in the first stage and the Kohonen's self-organization in the second stage (S.C.Amartur, et al. 1992). The algorithm can obtain the optimal solution by escaping from the local minimum. The algorithm is able to shorten the computation time without a burden of parameter tuning. The results of skin detecting phase are shown in Fig.4. A white color is assigned on skin aria.

Fig. 4.Result of Skin detecting phase

2.2. Making pattern Phase

In this phase, the facial candidate patterns are generated. These patterns are very simple. Because, these patterns contain hair, skin and background. Hair, skin and background are assigned to 0, 128 and 255 in gray scale, respectively. The result of skin detecting phase divided the input image into skin area and other arias. This phase segments skin area and divides other arias into the hair and the background by using threshold and normalizes the segments (16×12 pixels). We guess feature of the hair and the background as shown below.

- A temperature of background is lower than that of skin.
- A temperature of head is similar to skin.
- A color of hair is similar to black.

And this phase checks into limitations to reduce candidates. The limitations are (a) the ratio, height : width, of pattern, (b) the symmetry of pattern, (c) amount of hair and position from the center of gravity.

The facial candidate patterns are shown in Fig.5.

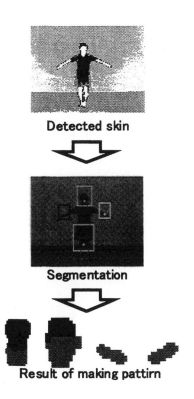

Detected skin

Segmentation

Result of making pattirn

Fig. 5.Process of making patterns phase

2.3 Recognition phase

CombNET consists two modules (Fig.6) (A.Iwata, et al.,1990). In the first module, kohonen's self-organizing algorithm is used as a stem network (T.kohonen,1993). In the second module back-propagation algorithm is used as branch network.

Training of CombNET

As training of the stem networks, the stem neurons

are generated and the synaptic weights were modified by the training procedure. After the stem network training, all input data is classified into several groups according to the best matching criteria to the synaptic weight vectors of neurons. Since the input data which contains similar patterns is assigned into a same group, the output data is assigned to a same branch module. In stem network module, group includes head and others (Fig.7). And branch network modules are trained for each sub-space to make a mapping function between the input and output data. backpropagation is utilized to train branch modules. Each branch neural network which is three layered hierarchical network doesn't have so complex mapping function that it is easy to train. The outputs of the networks are between 0 and 1. If the output is upper 0.6, it is correct head. And other is incorrect head. If the output is correct head, the system drew rectangle pointing face position(Fig.8).

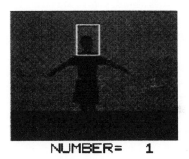

NUMBER= 1

Fig.8.Result of recognition phase

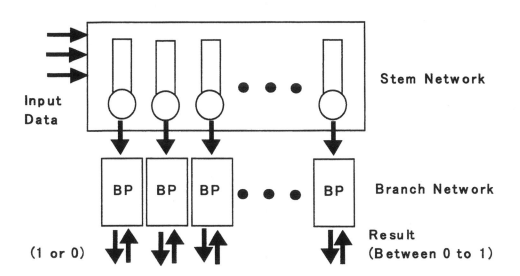

Fig. 6.CombNET for face detecting system

133

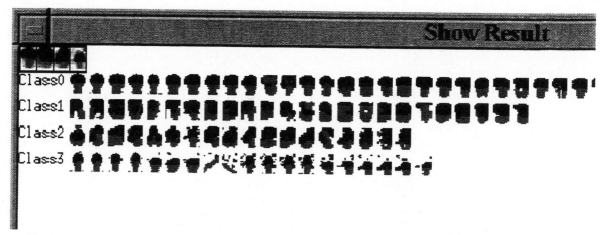

Fig. 7. Result of stem network (a part)

3.EXPERIMENTAL RESULTS

In our study, the experimental data is prepared considering varying illumination, background, distance between people and camera. The number of image samples is 940, consist of 100 training image and 840 testing image.

In the skin detecting phase, the size of input images is 128×96. In the recognition phase, the inputs of the stem network module are classified into four categories. The learning condition of each branch network module is as follows: the learning rate is 1.0, the training procedure is finished when the RMS error is less than 0.1, the size of training pattern is 16×12, the number of hidden unit is 30 and output units is 1. The test samples including many conditions described as follows.

- Table 1 shows the results under varying illumination. The average accuracy is 78.6% in the illuminations: $18lx$ (300 images), $58lx$ (240 images) and $500lx$ (300 images).
- Table 2 shows the results under varying background. The average accuracy is 78.2% in the background: type-B1 (210 images), type-B2 (210 images), type-B3 (210 images) and type-B4 (210 images).
- Table 3 shows the result under varying distance between people and camera. The average accuracy is 78.2% in the distance: $1m$ (280 images), $3m$ (280 images), $6m$ (280 images).

In the experimental results, the total average accuracy is 78.3%.

4.CONCLUSIONS

Our system achieved the demand for robust detection by Hyper-Spectrum images. We developed the hybrid self-organization classification neural model to detect skin from Hyper-Spectrum images and the CombNET to recognize human face. Our system does not too sensitive to varying illumination, background and distance between camera and human. According to the experimental result, our system is available to count the people who wait the elevator and so on. To realize such kind of recognition systems, not only preciseness but also flexibility are reuired.

Table 1 Illumination

	Test	Correct	[%]
18(lx)	300	237	79.0
58(lx)	240	201	83.8
500(lx)	300	219	73.0
Mean			78.6

Table 2 Background

	Test	Correct	[%]
B1	210	167	79.5
B2	210	180	85.7
B3	210	160	76.2
B4	210	150	71.4
Mean			78.2

Table 3 Distance

	Test	Correct	[%]
1(m)	280	248	88.6
3(m)	280	229	81.8
6(m)	280	180	64.3
Mean			78.2

ACKNOWLEDGEMENTS

The part of this study was the work of Advanced Software Enrichment Project produced by INFORMATION-TECHNOLOGY PROMOTION AGENCY, JAPAN. We thank Tetuo Kimura and Akihisa Kudoh for helpful discussions and Souichi Oka for helpful check english in this paper.

REFERLENCES

A.Iwata, T.Tohma, H.Matsuo and N.Suzumura(1990). ``A Large Scaled Neural Network ``CombNET'' and its application to Chinese Character Recognition'', *Proc. of INNS90*

S.C.Amartur, D.Piranio, and Y.Takefuji(1992). ``Optimization Neural Networks for the Segmentation of Magnetic Resonance Images'', *IEEE Trans. on Medical Imaging*, **vol.11, no.2**, pp.215-220,.

S.Oka, T.Ogawa, T.Oda and Y.Takefuji(1998). ``A New Self-Organization Classification Algorithm for Remote-Sensing Images'', *IEICE Transactions on Information and Systems*, **vol.E81-D, no.1**, p. 132-6

T.Kohonen(1993). ``Physiological interpretation of the selforganizing map algorithm'', *Neural Networks*, **vol.6**, pp.895-905.

T.Chashikawa, S.Oka, Y.Ajioka, Y.Takefuji(1998). ``Analyzing Human skin detection by Hyper-Spectrum Images'', *The Japanese Jurnal of Egonomics*, **vol.34 Supplement**, pp.454-455

Recognition of Human Shapes
by Deformable Template and Neural Network

Shunta Tate, Souichi Oka, Yoshiyasu Takefuji

Keio university Graduate School of Media & Governance
5233 Endo, Fujisawa Kanagawa 252-0816 Japan
tate@sfc.keio.ac.jp, takefuji@sfc.keio.ac.jp

Abstract: We propose a system which detects human shapes from video image. In this system, human contours are mainly extracted in order to recognize human shapes. The system has several human contour templates to compare the extracted contours with them based on deformable template matching model. After the template matching, the system detects human shapes from the results of the matching by the back-propagation neural network. We certain that the system is useful for various situations and objects. The experimental result shows that the system detects indoor targets with 93% accuracy and outdoor targets with 79% accuracy. Copyright©1998 IFAC

Keywords: Detection algorithms, Pattern identification, Backpropagation, Learning systems

1. Introduction

This paper addresses the problem of recognizing human shape from video images. The problem has wide related work in image processing and computer vision, including optical flow (Yauchida, 1990), active contour model (Kaas, 1987), and template matching (Kobayashi, 1996; Nagai, 1996; Ito, 1996). A number of studies and applications have been developed for detecting human shapes. However, there are few security camera systems that practically work well under various situations; in a room, outdoors, under an indefinite lighting, various target objects, and various object sizes.

The general additive noises to captured images are as follows:

1. Lighting variance.
2. Migratory light source, e.g. car headlight, flashlight.
3. Migratory shadow of object.
4. Objects except main target.

In addition to the noise, human appears with various contour shapes in an image. Since detecting human shapes under noise is one of ill-posed problems, proper expert knowledge of a target object and knowledge of noises are needed for developing a practical recognition system. Conventional detection methods are limited for a particular purpose or on an ideal condition.

In this system, human contours are mainly extracted in order to recognize human shapes. The system has several human contour templates in order to compare the extracted contours with them based on Deformable Template Model (Amit, 1991; Jain, 1996). The template matching process outputs values of matchings, and each of them is fed to the back-propagation neural network. The neural network sufficiently learns the data sets.

Our model realizes the flexible recognition by concerning various situations and objects.
The reason we proposed this system is that a human shape is indefinitely formed since man poses various postures and styles. If a system attempts to prepare templates of all the shapes rigidly, then mingled objects, additive noises, and the computing cost become crucial problems. Generally, a system cannot completely predict all the shapes of target and noises. In this paper, the proposed model can cover variety of

poses by using the neural network and the several partial deformable templates only for fundamental shapes.

The system is useful for detecting outdoor targets where a solution quality is improved and wrong detections are eliminated.

2. Outline of the Method

2.1 Deformable Template Matching

In this section, the algorithm proposed by Jain (1996) is described. The algorithm deforms a 2D template image as if the image is drawn on a 2D plastic sheet. We assume that the template image is drawn on a unit square $S = [0,1]^2$. Each point in the square is mapped into deformed point as $D_\xi(x,y) : S^2 \to S^2$ by the following function:

$$D_\xi(x,y) = (x,y) + \sum_{m=1}^{M}\sum_{n=1}^{N} \frac{\xi_{mn}^x \cdot e_{mn}^x + \xi_{mn}^y \cdot e_{mn}^y}{\lambda_{mn}} \quad (1)$$

$$\lambda_{mn} = \alpha\pi^2(n^2 + m^2), \quad (2)$$

where λ_{mn} are the normalizing constants and $e_{mn}^x(x,y), e_{mn}^y(x,y)$ are following orthogonal bases,

$$e_{mn}^x(x,y) = (2\sin(\pi m x)\cos(\pi m y), 0) \quad (3)$$

$$e_{mn}^y(x,y) = (0, 2\cos(\pi m x)\sin(\pi n y)). \quad (4)$$

The parameters $\xi = \{\xi_{mn}^x, \xi_{mn}^y | m,n = 1,2,...\}$. By choosing coefficients ξ the family of the functions defined in (1) represents complex smooth deformations. ξ are assumed to be independent of each other, and Gaussian distribution with mean zero as follows:

$$\Pr(\xi) = \prod_{m,n=1}^{M,N} \frac{1}{2\pi\sigma^2}\exp\left\{-\frac{\xi_{mn}^{x^2} + \xi_{mn}^{y^2}}{2\sigma^2}\right\}. \quad (5)$$

The algorithm estimates whether a template T_ξ is similar to an contour in the target image Y by repeating deformation process. The measurement of the similarity E is given as follows:

$$E(T_\xi, Y) = \frac{1}{n_T}\sum(1 + \Phi(x,y)|\cos(\beta(x,y))|) \quad (6)$$

$$\Phi(x,y) = -\exp\left\{-\rho(\delta_x^2 + \delta_y^2)^{\frac{1}{2}}\right\}, \quad (7)$$

where function defined in (7) is edge potential, (δ_x, δ_y) is the displacement from (x,y) to the nearest edge, ρ is a smoothing factor controlling precision of similarity, n_T is the number of pixels on the template, and $\beta(x,y)$ is the angle between the tangent of the nearest edge and the tangent direction of

the template at (x,y). The maximum similarity is achieved when E is 0, the minimum is 1.

Using Bayes rule, prior probability of the deform coefficients given in (5) and the similarity defined in (6) can be combined to obtain a posteriori probability density of the deformed template under the condition of given the input image:

$$\Pr(\xi|Y) = \frac{\Pr(Y|\xi)\Pr(\xi)}{\Pr(Y)}, \quad (8)$$

where $\Pr(Y|\xi)$ is as follows:

$$\Pr(Y|\xi) = \kappa\exp\{-E(T_\xi, Y)\}, \quad (9)$$

κ is a normalizing constant to ensure that the total values of the above function are integrated to 1. Our objective is computing maximum value of the posteriori probability given in (9), that is, it is equivalent to minimize the following function with respect to ξ:

$$L(T_\xi, Y) = E(T_\xi, Y) + \gamma\sum_{m=1}^{M}\sum_{n=1}^{N}(\xi_{mn}^{x^2} + \xi_{mn}^{y^2}), \quad (10)$$

where $\gamma = 1/2\sigma^2$ provides a relative weighting between deviation of the deformation and the fitness of the template to the input image. Since function defined in (10) cannot be minimized analytically, we compute partial differentiation numerically and use gradient method to update ξ as follows:

$$\Delta\xi = -\frac{\partial L(T_\xi, Y)}{\partial\xi}. \quad (11)$$

2.2 Learning and Discrimination by Back-propagation Neural Network

The proposed system has several template images. A template consists of contour image which is not necessarily closed as shown in fig. 1. The templates are classified into three types:

a). Contour of the target object.
b). A part of contour of the target object.
c). Contour of the non-target object.

Training images are divided into positive samples and negative samples. Positive sample images contain

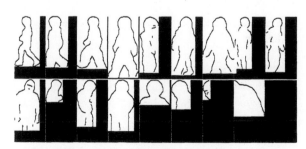

Fig. 1. The whole of the defined templates.(each template size is 67 x 133 pixel)

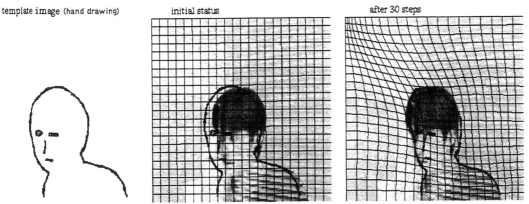

template image (hand drawing)   initial status   after 30 steps

Fig. 2. A experimental example which obtains convergence on a matching

target object(s). The probability density of the template is computed by applying template to the input image. The computation results over all the templates are given as input data to the neural network. Then the NN runs learning process with referring to given teacher value.

3. Experiment and Estimation

3.1 Preprocessing

We captured 430 frames by CCD camera at 16 scenes. Each frame is 160x120 pixel, 256 colored gray-scale, sampled every 500 msec.
Varing regions of two consecutive images are extracted as following algorithm:

1. A sequential n-frames are placed   .
2.   is the differences image between   and   .
3. Logical product image of   and   is generated.
4. Laplacian Gaussian operator   with Zero Cross method convert the image into the contour image.

Process 2-4 have each threshold to discard infinitesimal noise of the frames. The contour image is defined as an input image of the matching process.

3.2 Template matching

We prepare 17 hand-drawing template bitmaps as shown in fig. 1. Since the proposed deforming algorithm provides well fittings to divergence of partial deforms but not to excessive divergence of size or rotation, each template is initially affine transformed into plural templates with three different directions and four different scales: (1.0, 0.86, 0.7, 0.5).

The system descends the value of function by using gradient method. The average number of iteration steps which one template matching needs to converge is about from 10 to 50 . Fig. 2 shows example of sufficiently converging to input image.

Each matching value of all 17 templates are computed for an input image. Since the system has four differently sized template images for a template, the matching process means that an input image is mapped to the   dimensional feature space.

3.3 Learning by Back-propagation Neural Network

The neural network consists of as follows: it is three layered back-propagation network, where the number of the input layer neuron is 68, hidden layer 20, output layer 1, besides, the input layer and the hidden layer have a bias neuron respectively. The inputs of bias neurons are 1. Synaptic weights are initialized with random value (see fig. 3).
The NN is trained so that it outputs about 1 to a positive examples and 0 otherwise. To avoid over-learning, inadequate data including subtle distinction, or excessive small sized human shapes, should not be fed to the NN as training data. 100 images from 430 frames are used for training data, and the other test data. Learning process finishes when all the training examples are divided into two categories with the boundary 0.5, with 98% accuracy.

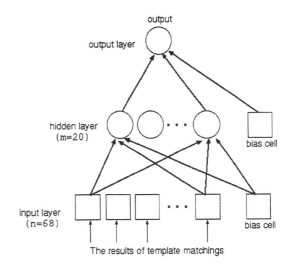

Fig. 3. The structure of the neural network

139

Fig. 4. Experiment 1 with outputs of NN

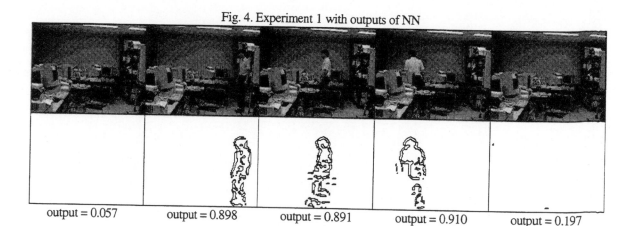

| output = 0.057 | output = 0.898 | output = 0.891 | output = 0.910 | output = 0.197 |

Fig. 5. Experiment 2 with outputs of NN

| output = 0.055 | output = 0.055 | output = 0.065 | output = 0.057 | output = 0.390 |

Fig. 5. Experiment 3 with outputs of NN

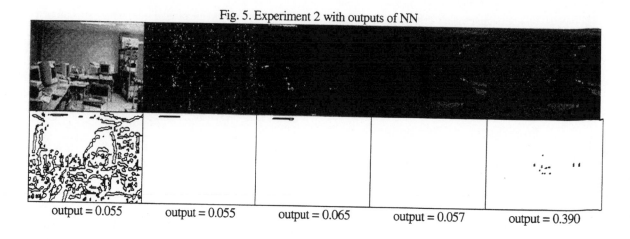

| output = 0.057 | output = 0.069 | output = 0.510 | output = 0.290 | output = 0.057 |

3.4 Detection tests and results

After the learning process finishes, the test set is given to NN for testing. Three results are shown in fig. 4, 5, and 6. Fig. 5 is frames under valiance lighting. In fig. 5, after turning off the lighting, we cast flashlight around the room. In fig. 6 the additive burst noise is given to the system artificially. Fig. 5 and 6 show the system successes in the leaning even though an input image contains noise.

The proposed system can be applied to both indoor and outdoor images. Fig. 7 and 8 show statistical results of indoor and outdoor situation. Table 1 shows the accuracy of the detection, where distinct border of the NN is 0.6. Table 2 shows the average run-time of a whole process.

Although outdoor images containing a number of the noises reduce the accuracy of the system, the model is robust to various situations; indoors, outdoors, mingling various objects. Some errors are observed when the size of the human contour is smaller than that of the template, or when the target object is excessively close to the camera scope.

Table 1 Accuracy of the detection with the indoor images and the outdoor images

indoor images

	actual result (expectation)	
	A	¬ A
NN output P	204 (110.0)	2 (96.0)
NN output ¬ P	25 (119.0)	198 (104.0)
accuracy(± error)	93.71% (± 1.17%)	

outdoor images

	actual result (expectation)	
	A	¬ A
NN output P	134 (71.5)	4 (66.5)
NN output ¬ P	81 (143.5)	196 (133.5)
accuracy(± error)	79.52% (± 1.98%)	

Table2 Average of the runtime

CPU: WS Sun-SS20	
preprocessing	1.2 sec
pattern matching	10.5 sec
neural network	0.01 sec
total	11.7 sec

Fig. 7. Distribution of outputs on indoor images

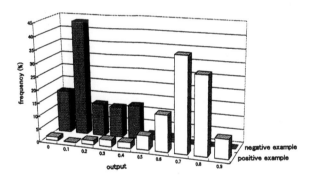

Fig. 8. Distribution of outputs on outdoor images

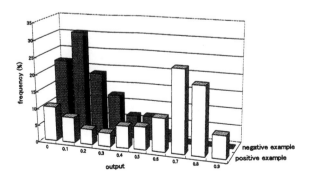

4. Conclusion and Future Works

We propose the model of the pattern detection problem. Since the model extracts human contour and uses deformed templates, it is robust to the variance of the forms and noises.

One of the critical problems for real use is computational time for the matching process. For future works, the system should be improved to the parallel processing system.

References

Amit, Y., U. Grender and M. Piccioni. (1991). Structural image restoration through deformable template. *Journal Am. Statisical Assn.*, **Vol.86, No.414,** 376-387.

Itoh, W., H. Ueda. (1996). A template matching method following object size change. *IEICE Techinical Report PRMU96,* **Vol.96, No.141,** 45-50.

Jain, A. K., Y. Zhong and S. Lakshmanan. (1996). Object matching using deformable templaes. *IEEE transactions on pattern analysis and machine intelligence,* **Vol.18, No.3,** 267-277.

Kaas, M., A. Witikin and D. Terzopoulos. (1987). SNAKES: Active contour models. *Proc. of 1st ICCV,* 259-268.

Kobayashi, N., J. Yamaguchi. (1996). Human tracking in outdoors using spatio-temporal information. *Proc. of the 2nd symposium on sensing via image information,* 173-176.

Nagai, A., Y. Kuno and Y. Shirai. (1996). Detection of intruders on changing background. *Proc. of the 2nd symposium on sensing via image information,* 177-182.

Yatauchi, M. (1990). *Animation processing and understanding.* Shokoudo, Tokyo.

Synthetic Optimization Approach of Combining Regional Guided Order Principle and Biological Evolutive Strategies

CHUANCAI   LIU
China Ship Scientific Research Center
CHANG-YUN SHEN
China Ship Research & Development Academy
Emial:shency@ship.cetin.net.cn

Abstract: A novel technique, called synthetic optimal approach of combining regional guided order principle and biological evolutive strategies (GOBE), is described in this paper. The GOBE approach adopts respective advantages of guided evolutionary simulated annealing (GESA) and Darwin & Boltzmann (D&B) strategies. It retains the local and global competitions of GESA and judges whether Metropolisian process is stable by using D&B strategies. We apply the technique to ore-burden and gasoline blending problems as examples. The results show that the convergence ratio of GOBE method is fast and the solutions are stable. It outperforms other optimization methods in engineering practices that need stable solutions. *Copyright © 1998 IFAC*

Keywords: evolutionary simulated annealing, biological evolution, optimization

1. Introduction

We put forward a new type of parallel and distributed approach to search for optimal solutions. It is neither gradient search nor local search technique. The new search technique is called synthetic optimization approach of combining regional guided order principle and biological evolutive strategies (GOBE). The GOBE approach integrates order principle with biological evolutive strategies and adopts respective advantages of GESA algorithm and D&B strategies. The local and global competitions and the regional guidance of GESA algorithm are retained, and the D&B strategies are in corporate in the local competitions for the

purpose of judging whether Metropolisian iterative process is stable. So GOBE approach could solve optimal problems rapidly and steadily. We demonstrate the GOBE technique with two examples: one is an ore-burden problem of an Iron & Steel Corporation, and the other is gasoline blending problem, which was often used in many papers as a benchmark.

The GOBE approach is described in the next section. Simulation results of ore-burden and gasoline problems are presented and compared with those from other techniques in the section 3. We then conclude our brief discussion with section 4.

2. Synthetic scheme of regional guide order principle & biological evolutive strategies

In nature and society all complex systems are some evolutive products. There are two kinds of basic evolutive processes: one is order principle and the other is biological evolutive strategies. The main idea of order principle is irreversible

thermodynamics process and the concrete form is Boltzmann strategies combined with annealing process. The biological evolutive strategies generally includes genetic (self-reproduction), variation (mutation) and natural selection etc.

We synthesized order principle and biological evolutive strategies and extract respective advantages of GESA algorithm and D&B strategies, and newly developed a synthetic optimal technique of combining regional guided order principle & biological evolutive strategies (GOBE). Generally, the optimal problems considered as follow:

$$\min F(\bar{x}) \qquad (1)$$

$$s \cdot t(\bar{x}) \in S \qquad (2)$$

where $\bar{X} = \{x_1, x_2, \ldots x_n\}$ is a set of system variable; S is likely solution sets. If S_{opt} is an optimal solution set, i.e.

$$S_{opt} = \left\{ \bar{X}_{opt} \in S \mid f(\bar{X}_{opt}) \leq f(\bar{X}); \bar{X} \in S \right\}$$

The GOBE search technique starts out with N likely candidates (solutions), which actually are chosen at random in the search space. These candidates are so-called parents. Initially, each patent can reproduce a number of children, say K. The GOBE algorithm is given in Fig. 1. $N(\bar{X}) \subset S$ is an adjacent domain of $\bar{X}$ in the GOBE algorithm.

Step 1. Initial state: in the solution space, select N likely candidates randomly as initial solutions $\bar{X}^j, j = 1, 2, \ldots N$; give an initial mutation coefficient M > 0.

Step 2. Local competition: produce K children for each parent $\bar{X}^j, \forall \bar{X}^k \in N(\bar{X}^j)$, k = 1,2,匡. Children generated from the same parent complete with each other and one with the lowest objective value will survive. That is the best child, $\bar{X}^p$ (p is one integer between 1 and K).

Step 3. Selection of parent for the next generation for each family: if $f(\bar{X}^p) \leq f(\bar{X}^j)$

or $\min\left\{ 1, \exp\left(\dfrac{f(\bar{x}^j) - f(\bar{X}^p)}{M} \right) \right\} > \eta$, then $\bar{X}^j = \bar{X}^p$, where $f(\bar{X}^p)$ is the

objective value of the best child; M is the mutation coefficient; η is a random number uniformly distributed between 0 and 1.

Step 4. Metropolisian process: for a mutation coefficient M, if Metropolisian iterative process is unstable (the stability of iterative process is that the value of the function is gradually close to a value), the mutation process does not end yet, then jump to step 2.

Step 5. Global competition: Find the number of children for each family that will be generated from the parents of the next generation. The details of this step are given in Fig.2.

Step 6. Change of the mutation coefficient: decrease the mutation coefficient M.

Step 7. Iterative process: repeat process from step 2 to step 6 until an acceptable solution is found or until a certain number of iterations is reached.

Fig. 1 The GOBE algorithm

It is necessary to explain the equilibrium state of the Metropolsian iterative process here. In the Metropolisian iterative process, theoratically, at temperature T, if the energy probability close to the value of eq. 3, the equilirium state just appears.

$$P(E_i) = \frac{\exp(-E_i / (KT))}{\sum_i \exp(-E_i / (KT))} \qquad (3)$$

where E_i is an energy value; $P(E_i)$ is an energy probability.

Step 1. For $\overline{X}^j$, repeat step 2 to step 5 for each family to find the acceptance number of the family, A_j; go step 6.

Step 2. Count $\leftarrow$ 0.

Step 3. Repeat step 4 for each child; go to step 5.

Step 4. If $f\left(\overline{X}^p\right) \leq f\left(\overline{X}^j\right)$ or $\min\left\{1, \exp\left(\dfrac{f\left(\overline{x}^j\right) - f\left(\overline{X}^{p}\right)}{M}\right)\right\} > \eta$, increase Count by 1,

where $f\left(\overline{X}^p\right)$ is the objective value of the child, $f\left(\overline{X}^j\right)$ is the lowest objective value ever found, M is the mutation coefficient, η is a random number uniformly distributed between 0 and 1.

Step 5. Acceptance number of the family (i.e. for $\overline{X}^j$, j=1, 2, N), A_j, is equal to Count.

Step 6. Sum up the acceptance numbers of all the families.

Step 7. For $\overline{X}^j$, j=1, 2, 匝, the number of children to be generated in the next generation can be calculated according to the formula: $m_j = \dfrac{T * A_j}{S}$, where m_j is the number of children that will be generated for the Jth family; T is the total number of children; A_j is the acceptance number of the Jth family; $S = \sum_{j=1}^{N} A_j$ is the sum of the acceptance numbers.

Fig. 2 Algorithm for determining the number of children in each family of the next generation

In practice, while training neural nets, some simple methods are adopted generally. For instance, after random disturbance of several steps, if the objective function's value has a little variation, we consider that the Matropolisian iteration reaches an equilirium state. Perhaps we think that the Metropolisian iteration is stable after undergoing several designated steps. A brief comparison of GESA to D&B is shown in the following table.

GESA	D&B
Some parents Some children	Single parent Single child
Local and global competitions	No competition
No	Stability check
Regional guidance	No

In the GOBE, there are two levels of competition. One is the local competition, the children in the same family (i.e. generated from the same parent) compete with each other and only the one with the lowest objective value survives. The best child then competes with its parent to find the parent for the next generation of its family. If the best child is better than its parent, then it is accepted and the parent for the next generation. If the best child is worse than its parent, we have a Boltzmann probability that the child be accepted. Next we judge whether the Metropolisian iterative process is stable. This local competition creates the parents for the next generation.

The global competition is the competition among the families. The number of children that should be self-reproduced in the next generation depends on the results of this level competition. The 2nd level competition actually provides the

145

measurement of the regional information. The different acceptance numbers actually show the information of how good the chances are in the different regions, and we put more focuses of search on the better regions.

3. Simulation results

Example 1: Ore-burden of an Iron & Steel Corporation

There are 49 composites for ore-burden. The ore-burden heavily affect the quality of product. This is a problem existing for several years. This is an optimal problem in a search space of 49 dimensions because of the affect of each composite on the quality of product. We build a Random-Vector Version Functional-Link net (RVFL) to learn the mapping from the inputs of composite to the outputs of quality indices of products with samples, which are selected randomly from all samples. The output of the RVFL net is as follow:

$$f = \sum_{i=1}^{N\_x} A_i X_i + \sum_{j=1}^{N\_r} \frac{B_j}{1+e^{-z}} \quad (4)$$

where

$$z = \sum_{i=1}^{N-x} rd\_vect[j*N\_x+i]*x_i*Const$$
$$+ rd\_vect[j*N\_x+N\_x+1];$$

N_x is the number of variables (i.e. the number of composites type); N_r is the number of enhancement nodes; "Const" is a constant. Here $N\_x=49$, and N_r equates some constant. "rd_vect[ ]" is a random vector produced by a random function. Fig. 3 indicates the scheme of a Random-Vector Version Functional-Link net.

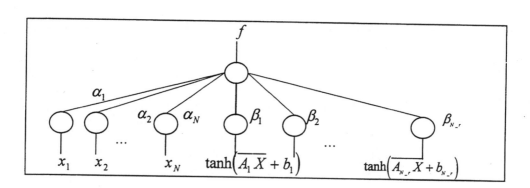

Fig. 3 Random-Vector Version Functional-Link net

In order to find out those ore composites which affect the quality and improve the quality of produce. The partial differential of eq. 5 is as follow:

$$\frac{\partial f}{\partial x_i} = A_i + \sum_{j=1}^{N\_r} \frac{e^{-z}*rd\_vect[j*N\_x+i]*Const}{(1+e^{-z})^2}$$

$$(5)$$

It would improve the quality of product if increase the contents of these composites which gradient values are negative. On the contrary, the quality of product would be improved if decrease the contents of those composites which gradient are positive. We had 286 samples in our experiment which were delivered by the Iron & Steel Corporation. 144 samples were selected randomly as training sets and the rest 142 as testing sets. The result was satisfactory.

Example 2: Gasoline blending problem

The training sets of the gasoline blending problem consist of 17 patterns which are listed in Table 1 and the testing sets of 9 patterns in Table 2. The results by using GESA and GOBE are also listed in Table 1 and Table 2. In our simulation, we adopted Random-Vector Version

Functional-Link net, and applied respectively GESA algorithm and GOBE technique to train the net. While using GESA algorithm and GOBE technique, we selected 10 families and a total of 100 children in each generation. Here, the hyperbolic tangent function is used as the activation function. The output of the net is as follow:

$$f = \sum_{i=1}^{s} \alpha_i x_i + \sum_{j=1}^{N-r} \beta_j \tanh\left(\overline{A_j} * \overline{x} + b_j\right) \qquad (6)$$

We terminated the training procedure when iterative number reached a fixed value.

Other examples:

4. Conclusion

The GOBE could apply to solving optimal solution set of complex system, combinatorial optimization problems, optimization of parameters of control systems as well as learning

Optimization of parameters of a ship in the phase of the conceptual design is one of multiple-parameter optimal problems. Speediness of the ship is an example of the subject. There are more than 14 parameters which have different effects on the speediness of the ship, for example. The objective function is defined from experimental formula. GESA and GOBE were applied to optimizing all parameters with accordance to the objective function of speediness of the ship in the conceptual design separately. The stability of the outputs of GOBE is higher.

of neural nets. The GOBE approach has been found to compare favorably with other optimal techniques in engineering practices. It could be apply to material selection, quality analysis and composites scheduling.

References

Berlinear, L. M. (1987) *"Bayesian Control in Mixture Models"*, Technometrics, **Vol. 29**, pp. 455 - 460.

Fogel, D. B., L. J. Fogel and V. W. Porto,(1990) *"Evolving Neural Networks"*, Biological Cybernetics, **Vol. 63**, pp. 487 - 493.

Percy P. C. Yip (1993) *"The Role of Regional Guidance in Optimization: The Guided Evolutionary Simulated Approach"* (PH. D. thesis), Dept. of Electrical Engineering and Applied Physics, Case Western Reserve University, USA.

Snee, R. D.(1981) *"Developing Blending Models for Gasoline and Other Blends"*, Technometrics, **Vol. 23**, pp. 119 - 130,.

Tian Peng (1993) *"Darwin & Boltzmann Mixed Strategy for Discrete Optimization: Theory and Application"* (PH. D. thesis), Dept. of Automatic Control, Northeastern University, Shenyang, China,.

Yoh-Han Pao and Boris Igelnik (1994) *"Mathematical Concepts Underlying the Functional-Link Approach"*, invited paper presented on June 6, 1994 at 1994 International Neural Networks Society Annual Meeting, San Diego, CA, USA.

Table 1. Results of the training data of the gasoline blending

x_1	x_2	x_3	x_4	x_5	target	output(GESA[1])	output(GESA[2])	output (GOBE)
0.000	0.300	0.000	0.100	0.600	100.000	100.020	100.018	100.009
0.150	0.150	0.100	0.600	0.000	97.300	97.291	97.305	97.304
0.000	0.300	0.000	0.489	0.211	97.000	96.991	97.008	97.005
0.150	0.000	0.311	0.539	0.000	99.700	99.676	99.683	99.692
0.000	0.300	0.285	0.415	0.000	99.800	99.850	99.845	99.837
0.000	0.080	0.350	0.570	0.000	100.000	99.918	100.081	100.002
0.150	0.150	0.266	0.434	0.000	99.500	99.492	99.493	99.503
0.150	0.150	0.082	0.018	0.600	101.900	101.974	101.971	101.941
0.000	0.000	0.300	0.461	0.239	100.900	100.822	100.890	100.895
0.150	0.034	0.116	0.100	0.600	101.200	101.204	101.210	101.194
0.068	0.121	0.0175	0.444	0.192	98.700	99.207	98.587	98.602
0.000	0.300	0.192	0.208	0.300	100.600	100.509	100.571	100.576
0.150	0.150	0.174	0.226	0.300	100.600	100.440	100.493	100.591
0.075	0.225	0.276	0.424	0.000	99.100	99.237	99.168	99.116
0.075	0.225	0.000	0.100	0.600	100.400	100.376	100.423	100.421
0.150	0.150	0.000	0.324	0.376	99.400	99.426	99.425	99.423
0.000	0.300	0.192	0.508	0.000	98.600	98.551	98.610	98.603

Table 2. Results of the testing data of the gasoline blending

x_1	x_2	x_3	x_4	x_5	target	output(GESA)	output(GESA)	output (GOBE)
0.000	0.000	0.350	0.600	0.060	100.000	100.248	100.049	100.047
0.000	0.300	0.100	0.000	0.600	101.000	102.224	101.457	101.450
0.150	0.000	0.150	0.600	0.100	97.800	98.123	97.906	97.816
0.000	0.300	0.049	0.600	0.051	96.700	97.543	97.331	97.031
0.150	0.127	0.023	0.600	0.100	97.300	97.082	97.256	97.276
0.000	0.158	0.142	0.100	0.600	100.700	101.618	100.813	100.806
0.067	0.098	0.234	0.332	0.270	100.500	100.121	100.321	100.328
0.000	0.126	0.174	0.600	0.100	98.400	98.133	98.312	98.368
0.075	0.000	0.225	0.600	0.100	98.200	98.851	98.799	98.798

---

[1] The output is produced by using the iterative technique.

[2] The output is produced by using the GESA optimization approach.

ADAPTIVE ESTIMATION USING MULTIPLE MODELS AND NEURAL NETWORKS

Tyrone L. Vincent * **Cecilia Galarza** **
Pramod P. Khargonekar **

* *Engineering Division, Colorado School of Mines, Golden, CO
80401, tvincent@mines.edu*
** *EECS Department, University of Michigan, Ann Arbor, MI
48109*

Abstract: A method is presented to combine multiple model estimation with a neural
network to obtain more accurate estimates. The key idea is to use the data from
the initial phase of the run for system identification, and then run a single estimator
designed for the identified model for the remainder of the run. The use of multiple
models and neural networks allows the on-line identification to take place extremely
quickly. The method is validated on actual data from an important estimation
problem in microelectronics manufacturing which is subject to model uncertainties:
determining end-point to an etch step using reflectometry data. *Copyright © 1998
IFAC*

Keywords: Adaptive Algorithms, Neural Networks, Estimation Algorithms

1. INTRODUCTION

Estimation is a key element of advanced sensing
and control of complex systems. Often a model
is available but may have unknown parameters,
requiring adaptive estimation techniques to be
applied. A useful approach to adaptive estimation
is the so called multiple model estimation. In
this case, the system dynamics are unknown, but
assumed to be one of a fixed number of known
possibilities. That is, consider the set of models

$$x(k+1) = f(k, x(k), u(k), \theta_j) + w(k) \quad (1a)$$
$$y(k) = g(k, x(k), u(k), \theta_j) + n(k) \quad (1b)$$

where $x \in \mathbb{R}^n$ is the state, $y \in \mathbb{R}^p$ is the output,
$w \in \mathbb{R}^n$ is a random disturbance and $n \in \mathbb{R}^p$ is
random measurement noise. The model is param-
eterized by $\theta_j \in \mathcal{S} \subset \mathbb{R}^d$, .where $\mathcal{S}$ is a finite set.
It is assumed that the true system captured by
some model j. This approach began with the work
of Magill (1965), who considered the problem under the usual linear/Gaussian assumptions. Magill

derived the optimal least squares estimator, which
consists of running several Kalman filters in parallel, the number of filters equal to the number
of possible modes. The minimum variance state
estimate is given by

$$\hat{x}(k|k) = \sum_j \hat{x}_j(k|k) P[\theta_j | Y^k] \quad (2)$$

where $\hat{x}(k|k)$ is the minimum variance estimate
for model j, and $P[\theta_j | Y^k]$ is the probability that
model j is the correct model given Y^k, the measured data up to time k. The linear/Gaussian
assumptions allow a recursive algorithm for the
calculation of $P[\theta_j | Y^k]$.

The multiple model approach has been used in
many applications, including flexible structures
(Griffin Jr. and Maybeck, 1995), polymer deposition (Krauss and Kamen, 1996) and oxide etching
(Vincent et al., 1997b) in electronics manufacturing, and unmanned flight control (Lane and Maybeck, 1994). Often, these applications do not exactly match the multiple model estimation struc-

ture. In particular the model uncertainty may be parametric, but the unknown parameter may vary continuously, rather than discretely. A possible approach to apply the multiple model method is to discretize the model parameter space. The choice of discretization is discussed by Sheldon and Maybeck (1993), who propose a design method to best choose a discrete model set to approximate the true uncertainty in order to minimize the estimation error. However, for computational reasons, it may not be possible to grid the parameter space sufficiently fine to obtain the true model parameter to the desired accuracy. Our approach is to use a neural network in conjunction with the multiple model estimators to interpolate between them. This method is related to that described in (Fisher and Rauch, 1994), where a neural network is used to extend the region of operation of an extended Kalman filter. However, there are many advantages to using a multiple model approach, and it is our objective to reduce the computational expense by enabling a coarser discretization of the uncertain parameter space.

This work is very strongly motivated by the following technological application. In microelectronics manufacturing, a common processing step involves reactive ion etching, where features are etched into previously deposited material. An important problem is estimating when to stop the etch, or in other words, when to call the process endpoint. One sensor which is used to determine film thickness is reflectometry, where by collimated light is directed onto the wafers surface and the reflected light intensity is monitored. As has been shown in (Vincent et al., 1996; Vincent et al., 1997b), the etching/reflectometry system can be modeled with linear dynamics and a nonlinear output equation, which maps the remaining film thickness to reflectance. This mapping is well defined if the underlying materials are known. On the other hand, if there is some uncertainty in the underlying stack then the nonlinear output equation is uncertain.

To deal with this uncertainty, one approach would be to define new state variables corresponding to the uncertain parameters with integrator dynamics. This approach is not successful in this application as there are severe observability problems in the resulting estimation problem. In addition, the etching step can be fairly short, and there is limited time for an adaptive estimator to converge. This suggests the use of the multiple model estimation technique where by a set of models is chosen which discretizes the uncertainty space. However, accuracy is critical, and errors in the stack model will contribute to errors in the estimated remaining thickness.

To improve the accuracy of the multiple model approach requires exploiting some of the specific requirements of the end-point estimation problem. The key task is to obtain accurate estimates of the remaining film thickness, but only for the period of time just before the desired endpoint is reached. If, before the endpoint is reached, an estimate of the true underlying stack can be obtained, an improved estimate of the remaining thickness could be calculated during the remainder of the etch using a single estimator that uses this newly estimated stack model. Restated, a possible procedure is to process data in preparation for a system identification until accurate estimates are required, and then perform the system identification and use an estimator designed for the identified model after this point. Thus we have explicitly separated the tasks of system identification and estimation which are usually combined in an adaptive estimation algorithm. However, the on-line nature of the estimation problem lends greater demands for fast convergence and low computational requirements than for an off-line system identification problem. The method presented here of combining multiple model estimation with a neural network addresses this need.

2. METHOD

In multiple model estimation, the a-posteriori probabilities $P[\theta_j|Y^k]$ for a discrete set of models are calculated (under linear/Gaussian assumptions) via

$$P[\theta_j|Y^k] = \frac{p(Y^k|\theta_j)P[\theta_j]}{\sum_i p(Y^k|\theta_j)P[\theta_j]}$$

where $P[\theta_j]$ is known a-priori and $p(Y^k|\theta_j)$ is the probability distribution function for the measurements Y^k given model $\theta_j \in S$ and is calculated on-line as

$$p(Y^k|\theta_j) = \frac{\exp\left(-\frac{1}{2}\sum_{i=1}^{k}(y_i - \hat{y}_i^j)'(C_i^j)^{-1}(y_i - \hat{y}_i^j)\right)}{(2\pi)^{\frac{kn}{2}}\prod_{i=1}^{k}\left|C_i^j\right|}$$

Where $\hat{y}_i^j$ is the Kalman Filter output estimate for model j and time i, and C_i^j is the output covariance matrix which can also be calculated by the Kalman Filter. Note that $p(Y^k|\theta_j)$ is simply a measure of the closeness of the observed data to that which would have been produced by model θ_j.

The optimal multiple-model estimate given in (2) is a mixture of the estimates of each model weighted by the a-posteriori probabilities $P[\theta_j|Y^k]$. The greatest weight is given to the model with the maximum a-posteriori probability. This model is

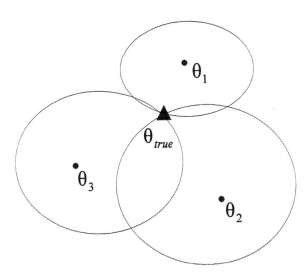

Fig. 1. Identification using distance from multiple models

termed the maximum a-posteriori (MAP) model estimate. Often, the weight corrisponding to a single model dominates the others. For example, in (Baram and Sandell Jr., 1978) general conditions, including the usual LTI/Gaussian assumptions, are given such that $\lim_{k\to\infty} P[\theta_j|Y^k] = 1$ for θ_j closest to the true model in an information distance metric, and $\lim_{k\to\infty} P[\theta_i|Y^k] = 0$ for all other models. This makes explicit the system identification which is implicitly performed by the multiple model adaptive estimation. In order to calculate the MAP estimate in the case of continuous parameter variation, one would like to calculate the probability density $p(\theta|Y^k)$ for arbitrary values of θ in order to perform a numerical search for the maximum. This would be given by

$$p(\theta|Y^k) = \frac{p(Y^k|\theta)p(\theta)}{\int p(Y^k|\theta)p(\theta)d\theta}$$

but it cannot be calculated as $p(Y^k|\theta)$ is only available for the discrete values of θ in S. Clearly, a numerical search for the maximum of $p(\theta|Y^k)$ would be very computationally expensive for a general, perhaps time varying or nonlinear, model.

As an alternative, it is proposed to use the information contained in $p(Y^k|\theta_j)$ as a measure of the distance of the model θ_j to the correct model. By comparing the relative distance to each model in S, an estimate of the true model parameter can be obtained. The idea is depicted in Figure 1 for the case when $\theta \in \mathbb{R}^2$. Suppose that we have calculated that $p(Y^k|\theta_j) = c_j$ for $\theta_j \in S$. The rings represent the set in $\mathbb{R}^2$ for which the expected value of $p(Y^k|\theta_j)$ is c_j given θ_{true} is the correct parameter. That is, each ring represents the set $\{\theta|E\left[p(Y^k|\theta_j)|\theta=\theta_{true}\right] = c_j\}$ for some j, where the expectation is taken over the data Y^k. If the intersection of these sets could be found, an estimate of the true parameter θ_{true} could be obtained.

In order to simplify calculations and to improve scaling, define

$$v_j(Y^k) := \sum_{i=1}^{k}(y_i - \hat{y}_i^j)'(C_i^j)^{-1}(y_i - \hat{y}_i^j)$$

and observe that $-\log p(Y^k|\theta_j) \propto v_j(Y^k)$ modulo a constant. Thus $v_j(Y^k)$ also qualifies as a measure of the goodness of fit, and will be used in what follows.

For clarity, first consider the case with the noise and disturbance equal to zero and the input and initial conditions fixed and known. Then the output Y^k is a function of θ only, and thus so is the measure of fit $v_j(Y^k)$. Let

$$\Gamma(\theta) = \begin{bmatrix} v_1(Y^k(\theta)) \\ v_2(Y^k(\theta)) \\ \vdots \\ v_N(Y^k(\theta)) \end{bmatrix}$$

denote the mapping from the true parameter θ to the vector of measures of fit for each estimator. This mapping can be (locally) inverted if it is (locally) one to one and continuous. Clearly, at minimum, we require $N \geq d$ where d is the dimension of θ. Because this mapping is an extremely complex function of θ, it is proposed take advantage of the well known universal approximation properties of a feed forward neural network to determine this inverse mapping $\Gamma^{-1}(v)$. Thus, we will have

$$\hat{\theta} = NN(v_1(Y^k), v_2(Y^k), ..., v_N(Y^k))$$

where $NN(\cdot)$ indicates a neural network. In this way, the neural network interpolates between the fixed values of θ_j chosen for the multiple model estimator.

The training of the neural network takes place off line, by simulating the system (1) with choices of θ distributed over the expected range. Note that this distribution can be much finer than that chosen for the multiple model estimators. Since this neural network is trained off line, the online computational cost is determined solely by the number of estimators in the multiple model estimator. Because of the interpolation which is afforded by the neural network, fewer estimators may be required, and the speed of the multiple model estimation can be improved.

The proposed estimation procedure as depicted in Figure 2 is thus as follows:

- N estimators are designed for different values of θ
- For a fixed amount of time (up to time k), these estimators are run in parallel to produce measure of fit $v_j(Y^k)$

151

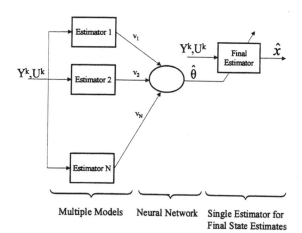

Fig. 2. Structure of combined estimator

- These measures of fit are use by the neural network to obtain estimate $\hat{\theta}$
- A single estimator designed for $\hat{\theta}$ is used in the remainder of the run. This estimator can be initialized using the states and error covariances of the estimator in the multiple model bank with the smallest value of $v_j(Y^k)$, or the data collected up to the current time can be run through the estimator.

In the more general case when input u and initial conditions x_0 are not known ahead of time, the trajectory Y^k must be considered a function of them as well. Let the input and initial conditions be parameterized in a suitable form, for example

$$u(t) = \sum_{i=1}^{M} \alpha_i u^i(t)$$

$$x_0 = \sum_{i=1}^{L} \beta_i x_0^i$$

where $u^i(t)$ and x_0^i are fixed and known. Let $\alpha = \begin{bmatrix} \alpha_1 & \cdots & \alpha_M \end{bmatrix}^T, \beta = \begin{bmatrix} \beta_1 & \cdots & \beta_L \end{bmatrix}^T$. Then the mapping becomes $\Gamma(\theta, \alpha, \beta)$ and $N \geq d + M + L$ for inversion.

3. RESULTS

This technique was applied to the reflectometry estimation problem described above. As described in (Vincent *et al.*, 1997b), the etching/reflectometry system can be modeled with simple linear dynamics and a nonlinear output equation:

$$\begin{bmatrix} er_{k+1} \\ \delta er_{k+1} \\ d_{k+1} \end{bmatrix} = \begin{bmatrix} a & 0 & 0 \\ 0 & 1 & 0 \\ \Delta T & \Delta T & 1 \end{bmatrix} \begin{bmatrix} er_k \\ \delta er_k \\ d_k \end{bmatrix}$$
$$+ \begin{bmatrix} 0 \\ w_1 \\ w_2 \end{bmatrix} + \begin{bmatrix} bu_k \\ 0 \\ 0 \end{bmatrix}$$
$$y_k = r(d_k, \theta, \phi) + n_k$$

where er is the nominal etch rate of the RIE, δer is the etch rate drift due to disturbances, and d is the film thickness. Note that the nominal etch rate has been modeled as a 1st order response. The input u is the forward power to the RIE, and is usually held fixed through the etch, thus the input will be a step. The constants a and b can be fit to match the nominal response of the RIE. The reflectometry output y is the amount of reflected light, which is a function of the material stack properties (θ, ϕ) and the top layer thickness d, with added measurement noise n. Those stack properties which are known are included in the vector ϕ, while those that are unknown are included in θ. Sensor gain and offset can also be included, but will not be considered here. The unknown inputs w and n are mutually independent white Gaussian noise. Because of the nonlinear output equation, the extended Kalman filter is applied.

Consider an etch of amorphous silicon (a-Si) on a silicon nitride/tantalum (SiN$_x$/Ta) stack with an initial a-Si thickness of 900 Å and a desired endpoint a-Si thickness of 500 Å (see Figure 3.) In this case, d is the thickness of the a-Si layer, while ϕ contains the refractive indices of a-Si, SiN$_x$ and Ta. The nominal thickness of silicon nitride is known but the actual thickness varies around the nominal, thus the system model is uncertain and θ will be the thickness of SiN$_x$. (Ta is highly absorbing at the wavelengths of interest and may be considered semi infinite.).

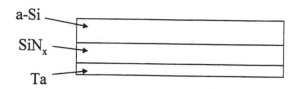

Fig. 3. Experimental stack to be etched

The operation of the multiple model estimator was as follows: since the goal was to achieve endpoint at 500 Å, a decision as to the true SiN$_x$ thickness was made with 600 Å remaining. The best estimate of remaining film thickness (the film thickness estimate from that estimator with the smallest value of $v_j(Y^k)$ was used to determine when 600 Å was left. To account for variations in etch rate, $v_j(Y^k)$ was calculated as follows:

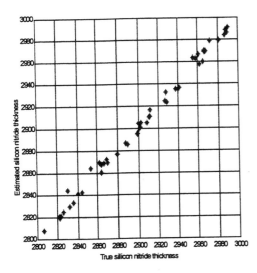

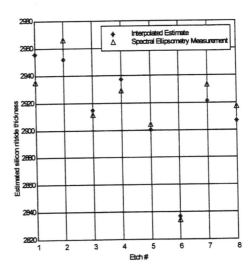

Fig. 4. Validation Results

$$v_j(Y^k) = \frac{1}{k_{600}-k_{800}} \sum_{i=k_{800}}^{k_{600}} (y_i - \hat{y}_i^j)'(C_i^j)^{-1}(y_i - \hat{y}_i^j)$$

where k_{800} is the sample point at which 800 Å is estimated to be remaining, and k_{600} is the sample point at which 600 Å is estimated to be remaining.

The multiple model estimator was applied with 3 models. The models had silicon nitride thickness of 2800, 2900, and 3000 Å respectively. A neural network was trained on 400 simulated etches. In the simulated etches the silicon nitride thickness was varied uniformly between 2800 and 3000 Å, the initial amorphous silicon thickness was varied between 875 and 975 Å, and the steady state etch rate was varied between 4.5 and 5.5 Å/s. Measurement noise was also added. More details on the simulation of in-situ reflectometry can be found in (Vincent et al., 1997b). The 3 input/single output neural network had 2 hidden layers of 2 nodes each. The key idea is that at 600 Å remaining, the true silicon nitride thickness is estimated, and this value will be used in the last part of the etch with a single-model estimator for accurate endpoint.

The neural network was validated using 50 additional simulated etches with similar variations in initial conditions, and the results are shown in Figure 4. Note that the neural network is able to recover the true silicon nitride thickness quite accurately for silicon nitride between the thicknesses of 2800 and 3000 Å, even though the multiple model estimator contained film stack models only for the discrete values of 2800 Å, 2900 Å and 3000 Å. The mean squared error was .6 Å and the maximum error was 13.8 Å.

This estimator was also validated using experimental data for the etch described above. This data is from experiments previously reported in (Vincent et al., 1997a) where a multiple model

Fig. 5. Experimental Results

estimator was used with 3 models with SiN$_x$ thicknesses of 2850, 2900 and 2950 Å, but without interpolation so the SiN$_x$ estimate could only be rounded to the nearest 50 Å. In Figure 5, the SiN$_x$ estimate using the same neural network interpolator is shown, along with an independent measurement of SiN$_x$ thickness obtained using spectral ellipsometry (SE). The results are quite good, with a mean squared error of 3.8 Å and a maximum error of 16.4 Å.

4. CONCLUSION

An adaptive estimation method combining multiple model estimators and a neural network was presented. This method used initial data and a multiple model estimator to obtain measures of fit which were input to a neural network. Because of the structure of the motivating problem, the adaptive estimation technique consisted of a system identification phase followed by an estimation phase. Thus the main contribution was developing an identification method which was compatible with the speed and computational requirements of on-line, real-time estimation.

A motivating problem was described and used to validate the method. This problem was to determine from reflectometry data the amount of remaining amorphous silicon film on a wafer during etching in the face of uncertainties in the silicon nitride/tantalum stack. The combined estimator performed quite well, estimating the uncertain silicon nitride thickness with a mean squared error of 3.8 Å. However, there were the following fortuitous aspects: only a single parameter was unknown, the input was fixed, and the state dimension was small. This allowed the use of only 3 estimators, and training could be accomplished with 400 simulated runs.

This paper only introduced this particular method of combining multiple model estimators with a neural network, and many open equations remain. In particular it would be useful to determine under what conditions the mapping Γ is locally invertible for a class of models. It would also be useful to determine those parameters to which Γ is insensitive. For example, if the true system is stable, then the initial conditions will have an effect which decreases in time.

5. REFERENCES

Baram, Yoram and Nils R. Sandell Jr. (1978). An information theoretic approach to dynamical systems modeling and identification. *IEEE Trans. Aut. Cont.* **AC-23**(1), 61–66.

Fisher, William A and Herbert E Rauch (1994). Augmentation of an extended Kalman filter with a neural network. In: *Proc. of IEEE Conf. on Neural Networks, Orlando FL.* pp. 1191–1196.

Griffin Jr., G. C. and P. Maybeck (1995). MMAE/MMAC techniques applied to large space structure bending with multiple uncertain parameters. In: *Proc. 34th Conference on Decision and Control.* pp. 1153–1158.

Krauss, A. F. and E. W. Kamen (1996). A multiple model approach to process control in electronics manufacturing. In: *Proc. IEEE/CPMT International Electronics Manufacturing Technology Symposium.* pp. 455–461.

Lane, D. W. and P. S. Maybeck (1994). Multiple model adaptive estimation applied to the lambda urv for failure detection and identification. In: *Proc. 33rd Conference on Decision and Control.* pp. 678–683.

Magill, D. T. (1965). Optimal adaptive estimation of sampled stochastic processes. *IEEE Trans. Aut. Cont.* **AC-10**(4), 434–439.

Sheldon, S. N. and P. S. Maybeck (1993). An optimizing design strategy for multiple model adaptive estimation and control. *IEEE Trans. Automat. Contr.* **38**(4), 651–654.

Vincent, T. L., P. I. Klemicky, W. Sun, P. P. Khargonekar and F. L. Terry Jr. (1997a). A highly accurate endpoint method for a tft back channel recess etch. In: *Proceedings of the 1997 International Display Research Conference.*

Vincent, T. L., P. P. Khargonekar and F. L. Terry, Jr. (1996). An Extended Kalman Filter based method for fast in-situ etch rate measurements. In: *Diagnostic Techniques for Semiconductor Materials Processing II: Symposium held November 27-30, 1995, Boston, MA* (S. W. Pang, O. J. Glembocki, F. H. Pollak, F. G. Celii and C. M. Sotomayor Tor-res, Eds.). Materials Research Society, Pittsburgh, PA. pp. 87–94.

Vincent, Tyrone L., Pramod P. Khargonekar and F. L. Terry, Jr. (1997b). End point and etch rate control using dual wavelength reflectometry with a nonlinear estimator. *J. Electrochem. Soc.* **144**(7), 2467–2472.

Computational Intelligence Procedures for Monitoring and Control of MBE Growth of Arbitrary Film Structures

Z. Meng[1], W.T. Taferner[2], Q. Yang[1], K.G. Eyink[2] and Y.-H. Pao[1]

[1]AI WARE, Division of Computer Associates International, Inc.
3659 Green Road, Beachwood, OH 44122
[2]Air Force Research Laboratory, Materials & Manufacturing Directorate
Wright-Patterson AFB, OH 45433

Abstract

This paper describes the use of computational intelligence procedures for real-time interpretation of multiwavelength ellipsometry data and the use of the results in monitoring and control of Molecular Beam Epitaxy (MBE) growth of thin films. Data reduction is carried out with the Guided Evolutionary Simulated Annealing (GESA) approach, which is a genetic algorithm like optimization method, and a neural net model of the available multiwavelength spectroscopic materials data. The smaller noise level in the Ψ parameter was taken advantage of to improve the accuracy of the results. Control of the film growth is accomplished through adjustments of crucible temperatures. Due to the constraints in data acquisition and processing, such adjustments may be in either feedback mode or feedforward mode depending on the profile of the structure and the switching between the two modes are carried out automatically. Results from two growth examples of different target profiles are reported for the system of $Al_xGa_{1-x}As$ film on GaAs substrate. In one case, there was a single step change in composition while linear and parabolic grades were included in the other. *Copyright © 1998 IFAC*

1. Introduction

The ability to monitor and control the growth of thin films of arbitrary vertical growth profiles using MBE offers new freedom in device structure design. The performance of many device structures could be improved by including vertically continuously varying regions. One frequently noted example is the heterojunction bipolar transistor [1]. Although other types of sensors such as Desorption Mass Spectrometry (DMS) have been used in monitoring and control of the growth process [2], the most popular type of sensors for such purposes is probably optical in nature based on the theory of ellipsometry [3]. This is due to the fact that ellipsometry has the ability to determine not only the optical properties of the top surface but also of film-substrate structures and the measurement process is non-destructive in nature.

Ellipsometry utilizes the fact that the polarization state of the reflected light will be a function involving the optical and structural properties of the surface and the underlying structure. Thus, such properties in principle can be obtained from ellipsometry measurements.

However, what can be directly obtained from the ellipsometry measurements are the change of the polarization state in terms of the basic parameters Ψ and Δ. The optical and structural property parameters of interest have to be computed from a model which is in the form of a set of transcendental equations. Given these parameters, Ψ and Δ can be computed from the model but numerical methods have to be used for computations in the reverse direction, except for the simple case of a bare substrate for which an analytical relationship exists. Thus, accuracy of the estimated values of the parameters of interest depends directly on the numerical method used and the validity of the assumptions made in deriving the model in the first place [4].

It is possible to directly use the basic parameters to achieve control of film growth by using empirical relationships and extensive calibration [5, 6]. The advantage of such an approach is that high precision can be achieved and the load for data processing is small. However, the calibration parameters and the empirical relationship obtained are usually specific to one material and one experimental setup and may also be specific to a particular wavelength and substrate temperature.

Direct monitoring and control of film growth can be achieved if the optical properties and the growth rate are computed first. Traditionally, descent search methods have been used to compute the film parameters [7]. These methods are very successful for the case of insulating films on a substrate, since the index of refraction is real. But for more complicated situations such as growth of conducting materials, no

descent search method so far has been shown to guarantee convergence to the correct result unless accurate initial estimates are given. Otherwise, the result tends to converge to a local minimum point, an undesired and misleading outcome. Urban et al. [8] described the use of neural networks for providing initial estimates to a descent search solver based on the Levenberg-Marquardt technique.

It is also possible to use neural networks to estimate the film parameters directly. Park et al. [9] described the use of random-vector functional-link nets [10] to predict the optical parameters and thickness of layers. Multiple models serving different ranges of growth were used to enhance the accuracy. However, even that method was still not robust enough to handle cases with a relatively large amount of noise.

Another approach is to use stochastic search methods. The advantage is that a stochastic search method does not easily get trapped to a local minimum as a method relying on descent direction alone would. The disadvantage is that it usually takes longer to converge, and that is undesirable for real-time data processing. Guided Evolutionary Simulated Annealing (GESA) [11, 12] is a particular form of evolutionary search method with a mechanism to provide regional guidance during the search process in order to reduce the time to convergence.

Inversion of the optical model is general and applicable to any material. Information specific to the materials being deposited can then be computed using the optical parameters determined in the inversion process. In this work, inversion results obtained with GESA for several different wavelengths are used to feed the neural net model based on materials data to compute the composition parameters of the film being grown.

In the MBE system of this particular study [13], arsenic is kept at saturation level and temperatures of gallium and aluminum crucibles determine the rates of gallium and aluminum growth on the substrate. Temperature adjustments to these two crucibles can thus control the composition and total rate of the film growth. Control of the growth process is thus the process of generating set points for these two crucibles. The mode of control can be either feedback or feedforward based on the thickness of the uniform layer. For a uniform layer of sufficient thickness, feedback control to the film growth can be implemented based on the monitoring results. Due to the relatively slow data acquisition rate of the current system setup and the inherent delay in using ellipsometry in monitoring, feedforward control must be used for thin layers when the target varies quickly. This is the approach adopted in this work.

2. The GESA method in data reduction

Due to the conductive nature of the material being grown in this study, multiple measurement points need to be taken before solution to the parameters can be computed. For the case of $Al_xGa_{1-x}As$ film on GaAs substrate, with the pseudo-substrate assumption, there are three unknowns, namely, the real index of refraction (n_2), the extinction coefficient (k_2) of the surface layer and

the growth rate (R). This determines that at least two measurement points need to be taken. However, in order to reduce the effect of noise, a few more points are also taken and the root-finding problem is converted into an optimization problem in the least-square sense with the following error function

$$E = \sum_i [(\Psi_{mi} - \Psi_{ci})^2 + (\Delta_{mi} - \Delta_{ci})^2]$$

(1)

where the subscript mi indicates the ith measured value, and ci the ith computed value based on a specific set of values for the optical constants and growth rate. The goal of the optimization is to find a set of values for n_2, k_2, and growth rate values at all measurement points so that Equation (1) is minimized. Since the composition estimation is based on the results from this optimization task, this task is of vital importance to the investigation. In this study, the Ψ_{ci} and Δ_{ci} are computed using the Drude equations [14]. One important assumption made in deriving these relations is that the film should be uniform over the whole period during which the data points used in optimization are taken.

Due to the likelihood that a descent search algorithm is getting trapped to a local minimum which is completely undesired [8], and the lack of enough data to build a reliable neural net model to provide initial estimates to a descent search, the population based search algorithm Guided Evolutionary Simulated Annealing (GESA) is used to carry out the optimization of Equation (1).

GESA tries to combine the advantages of genetic algorithm [15, 16] and simulated annealing [17]. In the search process, the quality of the proposed solutions are evaluated for the various search regions and the GESA algorithm automatically focuses the search onto regions with indications of higher chances of attaining the global minimum of an objective or error function. This is achieved by competition within the population which is the genetic part of the algorithm. The simulated annealing part of the algorithm helps in trying to prevent the search from being trapped to local minima by accepting temporary degradation of the value of the objective function with certain probabilities in accordance with the degree of degradation and also depending on the stage of the search.

This general approach has been described in more detail in a previous manuscript reporting earlier monitoring results [18].

Due to the higher sensitivity to the presence of noise [9], the results of growth rate from the optimization process can be of less accuracy than those for optical properties. Improvement to the estimation of the growth rate has been made since. It took advantage of the fact that for this particular experimental setup, the noise level in the Δ parameter is much higher than that for the Ψ parameter, as illustrated in Figure 1. The reason is that the value of the Δ parameter frequently goes close to 180° when the intensity of the reflected light that passes through the polarizer goes to zero. Under such circumstances, any background noise becomes significant.

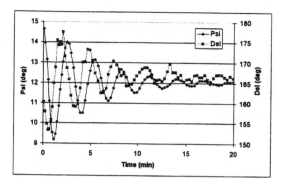

Figure 1: Example of experimental data of Ψ and Δ at $\lambda = 5286$ Å. It can be seen that the noise level in Ψ is much lower.

It can be seen from Figure 1 that both Ψ and Δ are of sinusoidal form with decreasing amplitudes. The "period" of the signals corresponds to a 2π change in the film phase thickness which can be converted to actual thickness. Thus, by measuring the period of the Ψ signal, the growth rate of the film can be obtained directly if the growth can be considered uniform during that period.

A peak finding algorithm was used to measure the period of the Ψ signal. When two peaks of the opposite type are detected, the period is twice the time elapsed between the two peaks. If two peaks of the same type are detected due to presence of noise, the algorithm will reset itself. Since growth rate estimates using this method will be available only once in a while, it is used to provide guidance to the optimization process by setting the search space around it.

3. Estimation of Composition with Neural Net Model of Materials Data

The availability of new and expanded set of multiwavelength spectroscopic materials data enabled the use of a neural net model to estimate the composition of the film from its optical characteristic parameters obtained during the inversion of the optical relationship equations using optimization. This not only saves time and storage space as compared to table lookup and local curve fitting procedures, but also may reduce the random noise that may present in the database due to the smoothing ability of the neural net.

Data required for training the neural net are obtained from the "AlGaAs3f" database which contains experimental results for relationship between wavelength (λ), real index of refraction (n_2), extinction coefficient (k_2), substrate temperature (T) and the composition (x) at various combinations of these parameters. The neural net model was established using Process Advisor[TM], a neural net modeling and optimization package from AI WARE. In training the neural net model of the relationship, the first four parameters are used as inputs to the net and x is used as output. The net architecture used in learning the model is a hidden layer net with one hidden layer of 10 nodes.

Logistic functions were used as activation functions for both hidden and output nodes. Accelerated back-propagation of error algorithm was used in training the net. Fifteen percent of the patterns were picked out randomly and were used as test patterns. The training and testing errors showed that the model could estimate composition to an accuracy better than 5‰, on average. Figure 2 shows the output of the net model for $\lambda = 5126\text{Å}$ and $T = 600°C$ for n_2 and k_2 values in the range shown. The seemingly flat region in the bottom portion of the surface shown in Figure 2

is outside valid range of combinations of n_2 and k_2, as is a smaller portion on the top. Unrealistic composition values (smaller than 0 or larger than 1) are shown for those regions due to non-physical extrapolation of the model beyond the ranges of validity.

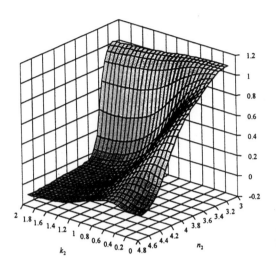

Figure 2: Model estimates of composition for various n_2 and k_2 values for given wavelength (5126Å) and substrate temperature (600°C). Results are not meaningful for certain ranges of values, such as for $x < 0$ and $x > 1$.

4. Integrated Control for Arbitrary Growth Profiles

The task of monitoring is achieved when the optical parameters and growth rate are computed by the GESA optimization algorithm and the composition parameter by the neural net model of materials data. The next task is to control the growth process by computing the set points for the crucibles. For that it is necessary to know the response of the crucible, i.e. the change of the deposition rate of the materials evaporated out of the crucible, for a given set point.

In principle, such a model can be established with a neural net. Given a set of experimental data, it is possible to learn a net to predict the response for the given input. In general, this is the method to take if nothing is known about the system except the experimental data. Fortunately, for this particular investigation, there exists a theoretical exponential relationship between the set point temperature and the deposition rate. From the experience of the users, the system seems to conform to that theoretical relationship reasonably well in the whole range of the set points of interest.

The theoretical relationship is given by Equation (2).

$$R = A\exp(-\frac{\Delta H}{kT})$$

(2)

where A and ΔH are characteristic constants of the crucible, k is Boltzmann constant and R is the deposition rate of the material in the crucible measured in Angstrom per second. This equation can be rewritten into the following linear form by taking logarithm of both sides of the equation.

$$\ln R = -\frac{m}{T} + b$$

(3)

where $m = \Delta H/k$ and $b = \ln A$. The values of the constants in this equation are easily obtained from the experimental results.

There is a simple relationship between the deposition rates of the aluminum and gallium and the composition of the film. Assuming uniform sticking coefficients of both aluminum and gallium, which is approximately true in normal range of substrate temperature with arsenic kept

at saturation level, the total growth rate will be the sum of the deposition rates of the two materials and the composition of the film is determined by the ratio of the two deposition rates. Since the lattice constants of both GaAs and AlAs are very close, the difference between the ratio of the deposition rate and the ratio of the number of atoms can be ignored. Thus, the relationship between the composition and the deposition rates R_{Al} and R_{Ga} can be written as

$$x = \frac{R_{Al}}{R_{Al} + R_{Ga}} = \frac{R_{Al}}{R_{Total}}$$

(4)

where R_{Al} and R_{Ga} can be evaluated using Equation (2) or (3) and R_{Total} is the growth rate.

In actual growth control, a frequently used method for determining the crucible temperatures is to set one crucible temperature constant and to compute the temperature of the other crucible and the expected total growth rate from the target composition x using above equations.

In principle, if the calibration constants are very accurate and there has been no drift in the system since the last calibration, temperature set point values obtained using such a method will yield good results. However, it is rarely true that the system conforms to the idealized theoretical relationships precisely. This may be because the system configuration is not quite like that assumed in the derivation of the theoretical relationships. Drift is also unavoidable due to the depleting of materials in the crucibles. Feedback can be used to remedy this problem.

Feedback control could be achieved using the PID model with proper adjustment of the gain parameters. However, such an approach fails to take advantage of the fact that there is a theoretical model available which is reasonably accurate. Therefore, the approach shown in Figure 3 was adopted to achieve the feedback control.

In this approach, the control actions are generated by the theoretical model of the relationship between the set points and the resulting composition, which are given in Equations (2)-(4). The detected difference between the target and the measurement is used to either update the model or shift the desired input value to the model for compensation.

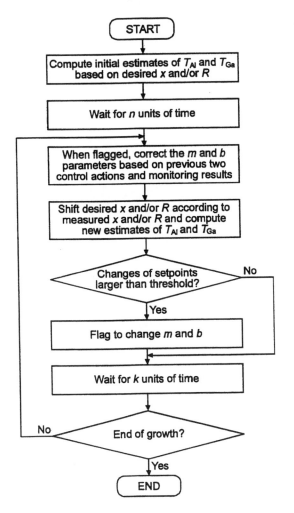

Figure 3: Flow chart for feedback control based on the theoretical model.

When the difference is small, that is, when the change in set points is smaller than a threshold value, the desired inputs to the model will be shifted an amount equal to the detected difference to compute the new set points. The assumption is that the difference is mostly due to normal fluctuations in sensor outputs and/or the imperfect agreement between the actual system and the theoretical model. If the difference is large, it may indicate that the previously calibrated model parameters are not accurate enough now due to drift and/or experimental errors in calibration. In this case, the results from previous two actual measurements will be used to update the model parameters. Subsequent set points will be computed using the updated model.

Due to the larger variation in the beginning of the growth and the larger experimental error of the Δ parameter due to the larger size of the spiral, the composition values obtained during this period have larger errors. Thus, an initial wait is introduced to skip this period before control actions are generated based on measured values. Due to delays in system response immediately after one control action, another wait is introduced in between the control actions to prevent incorrect control actions based on the transients. This delay includes the natural delay in the electric heating elements and the requirement of constant growth rate during the multiple measurement points for one optimization in order to get stabilized results.

Since control actions are generated based on detected difference from target values, feedback control in general requires accurate estimation of the system state with as little delay as possible. However, one measurement point is not enough to solve the current ellipsometry problem which means that the monitoring result has an inherent delay. The use of multiple measurement points only exacerbates this problem. Due to the slow data acquisition rate for the current system, which is typically every 9 seconds, the amount of materials grown during this period can be more than one hundred angstroms thick. For uniform layers of sufficient thickness, this will translate to slower convergence but will still work. However, the presence of this delay and the wait time introduced to enhance stability will prevent this feedback approach from being used for cases where the target varies quickly with respect to the thickness of the film. Feedforward control has to be incorporated to handle such cases. Given a system model, a schedule of control actions can be computed from an arbitrary growth profile. Since the monitoring results are irrelevant for feedforward control, control actions can be generated at any time and as frequently as desired.

In order to use feedforward control, an accurate model of the system is required. In general, a neural net model of the system can be established to serve this purpose. Fortunately, for this investigation, existence of a theoretical model obviates the need for learning a model of the operation of the crucibles. The few parameters present in the theoretical model also make the work of calibration simpler.

However, the theoretical model can not capture all details of the actual experimental setup. It is just a description of the behaviors of an idealized system. Thus, even a good calibration of the model will still introduce errors in the control system. In addition to that, drift of the system parameters is also unavoidable due to depletion of materials in the crucibles. Thus, the quality of the system model degrades over time. This means that feedforward control will be suitable for only a short period after calibration.

The fact that both feedforward and feedback control utilize the same theoretical model of the system enables the formation of integrated control which takes advantage of both modes of control for an arbitrary growth profile. For this to work, the growth profile is divided into uniform layers. The thickness of the layers will be determined by the variation of the profile. The slower the variation is, the thicker the layer will be. During the growth, automatic switching between feedback and feedforward modes of control will be carried out based on the thickness of the current target layer. If the thickness of the current layer is larger than a set threshold, feedback mode of control will be used. This will provide better tracking of the target value and automatic correction of the system model if necessary. For thinner layers, feedforward control will be used based on the current system model to achieve the desired profile.

5. Results and discussions

Two examples of target growth profiles are reported for the system of

Al$_x$Ga$_{1-x}$As film on GaAs substrate. For all of the results in this section, the wavelengths used are 4661Å, 4966Å, 5126Å, 5286Å and 5446Å. The reduction of the five sets of data at each measurement point takes less than half a second on a 200 MHz Pentium PC.

The first example is a simple step in between two thick layers of uniform composition at 0.25 and 0.35 respectively. The results are shown in Figure 4. In this case, the aluminum crucible temperature was kept constant at 1133°C and the gallium crucible temperature was adjusted to achieve the desired composition. The initial variation was likely due to the transient when the shutter was opened and sensor artifact in the Δ parameter. It can be seen that the initial guess of the gallium crucible temperature based on previous calibration was off by approximately 10°C compared with the steady state value and that the control algorithm was able to bring the composition back to the desired value. The model parameters was also adjusted in the meantime so that when the target composition value was changed in the middle of the run, the guess for the gallium crucible temperature corresponding to the new desired composition value was much closer to the steady state value. The steady state composition value for time between 20 and 48 minutes is 0.250 with a standard deviation of 0.004 and for time after 51 minutes is 0.350 with a standard deviation of 0.005 for the target values of 0.25 and 0.35 respectively.

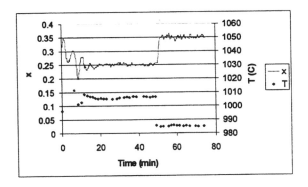

Figure 4: Control result of composition x and temperature adjustment to the gallium crucible with the alluminum crucible temperature kept constant at 1133°C.

It is worth mentioning that computational intelligence control is not feedback control. Instead, ideally, one learns a model of the composition and growth rate as functions of crucible temperature settings and determines the appropriate temperature schedule for any desired film structure. The experiment is then run in feedforward open-loop manner. There is no time to operate by trial and error because that would result in incorrect layers of film included in an otherwise correct film structure. For this particular problem, a simple exponential relationship between the crucible temperature and growth rate exists for steady state situations. Therefore, control of growth of layers of uniform film can be achieved with the exponential relationship model and feedback can be a handy tool to make small real-time corrections to the calibrations of model parameters. However, for complicated film structures, strict feedforward open-loop control must be used.

There are two aspects of the feedforward open-loop control. One part is concerned with demonstrating that control actions do result in desired results. This can be achieved by adjusting model parameters dynamically through feedback during growth of buffer layers. Once that is established, the model can be inverted, with an optimization procedure if necessary, so that the optimal control path can be obtained for any desired outcome trajectory. For this specific problem, given the desired profile of a film structure, one can compute the desired control temperature sequence.

In testing the ability of the program to control the growth of varying targets over thickness of the film, linear and parabolic grades are included. This profile contains two linear grades of 500 Å each from x = 0.3 to 0.003 and back. Another varying target is a parabola of 500 Å wide and again changes from x = 0.3 to 0.003 and back to 0.3.

During the growth, the crucible temperature for gallium was kept at 1000°C. The set points for the aluminum crucible was varied based on the profile to achieve the desired composition. The growth started on pure GaAs and a thin GaAs cap layer was also grown on top of the structure afterwards. Due to a preset low temperature limit to the aluminum crucible, the temperature was held at that value when composition targets dropped below approximately 0.03.

The resulting sample was analyzed with secondary-ion mass spectrometry (SIMS). The results from this analysis versus the target profile are shown in Figure 5. It is not clear yet whether the slight drifts for the flat regions occurred during growth time or were just artifacts of the SIMS. However, it still can be seen that the results followed the target profile very well. The measured composition at the bottom of the two

wells also agrees to the expected values due to the preset low temperature limit.

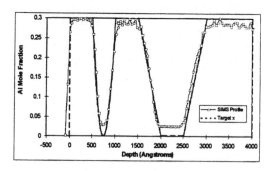

Figure 5: SIMS results versus target profile which contains two linear grades and one parabolic grade.

6. Conclusions

The use of a combined genetic algorithm and simulated annealing optimization process, together with optical and neural net models for real-time interpreting of ellipsometry data is described. Optical relations for ellipsometry are used in an optimization which gives film optical parameters n_2 and k_2 at several wavelengths and also an estimate of the growth rate. The film composition parameter x is obtained by consulting a neural net model which is based on previously obtained materials spectroscopic data. In all of this, the use of a guided stochastic search optimization procedure and the neural net model is the key to performance.

Control of film growth can be achieved in either the feedforward or the feedback mode. Feedback control may be used for cases of uniform layers of sufficient thickness to achieve better steady state results and for providing dynamic adjustment to model parameters in preparation for subsequent feedforward control. Feedforward control may be used to handle cases of complicated film structure when trial and error can not be tolerated. Integration of the two control modes will provide robust control for growth of arbitrary profiles.

The general approach to monitoring and control can be applied to MBE and related procedures such as CVD and CBE for growth of thin film structures.

Results obtained for the case of monitoring and control of the MBE growth of $Al_xGa_{1-x}As$ film on GaAs substrate are reported.

At the time of writing, it is known that certain crystalline rugate filter structures [19] have been fabricated successfully with the use of this technology. Further analysis and characterization is underway. This is encouraging and suggests that such technology might find ready use in other materials research and device fabrication tasks.

Acknowledgments

The authors would like to thank Dr. Steven R. LeClair, director of the Materials Directorate at the Wright Laboratory of the Wright-Patterson Air Force Base, for his guidance and support throughout the course of this investigation. This work was performed while one of the authors (W. T. T.) held a National Research Council AFRL/MLPO Research Associateship.

References:

[1] Hayes, J. R., Capasso, F., Malik, R. J., Gossard, A. C. and Wiegmann, W. (1983) Optimum emitter grading for heterojunction bipolar transistors, *Applied Physics Letters*, **43**(10), 949-951.

[2] Evans, K. R., Kaspi, R., Cooley, W. T., Jones, C. R. and Solomon, J. S. (1993) Arbitrary composition profiles by MBE using desorption mass spectrometry, presented at MRS Spring 1993 Meeting, San Francisco, CA.

[3] Passaglia, E., Stromberg, R. R. and Kruger, J. Eds. (1964) *Ellipsometry in the Measurement of Surfaces and Thin Films*, National Bureau of Standards Miscellaneous Publication 256, U.S. Government Printing Office, Washington D.C.

[4] Tompkins, H. G. (1993) *A User's Guide to Ellipsometry*, Academic Press, San Diego, CA.

[5] Aspnes, D. E., Quinn, W. E. and Gregory, S. (1990) Optical control of growth of $Al_xGa_{1-x}As$ by organometallic molecular beam epitaxy, *Applied Physics Letters*, **57**(25), 2707-2709.

[6] Aspnes, D. E., Quinn, W. E., Tamargo, M. C. (1992) Closed-loop control of growth of semiconductor materials and structures by spectroellipsometry, *Journal of Vacuum Science and Technology A*, **10**(4), 1840-1842.

[7] McCrackin, F. L. (1969) *A Fortran Program for Analysis of Ellipsometer Measurements*, National Bureau of Standards Technical Note 479, U.S. Government Printing Office, Washington D.C.

[8] Urban III, F. K., Park, D. C. and Tabet, M. F. (1992) Development of

artificial neural networks for real time *in situ* ellipsometry data reduction, *Thin Solid Films*, **220**, 247-253.

[9] Park, G. H., Pao, Y.-H., Igelnik, B., Eyink, K. G. and LeClair, S. R. (1996) Neural-net computing for interpretation for semiconductor film optical ellipsometry parameters, *IEEE Transactions on Neural Networks*, **7**, 816-829.

[10] Pao, Y.-H., Park, G. H. and Sobajic, D. J. (1994) Learning and generalization characteristics of the random vector functional-link net, *Neurocomputing*, **6**, 163-180.

[11] Yip, P. C. (1993) *The Role of Regional Guidance in Optimization: The Guided Evolutionary Simulated Annealing Approach*, Ph.D. Dissertation, Case Western Reserve University.

[12] Yip, P. C. and Pao, Y.-H. (1994) A guided evolutionary simulated annealing approach to the quadratic assignment problem, *IEEE Transactions on Systems, Man and Cybernetics*, **24**(9), 1383-1387.

[13] Taferner, W. T., Eyink, K. G., Brown, G. J., Szmulowicz, F., Hegde, S. M., Wheeler, B. Mahalingam, K. Walck. S. D., Solomon, J. S. and Lampert, W. V. (1998) Determination of the quality of in-situ monitoring of composition and thickness of AlGaAs by spectroscopic ellipsometry for the formation of p-type quantum well infrared photodetectors, submitted for publication to *Journal of Electronic Materials*.

[14] Drude, P. (1889) Ueber Oberflächenschichten, *Annalen der Physik und Chemie*, **36**, 865-897.

[15] Fogel, L. J., Owens, A. J. and Walsh, M. J. (1966) *Artificial Intelligence through Simulated Evolution*, Wiley, New York.

[16] Goldberg, D. E. (1989) *Genetic Algorithms in Search, Optimization and Machine Learning*, Addison-Wesley, Reading, MA.

[17] Kirkpatrick, S., Gelatt Jr., C. D. and Vecchi, M. P. (1983) Optimization by simulated annealing, *Science*, **220**, 671-680.

[18] Meng, Z., Yang, Q., Yip, P. C., Eyink, K. G., Taferner, W. T. and Igelnik, B. (1997). Combined use of computational intelligence and materials data for on-line monitoring of MBE experiments, presented at the Australiasia-Pacific Forum on Intelligent Processing and Manufacturing of Materials (IPMM'97), Gold Coast, Australia.

[19] Bovard, B. G. (1993) Rugate Filter Theory: An Overview. *Applied Optics*, **32**(28), 5427-5442.

DATABASE RETRIEVAL ENHANCEMENT USING THE BOXES METHODOLOGY

David W. Russell [a],

Dennis L. Wadsworth [b].

[a] *Engineering Division*
Penn State Great Valley
Malvern. PA 19355, USA.
drussell@psu.edu

[b] *Natural Language Understanding*
Unisys Corporation[1]
Malvern. PA 19355, USA
dennis.wadsworth@unisys.com

Abstract

The problem in temporal sequenced data has always been in identifying and defining a process that will describe anticipated changes in patterns of use in the application domain, and to then act on the projected changes prior to their need. This paper presents the concept and results from modeling a database as it learns to identify anticipated temporal sequenced data and accessing it using an adaptation of the BOXES paradigm as an intelligent agent. The database and data request streams are modeled, and the intelligent agent acts as a pre-processor to the internal data access caching mechanism. As the agent learns, improvements in access latency for the simulated application environment are clearly seen. *Copyright © 1998 IFAC*

Keywords - Database, Machine learning, Temporal logic, Problem identification, BOXES methodology.

1. INTRODUCTION

Jefferey Elman [6] writes, "[Time] is inextricably bound up with many behaviors which express themselves as temporal sequences. Indeed, it is difficult to know how one might deal with such basic problems as goal-directed behavior, planning, or causation without some way of representing time." The Mayas may have been the first *keepers of time* setting to record the temporal patterns of the seasons directing their food production (Tedlock [17]). The advent of computer technology has placed much of the valuable commercial information in databases and because of the intrinsic value of such data, "databases have become the cornerstone of corporate software development for more than two decades" (Bloor [2]).

The majority of research in the use of applied artificial intelligence (AI) in database technology seems to have focused on designing data layouts or in data retrieval (Bloor [2], Choobineh [5]). This paper investigates the use of an AI agent to non-invasively enhance the access of data in a real-time commercial application and uses a model developed during a research course at Penn State Great Valley (a graduate campus of The Pennsylvania State University in the Philadelphia suburbs) to support the findings. This paper explores time-oriented data whose need and duration is sensitive to and driven by

time (Wadsworth [18]) as the trigger in intelligent caching. The AI agent presented in this paper uses an adaptation of the BOXES methodology (Michie and Chambers [12]) to intelligently cache the required database record.

The BOXES methodology is used to forecast when requests for time-oriented data records are most likely to be made by the application and consequently decrease its access time by intelligently caching the record just prior to, and in anticipation of, the application's request. The major strength of BOXES is its unique ability to control a system without a detailed mathematical model (Russell [15]). The BOXES methodology does not employ a forecast model like the Box-Jenkins ARIMA methodology (Reynolds, *et al* [13]) to control a system. The BOXES methodology achieves control of a system through statistical inferencing based on a process of rewards and penalties applied to a Signature Table. For example, a reward is applied when advancement towards a pre-defined goal is made and vice versa; this is analogous to learning to play a board-game such as chess -- BOXES handles the task as a "game."

The paper briefly discusses the BOXES method, the database problem domain, the adaptation of the BOXES paradigm, and includes test results from simulations. The paper concludes with some

---

[1] The paper is for educational purposes and its views should not necessarily be attributed to Unisys.

discussion of the startling improvement in data access time that the intelligent enhancement agent has on the model of the database management systems (DBMS), and future research at Penn State Great Valley.

2. THE BOXES METHOD.

The BOXES paradigm (as shown in figure 1) operates and learns in real-time and requires the use of statistical data accumulated during a series of finite time intervals. The method partitions the system state space into a set of discrete cells or "boxes." These cells are defined by assigning a unique state number to each variable, based on some quantization algorithm, and then combining the individual state numbers into a system integer. For example, a variable may be divided into ten equal regions that allows that variable to be described by an integer I, such that $I \in \{1..10\}$. If for any reason the variable drifts outside of its pre-defined range, it is assigned a zero (0) state number and the system exits. If there are many variables in the system, the effect of quantization is to produce a grid-like structure of hyper-cubes which are defined by a set of integer state numbers as stylized in figure 2. This set of state numbers can be easily combined to form a "cell" index that is unique for each hyper-cube.

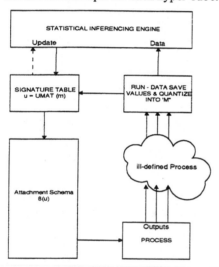

Fig. 1. A Typical BOXES Configuration

This index, which now completely defines the region or zone that the system is populating at any given instant, is linked to a specified control mechanism that spawns an array of control actions, all of which are pointed to by the cell index. This structure is sometimes called a Signature Table (Canudas de Wit [3]), as it contains detailed control signatures for system operations across the controllable domain. The Signature Table used in this research presents each cell as a structure that contains statistical data used to determine the cell's worth. A typical BOXES system presents signature data as bi-valued (e.g., left|right etc.). As the system operates, data is collected for all cells that are entered; typically data may include a running count of how often the cell was entered, or how long the system resided in the cell, or how long the system operated without failure after the influence of a cell, etc. After some arbitrary time period or on detection of a system failure (i.e. a

zero state index), the BOXES algorithm triggers the a learning procedure that manipulates and merges the recently acquired data with all previously collected information on a cell-by-cell basis. During this post-processing function, statistical rewards and penalties are applied to all cells some which are modified by the immediately previous run. If the last run was deemed an improvement to prior operations, the contributing cell strengths are increased to reinforce that the control value for that cell should probably be retained for future use. If not, the cell strength is weakened possibly causing the control value implanted into the Signature Table for that cell index to be changed.

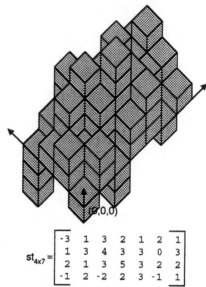

$$st_{4x7} = \begin{bmatrix} -3 & 1 & 3 & 2 & 1 & 2 & 1 \\ 1 & 3 & 4 & 3 & 3 & 0 & 3 \\ 2 & 1 & 3 & 5 & 3 & 2 & 2 \\ -1 & 2 & -2 & 2 & 3 & -1 & 1 \end{bmatrix}$$

Fig. 2. The Signature Table as a structure of hyper-cubes

Each cell structure within the Signature Table (ST) holds the "statistical data," which constitutes the knowledge-base of the method (Russell [14]). A forgetfulness factor ($ST[n,t] = ST[n,t] * \psi$)is applied to the statistical data to refine and smooth the switching surface (Kubat and Krizakova [10]) and to suppress any statistical data skewing inherited from early excursions of the program and random initial seeding of the control matrix.

3. THE DATABASE PROBLEM DOMAIN

3.1 Definitions

A database is a centralized collection of data organized into one or more data sets of various dimensions and complexity. A data set is a structure contained within a file system. The database system allows one or more user applications access to data tuples within the structure. Applications manipulate the data contained within these data sets through some well-defined accessing methodology.

This access methodology provides a bi-directional path between the logical and physical data (Inmon [9], Martin [11]) contained within the data set. An application typically has no direct knowledge of the accessing methodology, nor of its supporting systems. In larger systems, in fact, a disc processing unit (DPU) may schedule and buffer data transfers according to an independent agenda. Figure 3

(Martin [11]) illustrates a typical database configuration that contains all three parts; a physical database, the application domain and an access methodology which forms the basis, for the focus on the accessing methodology to be enhanced by the AI agent.

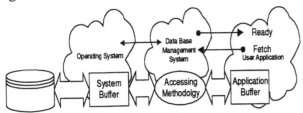

Fig. 3. A Typical Database Schematic

3.2 Data Retrieval

Any data retrieval methodology consists of a set of configurable generalized drivers that the host operating system spawns in order to service requests from the application domain. These data "reads" and "writes" use internal buffers contained either in local RAM or in a DPU as temporary storage locations. Each application is assigned pointers to internal buffers for each physical structure (data set, set, or sub-set) that is described within the database. The operating system then maps the data from these buffers back and forth into the user's own set of symbol table elements for local use.

4. ADAPTATION OF BOXES

Intelligent accessing, as discussed in the paper, addresses the internal buffering of data within the accessing methodology and is transparent to the user beyond a measurable performance upgrade. The paper presents a single user application and a non-distributive database environment so that the concept of intelligent accessing is not obscured by issues specific to complex system environments.

Figure 4 shows how the modified BOXES method may be configured to augment the performance of the DB's accessing methodology. The example used is a simple database that comprises three data sets $\{D_0, D_1, \text{ and } D_2\}$ and uses $\{R_n\}$ as the record index.

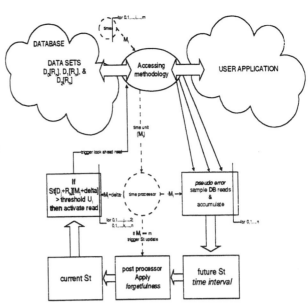

Fig. 4. BOXES' Augmentation of a Database Access Methodology

A decrease in retrieval time is achieved by a BOXES augmentation of the database's own accessing methodology. The augmentation of the accessing methodology is non-intrusive in nature since the method is contained in the database's own accessing routines and not in the user application or in the operating system. The AI enhancement to the accessing methodology is illustrated in the event trace taken from database modeling shown in figure 5.

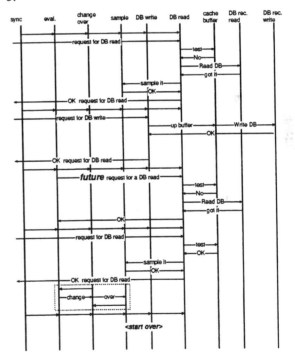

Fig. 5. An OMT event trace of the AI enhancement to the database.

The Signature Table is used to prompt the database to read the forecasted record updating the internal buffer maintained by the database on a per-structure per-application basis (just) prior to the application's request for the data.

The time oriented nature of the data is developed as the data driven user application accesses the database

169

though the AI enhanced accessing methods over multiple finite time intervals. The pattern reflected in the Signature Table (ST) may be thought of as a switching surface (Russell and Rees [16]) in that it is used to influence future requests.

A point on the surface may be described by (M_i, D'_R, U_i), where: D'_R is the unique identifier of the data set D_j, connected to a data set record index R_k; and M_i is some unit of time within the finite time interval, and U_i is the value of a statistical inference factor that rates the data set's use at that point in time. The switching surface is non-Markovian and akin to a "rubber sheet" in that there are *hills* (maximums) and *valleys* (minimums), and like a rubber sheet the switching surface's maximums and minimums migrate over the life of the database operation. A typical rule might be: *if a hill exceeds some pre-set value, then that data set should be considered likely for retrieval and cached forward*. This threshold can be visualized as a simple plane above the switching surface as seen in figure 6.

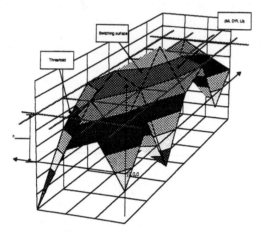

Fig. 6. A Signature Table presented as a surface with the threshold as a plane.

The threshold can also be described as a surface with the interaction to the AI triggering algorithm controlled through another probabilistic AI agent. One such control scheme is a Bayesian belief network.

"A Bayesian belief network is used to model a domain containing uncertainty in some manner" (HUGIN [8]). The belief network could provide a directed acyclic graph (DAG) (Castillo, *et al* [4]) of BOXES' switching surface on a cell to cell relationship with BOXES' ST. Each node of the belief network has the states of the random variable it represents and a conditional probability table. The conditional probability table of a node contains probabilities of the node being in a specific state given the states of its parents. The idea behind this additional AI agent is, as the BOXES surface peaks the belief network provides a causal dependency between peaking nodes to decide of caching is needed for a record. Figure 7 shows the relationship between the DAG and the cells within the BOXES' ST where X1 is the goal -- read the record based on the dynamic probabilistic relationships between X2, X3, and X4 as they relate to X1.

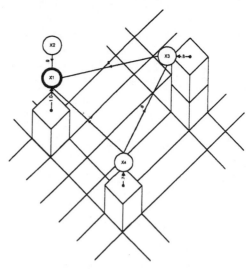

Fig. 7. The ST cell's relationship with a DAG.

The belief network coupled with the BOXES surface would provide a finer more flexible *cache forward* control through the intersection of the two surfaces. Other AI real-time control paradigm could also be used to interact with BOXES, for example, an ART2 network (Jambunathan, *et al* [7]) or a fuzzy adaptive control (Baglio, *et al* [1]).

4.2 *Learning*

At the end of some finite time interval, the adaptive BOXES post-processing routine re-computes all statistical inference values and updates the Signature Table, using a schema of "rewards" or "penalties" as shown in figure 8. The Signature Table values represent the probability that a data set will be used during the next time interval. This learning procedure is performed without interfering with normal database traffic. Figure 9 shows the progress of data in three contiguous finite time intervals. Each interval contains two equally spaced time units M_1 and M_2 within the interval and connected to data identifiers D'_1, D'_2, and D'_3 at record zero (Wadsworth [19]).

```
for (i=0; i < MAX_STRUCTURES; i++) {
  for (k=0; k < MAX_RECORDS; k++) {
    irx = DataIdentifier[i].rec + (i*MAX_RECORDS);
    if (DataIdentifier[i].fire == READ && k == DataIdentifier[i].rec) {
      if (DataIdentifier[i].incentive == REWARD)
      /*
       * REWARD
       */
      OneTimeInterval[irx][clock].probility += float(POINT_GAIN);
      else
      /*
       * PENALTY
       */
      OneTimeInterval[irx][clock].probility -= float(POINT_PAIN);
      OneTimeInterval[irx][clock].record = DataIdentifier[i].rec;
      OneTimeInterval[irx][clock].dataset = I;
      DataIdentifier[i].fire = 0;
    }
  }
}
```

Fig. 8. Reward Penalty algorithm used in the simulation.

The prompt to the database "read" is triggered when the statistical inference factor within a the cell structure is estimated as likely to break a controlled *threshold* (U_i) within plus δ (delta) time units. The threshold is described as a simple or complex surface at some height (h) above the switching curve as illustrated in figure 6. Delta (δ) is the look-ahead time-factor used by BOXES to forecast an application's need for that data and issue the corresponding *load cache from record* -- so that the database's internal buffer is loaded and waiting for the applications probable use. Figure 9 shows how one can follow the activity of data D'_1 at M_1 over three time intervals. As each time interval is post-processed the Signature Table cell at (D'_1, M_1) reflects the change in the statistical inference indicator, U_1, over time. The modified BOXES process updates the switching surface defined by the Signature Table (ST) to achieve the "look ahead" goal.

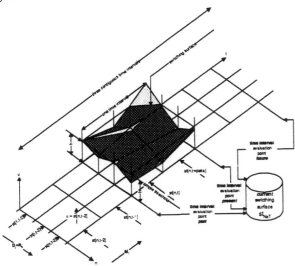

Fig. 9. Three Signature Tables at their processing intervals.

4.3 *Forecasting*

Forecasting is defined by the following algorithm:

if *$(U_i = ST[(D_j + R_k)][(\delta + M_i)].probability)$* **>=**
{the "need" defined by the rule}
then
{load the cache from the record indicated by $(D_j + R_k)$}

The Signature Table (ST) is indexed by some time unit (M_i) plus some look ahead value (δ), (see figures 4 & 5) testing the ST for candidates. The look-ahead, or offset factor, as shown in figure 9, allows the modified BOXES algorithm to forecast the need (U_i) for any data set record specified within the $\delta + M_i$ time window of some finite time interval, based on the curve contained in the Signature Table.

5. SIMULATION

The goal of the simulation was to model a controllable functionally correct database (DB) that uses logical and physical records, where the physical records are located in a simulated secondary storage. The modified BOXES methodology is used as the AI agent to learn the DB's temporal patterns through its accessing methodology. The AI agent is designed to assist the DB at the record level (R_k). This level of control requires that the Signature Table contains a unique cell structure for each record in the DB. The cells within the ST are indexed by unique record identifier ($D_j + R_k$) and some time unit index. In the

simulation, the user access demands are driven by a data generator.

5.1 *Simulation Results*

Figures 10 and 11 are representative of the results from the simulation and clearly demonstrate the improvement in switching surface produced by the adpative BOXES enhancement. This change in the switching surface may be attributed to three factors:

- The maturing of the switching surface as the statistical inference procedure takes effect.

- The use of a *forgetfulness factor* that is applied to each cell structure in the Signature Table, so that poor, naive, initial decisions are made progressively less significant over time.

- The different lengths of time interval used in a simulation.

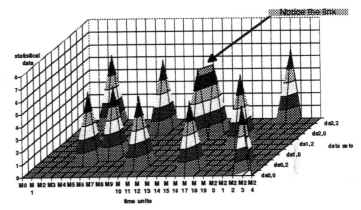

Fig.10. Switching Surface early in BOXES learning process.

In each simulation the length of time interval is selected and held constant. The Signature Table is initially populated with zeroes and data is accrued as real time progresses. It is apparent from figures 10 and 11 that the longer the activity is sustained, the more defined the statistical peaks become. This indicates that the caching process would proceed with a significant reduction in retrieval time latency.

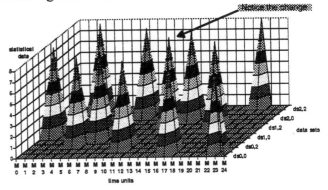

Fig. 11. Switching Surface with mature BOXES Learning.

6. CONCLUSION

The object of the paper has been to introduce the concept of enhancing the access methodology of a database system by attaching an AI agent. The selection of BOXES as a suitable AI agent is due to its statistical and learning capabilities. The paper has shown that the Signature Table maps database activity in such a manner that access predictions decrease retrieval time latency over sustainable finite time periods. Ongoing research at Penn State Great Valley is focusing on larger database models and heuristics within the firmware in modern DPU architectures.

REFERENCES

[1]   Bagilo, S., L. Fortuna, G. Giudice and G. Manganaro. 1997. Fuzzy Cellular System for a New Computational Paradigm, *Engineering Applications of Artificial Intelligence* **10**,   no. 1, February : 47-52.

[2]   Bloor, Robin. 1995. Metamorphosis of the Database, *DBMS* **8**, no. 2: 12.

[3]   Canudas de Wit, Carlos. 1988. *Adaptive control for partially known systems*. New York,N.Y.:Elsevier: 10-21.

[4]   Castillo, E., J. M. Gutierrez, A. S. Hadi and C. Solares. 1997. Symbolic propagation and sensitivity analysis Gaussian Bayesian networks with application to damage assessment, *Artificial Intelligence in Engineering* **11**, no. 2, April : 173-181.

[5]   Choobineh, Joobin. 1988. Expert Database Design System Based on Analysis of Forms, *IEEE Transactions on Software Engineering* **1**, no. 2 : 242-253.

[6]   Elman, Jefferey L. 1990. Finding Structure in Time. *Cognitive Science*, no. 14 :179-211.

[7]   Jambunathan, J., V. N. Fontama, S. L. Hartle and Ashforth-Frost. 1997. Using ART2 Networks to deduce flow velocities, *Artificial Intelligence in Engineering* **11**, no. 2,   April :135-141.

[8]   ·   HUGIN. 1997. [On-line] Available: http://www.hugin.dk/

[9]   Inmon, W. H. 1996. *Building the Data Warehouse*. 2d ed. New York, N.Y.: Wiley:25-26, 93-96.

[10]   Kubat, M and Krizakova. 1992. Forgetting and Aging of Knowledge in Concept Formation. *Applied Artificial Intelligence*. **6**, no. 2:195-206.

[11]   Martin, James. 1977. *Computer Data-Base Organization*. 2d ed. Englewood Cliffs, N.J.:Prentice-Hall:10-30, 41.

[12]   Michie, D., and R. A. Chambers. 1968. BOXES - an experiment in adaptive control. In *Machine intelligence* II, no. 2, 137-152. London: Oliver and Boyd.

[13]   Reynolds, Steven B., J. M. Mellichamp and R. E. Smith. 1995. Box-Jenkins Forecast Model Identification, *AI Expert* **10**, no. 6: 15-28.

[14]   Russell, David W. 1994. Failure-Driven Learning in the Control of Ill-defined, Continuous Systems, *Cybernetics and Systems* **25**, no. 4: 555-566.

[15]   Russell, David W. 1993. A Critical Assessment of the Boxes Paradigm, *Applied Artificial Intelligence* **7**, no. 4, 383-394. Washington DC: Taylor & Francis.

[16]   Russell David W. and S.J. Rees. 1975. System Control - a case study of a statistical learning automation, *Progress in Cybernetics and Systems Research* **2** : 114-120.

[17]   Tedlock, Dennis. 1993. *Breath on the Mirror*. San Francisco, CA.: Harper.

[18]   Wadsworth, Dennis L. 1995. Intelligent Database Accessing. Paper presented in *CMPEN 597: AI Applications to Engineering* Penn State Great Valley campus, PA, USA.

[19]   Wadsworth, Dennis L. 1997. *Adaptive BOXES: a learning paradigm*. Masters Thesis. Department of Engineering, Penn State Great Valley campus, PA, USA.

FUZZY OPTIMAL CONTROL IN L^2 SPACE

Takashi Mitsuishi*, Katsumi Wasaki*, Jun Kawabe**,
N. Pauline Kawamoto* and Yasunari Shidama*

* *Department of Information Engineering, Faculty of Engineering, Shinshu*
University, 500 Wakasato, Nagano 380-8553 Japan
** *Department of Applied Science, Faculty of Engineering, Shinshu University,*
500 Wakasato, Nagano 380-8553 Japan

Abstract: This article presents a framework for studying the existence of optimal
control based on fuzzy rules. The framework consists of two propositions: (1)
A set of fuzzy membership functions which are selected out of $L^2[a, b]$ is convex
and compact metrizable for the weak topology. (2) The defuzzification function
is continuous on the set above. Then, the existence of fuzzy optimal control is
proved. *Copyright ©1998 IFAC*

Keywords: fuzzy logic, nonlinear control systems, fuzzy control, fuzzy sets, optimal control.

1. INTRODUCTION

Since the notion of fuzziness and fuzzy sets was introduced by Zadeh (1968) in 1965, it has been applied to various fields of engineering. The emphasis of these applications is in the field of control,
and recently various practical research has been
conducted in the area of fuzzy control. However, unlike the theory of classical control and modern control, systematized considerations have not
been discussed sufficiently.

In Shidama *et al.* (1996), the authors consider optimal control problems as the problem of finding
the minimum (maximum) values of functions, by
treating control laws as problems of function approximation. The feedback laws of the control
which are defined by nonlinear functions are expressed as if-then type rules. The authors have
discussed the compactness and the application of
a membership function set F_Δ representing the
class of triangular, trapezoidal, and bell-shaped
functions with gradients less than some positive
value Δ.

This paper is an extension of that result. Instead
of using function with strict conditions, such as
those with gradients less than some positive value, the membership functions are considered as
elements of L^2 space. The compactness of a set
of such functions is proved, and the continuity of
fuzzy controller is shown for the weak topology in
L^2 space.

Finally, it should be noted that the book of Dunford and Schwartz (1988) will help the reader who
is not familiar with the mathematical definitions
and results used in this paper.

2. COMPACTNESS OF A SET OF MEMBERSHIP FUNCTIONS

Let $L^2[a, b]$ be the real Hilbert space of all square
Lebesgue integrable real functions on $[a, b]$. Consider the set $\mathcal{L}$ of membership functions defined
by

$$\mathcal{L} \triangleq \{\mu \in L^2[a, b] : 0 \le \mu(x) \le 1 \text{ a.e. } x \in [a, b]\}.$$

Proposition 1. (a) $\mathcal{L}$ *is closed under the lattice op-*

erations $\vee$ and $\wedge$. (b) If $0 \leq c \leq 1$ and $\mu \in \mathcal{L}$, then $c \wedge \mu, c \vee \mu, c\mu \in \mathcal{L}$. (c) $\mathcal{L}$ is a convex subset of $L^2[a,b]$. (d) $\mathcal{L}$ is compact metrizable for the weak topology on $L^2[a,b]$.

Proof. (a), (b) and (c) are obvious. (d) Since every convex, bounded, and norm closed subset of $L^2[a,b]$ is compact metrizable for the weak topology, it is sufficient to show that $\mathcal{L}$ is bounded and norm closed in $L^2[a,b]$. The boundedness of $\mathcal{L}$ can be easily proved. In order to prove that $\mathcal{L}$ is closed, assume that a sequence $\{\mu_n\} \subset \mathcal{L}$ converges to $\mu \in L^2[a,b]$ for the norm topology on $L^2[a,b]$. Then there exists a subsequence $\{\mu_{n_k}\}$ of $\{\mu_n\}$ which converges to μ almost everywhere on $[a,b]$. Hence it is easy to show that $0 \leq \mu(x) \leq 1$ a.e. $x \in [a,b]$, and this implies that $\mu \in \mathcal{L}$. Thus $\mathcal{L}$ is norm closed in $L^2[a,b]$. $\square$

3. CONTINUITY OF THE MOMENT CALCULATION

In fuzzy control, the control variable is generated by a family of fuzzy membership functions through the moment calculation. In this section, the moment calculation and its continuity are considered. The moment calculation M is defined by

$$M : \mu \mapsto M_\mu \triangleq \frac{\int_a^b x\mu(x)dx}{\int_a^b \mu(x)dx}.$$

To avoid making the denominator of the expression above equal to 0, we consider

$$\mathcal{L}_\delta \triangleq \left\{ \mu \in \mathcal{L} : \int_a^b \mu(x)dx \geq \delta \right\} \quad (\delta > 0),$$

which is a slight modification of $\mathcal{L}$. If δ is taken small enough, it is possible to consider $\mathcal{L} = \mathcal{L}_\delta$ for practical applications. Then, the following corollary is obtained.

Corollary 2. (a) $\mathcal{L}_\delta$ is closed under the lattice operation $\vee$. (b) $\mathcal{L}_\delta$ is a convex subset of $L^2[a,b]$. (c) $\mathcal{L}_\delta$ is compact metrizable for the weak topology. (d) The moment calculation M is weakly continuous on $\mathcal{L}_\delta$.

Proof. (a) is obvious. (b) Let $\mu, \mu' \in \mathcal{L}_\delta$ and $0 \leq \lambda \leq 1$. Then

$$\int_a^b (\lambda\mu(x) + (1-\lambda)\mu'(x))dx$$

$$= \lambda \int_a^b \mu(x)dx + (1-\lambda) \int_a^b \mu'(x)dx$$

$$\geq \lambda\delta + (1-\lambda)\delta = \delta,$$

and hence $\lambda\mu + (1-\lambda)\mu' \in \mathcal{L}_\delta$. Therefore $\mathcal{L}_\delta$ is convex.

(c) Let $\{\mu_n\}$ be a sequence in $\mathcal{L}_\delta$. Assume that $\mu_n \to \mu \in \mathcal{L}$ for the weak topology. Since $\mu_n \in \mathcal{L}_\delta$ for all $n \geq 1$,

$$\int_a^b \mu(x)dx = \lim_{n \to \infty} \int_a^b \mu_n(x)dx \geq \delta,$$

and this implies that $\mu \in \mathcal{L}_\delta$. Therefore, $\mathcal{L}_\delta$ is weakly closed in $\mathcal{L}$. Since $\mathcal{L}$ is compact metrizable for the weak topology by (d) of Proposition 1, the same can be said of $\mathcal{L}_\delta$.

(d) Assume that a sequence $\{\mu_n\} \subset \mathcal{L}_\delta$ converges weakly to $\mu \in \mathcal{L}_\delta$. Then, the weak continuity of M follows from the fact that

$$\int_a^b x\mu_n(x)dx \to \int_a^b x\mu(x)dx \quad (n \to \infty)$$

and

$$\int_a^b \mu_n(x)dx \to \int_a^b \mu(x)dx \quad (n \to \infty).$$

$\square$

4. AN APPLICATION TO FUZZY OPTIMAL CONTROL

In this section, using an idea and framework mentioned in the previous section, the existence of optimal control based on fuzzy rules will be established. $\boldsymbol{R}^n$ denotes the n-dimensional Euclidean space with the usual norm $\|\cdot\|$. Let $f(y,v) : \boldsymbol{R}^n \times \boldsymbol{R} \to \boldsymbol{R}^n$ be a (nonlinear) vector valued function which is Lipschitz continuous. In addition, assume that there exists a constant $M_f > 0$ such that

$$\|f(y,v)\| \leq M_f (\|y\| + |v| + 1) \tag{1}$$

for all $(y,v) \in \boldsymbol{R}^n \times \boldsymbol{R}$.

Consider a system given by the following state equation:

$$\dot{x}(t) = f(x(t), u(t)), \tag{2}$$

where $x(t)$ is the state and the control input $u(t)$ of the system is given by the state feedback

$$u(t) = \rho(x(t)).$$

For a sufficiently large $r > 0$,

$$B_r \triangleq \{x \in \boldsymbol{R}^n : \|x\| \leq r\}$$

denotes a bounded set containing all possible initial states x_0 of the system. Let T be a sufficiently large final time.

Proposition 3. Let $\rho : \boldsymbol{R}^n \to \boldsymbol{R}$ be a Lipschitz function and $x_0 \in B_r$. Then, the state equation

$$\dot{x}(t) = f(x(t), \rho(x(t))) \tag{3}$$

has a unique solution $x(t, x_0, \rho)$ on $[0, T]$ with the initial condition $x(0) = x_0$ such that the mapping

$$(t, x_0) \in [0, T] \times B_r \mapsto x(t, x_0, \rho)$$

is continuous.

For any $r_2 > 0$, denote by Φ the set of Lipschitz functions $\rho : \mathbf{R}^n \to \mathbf{R}$ satisfying

$$\sup_{u \in \mathbf{R}^n} |\rho(u)| \le r_2. \tag{4}$$

Then, the following (a) and (b) hold.

(a) For any $t \in [0, T]$, $x_0 \in B_r$ and $\rho \in \Phi$,

$$\|x(t, x_0, \rho)\| \le r_1, \tag{5}$$

where

$$r_1 \equiv e^{M_f T} r + (e^{M_f T} - 1)(r_2 + 1). \tag{6}$$

(b) Let $\rho_1, \rho_2 \in \Phi$. Then, for any $t \in [0, T]$ and $x_0 \in B_r$,

$$\|x(t, x_0, \rho_1) - x(t, x_0, \rho_2)\|$$

$$\le \frac{e^{L_f (1 + L_{\rho_1}) t} - 1}{1 + L_{\rho_1}} \sup_{u \in [-r_1, r_1]^n} |\rho_1(u) - \rho_2(u)|, \tag{7}$$

where L_f and L_{ρ_1} are the Lipschitz constants of f and ρ_1.

Proof. See Miller and Michel (1982) for the existence and the uniqueness of the solution. The joint continuity of the solution can be readily proved.

(a) Let $x(t) = x(t, x_0, \rho)$ be the solution of (3) on $[0, T]$ with the initial condition $x(0) = x_0$. Then

$$x(t) = x_0 + \int_0^t f(x(s), \rho(x(s)))ds, \quad 0 \le \forall t \le T.$$

By (1) and (4),

$$\|x(t)\| \le \|x_0\| + \int_0^t \|f(x(s), \rho(x(s)))\| ds$$

$$\le r + \int_0^t M_f(\|x(s)\| + |\rho(x(s))| + 1)ds$$

$$\le r + \int_0^t M_f(\|x(s)\| + r_2 + 1)ds,$$

and hence

$$\|x(t)\| + r_2 + 1$$

$$\le (r + r_2 + 1) + \int_0^t M_f(\|x(s)\| + r_2 + 1)ds$$

for all $0 \le t \le T$. By the Grownwall inequality (Miller and Michel (1982)),

$$\|x(t)\| + r_2 + 1 \le (r + r_2 + 1)e^{M_f T}.$$

Thus

$$\|x(t)\| \le r e^{M_f T} + (r_2 + 1)(e^{M_f T} - 1),$$

and the proof of (a) is complete.

(b) Let $\rho_1, \rho_2 \in \Phi$. Let $x_1(t) = x(t, x_0, \rho_1)$ and $x_2(t) = x(t, x_0, \rho_2)$ be the solutions of (3) for the feedback ρ_1 and ρ_2, respectively. Then,

$$|x_1(t) - x_2(t)\|$$

$$\le \int_0^t \|f(x_1(s), \rho_1(x_1(s))) - f(x_2(s), \rho_2(x_2(s)))\| ds.$$

Since $\|x_2(s)\| \le r_1$ by (a) of this proposition,

$$\|f(x_1(s), \rho_1(x_1(s))) - f(x_2(s), \rho_2(x_2(s)))\|$$

$$\le L_f(1 + L_{\rho_1})\|x_1(s) - x_2(s)\|$$

$$+ L_f \sup_{u \in [-r_1, r_1]^n} |\rho_1(u) - \rho_2(u)|,$$

where L_f and L_{ρ_1} are Lipschitz constants of f and ρ_1. Put

$$\alpha = \sup_{u \in [-r_1, r_1]^n} |\rho_1(u) - \rho_2(u)|,$$

then

$$\|x_1(t) - x_2(t)\|$$

$$\le \int_0^t L_f(1 + L_{\rho_1}) \left\{ \|x_1(s) - x_2(s)\| + \frac{\alpha}{1 + L_{\rho_1}} \right\} ds.$$

Adding $\alpha/(1 + L_{\rho_1})$ to both sides and using the Grownwall inequality again results in

$$\|x_1(t) - x_2(t)\| + \frac{\alpha}{1 + L_{\rho_1}}$$

$$\le \frac{\alpha}{1 + L_{\rho_1}} e^{L_f(1 + L_{\rho_1})t}.$$

Hence, the inequality (7) follows. □

Assume the feedback law ρ consists of the following l fuzzy control rules:

if x_1 is A_{i1} and ... and x_n is A_{in}
then u is B_i $(i = 1, \cdots, l)$.

For $\Delta_{ij} > 0$ $(i = 1, \cdots, l; j = 1, \cdots, n)$ and a sufficiently large $r_2 > 0$,

$$F_{\Delta_{ij}} \triangleq \{\mu : [-r_1, r_1] \to [0, 1] : |\mu(x) - \mu(x')|$$

$$\le \Delta_{ij}|x - x'| \text{ for } \forall x, x' \in [-r_1, r_1]\}, \tag{8}$$

where r_1 is the constant determined by (6).

$$\mathcal{L} \triangleq \{\mu \in L^2[-r_2, r_2] : 0 \le \mu(y) \le 1$$

$$\text{a.e. } y \in [-r_2, r_2]\}. \tag{9}$$

$$\mathcal{F} \triangleq \prod_{i=1}^{l} \left\{ \prod_{j=1}^{n} F_{\Delta_{ij}} \right\} \times \mathcal{L}^l. \tag{10}$$

The membership functions $\mu_{A_{ij}}$ of the fuzzy sets A_{ij} and μ_{B_i} of the fuzzy sets B_i ($i = 1, \cdots, l; j = 1, \cdots, n$) are selected from $F_{\Delta_{ij}}$ and $\mathcal{L}$, respectively. For simplicity, write "if" and "then" parts in the rules by the following notation:

$$\mathcal{A}_i \overset{\triangle}{=} (\mu_{A_{i1}}, \cdots, \mu_{A_{in}}) \quad (i = 1, \cdots, l),$$

$$\mathcal{A} \overset{\triangle}{=} (\mathcal{A}_1, \cdots, \mathcal{A}_l), \ \mathcal{B} \overset{\triangle}{=} (\mu_{B_1}, \cdots, \mu_{B_l}).$$

For $\delta > 0$, put

$$\mathcal{F}_\delta \overset{\triangle}{=} \left\{ (\mathcal{A}, \mathcal{B}) \in \mathcal{F} : \int_{-r_2}^{r_2} \beta_{B_i A_i}(x, y) dy \geq \delta \right.$$

$$\left. \text{for all } x \in [-r_1, r_1]^n \text{ and all } i = 1, \cdots, l \right\}. \quad (11)$$

Then, for any pair $(\mathcal{A}, \mathcal{B}) \in \mathcal{F}_\delta$, the feedback function

$$\rho_{AB}(x) = \rho_{AB}(x_1, \cdots, x_n) : [-r_1, r_1]^n \to \mathbf{R}$$

on the basis of the rules can be defined by

$$\rho_{AB}(x) \overset{\triangle}{=} \frac{\sum_{i=1}^{l} \int_{-r_2}^{r_2} y \beta_{B_i A_i}(x, y) dy}{\sum_{i=1}^{l} \int_{-r_2}^{r_2} \beta_{B_i A_i}(x, y) dy}, \quad (12)$$

where

$$\alpha_{A_i}(x) \overset{\triangle}{=} \bigwedge_{j=1}^{n} \mu_{A_{ij}}(x_j) \quad (i = 1, \cdots, l), \quad (13)$$

$$\beta_{B_i A_i}(x, y) \overset{\triangle}{=} \alpha_{A_i}(x) \mu_{B_i}(y) \quad (i = 1, \cdots, l). \quad (14)$$

See Mamdani (1974).

Proposition 4. *Let $(\mathcal{A}, \mathcal{B}) \in \mathcal{F}_\delta$. Then, the following (a) and (b) hold.*
(a) ρ_{AB} is Lipschitz continuous on $[-r_1, r_1]^n$.
(b) $|\rho_{AB}(x)| \leq r_2$ for all $x \in [-r_1, r_1]^n$.

Proof. (a) By Shidama et al. (1996), for each $i = 1, \cdots, l$, the mapping α_{A_i} is Lipschitz continuous on $[-r_1, r_1]^n$. Put

$$g(x) = \sum_{i=1}^{l} \int_{-r_2}^{r_2} y \beta_{B_i A_i}(x, y) dy,$$

$$h(x) = \sum_{i=1}^{l} \int_{-r_2}^{r_2} \beta_{B_i A_i}(x, y) dy.$$

Then for any $x_1, x_2 \in [-r_1, r_1]^n$,

$$|g(x_1) - g(x_2)| \leq r_2^2 lL \|x_1 - x_2\| \quad (15)$$

and

$$|h(x_1) - h(x_2)| \leq 2r_2 lL \|x_1 - x_2\|, \quad (16)$$

where L is the maximum of Lipschitz constants of the functions α_{A_i} ($i = 1, \cdots, l$).

Noting that $h(x) \geq l\delta$, $|g(x)| \leq r_2^2 l$ and $|h(x)| \leq 2r_2 l$ for all $x \in [-r_1, r_1]^n$, by (15) and (16)

$$|\rho_{AB}(x_1) - \rho_{AB}(x_2)|$$

$$\leq \frac{|g(x_1) - g(x_2)||h(x_2)| + |h(x_1) - h(x_2)||g(x_2)|}{(l\delta)^2}$$

$$\leq \frac{(2r_2 + r_2^2)^2}{\delta^2} \|x_1 - x_2\|,$$

and the Lipschitz continuity of ρ_{AB} is proved.

(b) It follows from an elementary inequality

$$g(x) \leq r_2 h(x)$$

that

$$|\rho_{AB}(x)| = \left| \frac{g(x)}{h(x)} \right| \leq r_2$$

for all $x \in [-r_1, r_1]^n$. $\square$

It is easily seen that every bounded Lipschitz function $\rho : [-r_1, r_1]^n \to \mathbf{R}$ can be extended to a bounded Lipschitz function $\tilde{\rho}$ on $\mathbf{R}^n$ without increasing its Lipschitz constant and bound. In fact, define $\tilde{\rho} : \mathbf{R}^n \to \mathbf{R}$ by

$$\tilde{\rho}(x) = \tilde{\rho}(x_1, \cdots, x_n)$$

$$= \begin{cases} \rho(x_1, \cdots, x_n), & \text{if } x \in [-r_1, r_1]^n \\ \rho(\varepsilon(x_1) r_1, \cdots, \varepsilon(x_n) r_1), & \text{if } x \notin [-r_1, r_1]^n, \end{cases}$$

where

$$\varepsilon(u) = \begin{cases} 1, & \text{if } u > r_1 \\ -1, & \text{if } u < -r_1. \end{cases}$$

Let $(\mathcal{A}, \mathcal{B}) \in \mathcal{F}_\delta$. Then it follows from Proposition 4 and the fact above that the extension $\tilde{\rho}_{AB}$ of ρ_{AB} is Lipschitz continuous on $\mathbf{R}^n$ with the same Lipschitz constant of ρ_{AB} and satisfies

$$\sup_{u \in \mathbf{R}^n} |\tilde{\rho}_{AB}(u)| \leq r_2.$$

Therefore, by Proposition 3 the state equation (3) for the feedback law $\tilde{\rho}_{AB}$ has a unique solution $x(t, x_0, \tilde{\rho}_{AB})$ with the initial condition $x(0) = x_0$. Though the extension $\tilde{\rho}_{AB}$ of ρ_{AB} is not unique in general, the solution $x(t, x_0, \tilde{\rho}_{AB})$ is uniquely determined by ρ_{AB} using the inequality (7) of Proposition 3. Consequently, in the following the extension $\tilde{\rho}_{AB}$ is written as ρ_{AB} without confusion.

The performance index of this control system for the feedback law ρ is evaluated with the following integral cost function:

$$J = \int_{B_r} \int_0^T w(x(t, \zeta, \rho), \rho(x(t, \zeta, \rho))) dt d\zeta, \quad (17)$$

where $w : \mathbf{R}^n \times \mathbf{R} \to \mathbf{R}$ is a positive continuous function. The following theorem guarantees the existence of a rule set which minimizes the cost function (17).

Theorem 5. *The mapping*

$$(\mathcal{A}, \mathcal{B}) \in \mathcal{F}_\delta \mapsto$$

$$\int_{B_r} \int_0^T w(x(t, \zeta, \rho_{\mathcal{AB}}), \rho_{\mathcal{AB}}(x(t, \zeta, \rho_{\mathcal{AB}}))) dt d\zeta$$

has a minimum (maximum) value on the compact metric space $\mathcal{F}_\delta$ defined by (11).

Proof. The proof is divided into several parts.
(i) *Compactness of $\mathcal{F}$*: By the Ascoli theorem, for each $i = 1, \cdots, l$ and $j = 1, \cdots, n$, $F_{\Delta_{ij}}$ is a compact subset of the Banach space $C[-r_1, r_1]$ of real continuous functions on $[-r_1, r_1]$ with the supremum norm $\|\cdot\|_\infty$. On the other hand, by (d) of Proposition 1, $\mathcal{L}$ is compact metrizable for the weak topology on $L^2[-r_2, r_2]$. Therefore, by the Tychonoff theorem

$$\mathcal{F} = \prod_{i=1}^{l} \left\{ \prod_{j=1}^{n} F_{\Delta_{ij}} \right\} \times \mathcal{L}^l$$

is compact metrizable for the product topology. It should be noted here that a sequence $\{(\mathcal{A}^k, \mathcal{B}^k)\} \subset \mathcal{F}$ converges to $(\mathcal{A}, \mathcal{B})$ for the product topology if and only if, for each $i = 1, \cdots, l$,

$$\|\alpha_{\mathcal{A}_i^k} - \alpha_{\mathcal{A}_i}\|_\infty \overset{\triangle}{=} \sup_{x \in [-r_1, r_1]^n} |\alpha_{\mathcal{A}_i^k}(x) - \alpha_{\mathcal{A}_i}(x)| \to 0$$

and

$$\mu_{\mathcal{B}_i^k} \to \mu_{\mathcal{B}_i} \text{ weakly on } L^2[-r_2, r_2].$$

(ii) *Compactness of $\mathcal{F}_\delta$*: Assume that a sequence $\{(\mathcal{A}^k, \mathcal{B}^k)\}$ in $\mathcal{F}_\delta$ converges to $(\mathcal{A}, \mathcal{B}) \in \mathcal{F}$. Fix $x \in [-r_1, r_1]^n$. Then, it is easy to show that for each $i = 1, \cdots, l$,

$$\int_{-r_2}^{r_2} \beta_{\mathcal{B}_i \mathcal{A}_i}(x, y) dy = \lim_{k \to \infty} \int_{-r_2}^{r_2} \beta_{\mathcal{B}_i^k \mathcal{A}_i^k}(x, y) dy \geq \delta,$$

and this implies that $(\mathcal{A}, \mathcal{B}) \in \mathcal{F}_\delta$. Therefore, $\mathcal{F}_\delta$ is a closed subset of $\mathcal{F}$, and hence it is compact metrizable.

(iii) Routine calculation gives the estimate

$$\sup_{x \in [-r_1, r_1]^n} |\rho_{\mathcal{A}^k \mathcal{B}^k}(x) - \rho_{\mathcal{AB}}(x)|$$

$$\leq \frac{r_2}{l\delta^2} \left\{ 2 \sum_{i=1}^{l} \left| \int_{-r_2}^{r_2} y\mu_{\mathcal{B}_i^k}(y) dy - \int_{-r_2}^{r_2} y\mu_{\mathcal{B}_i}(y) dy \right| \right.$$

$$\left. + r_2 \sum_{i=1}^{l} \left| \int_{-r_2}^{r_2} \mu_{\mathcal{B}_i^k}(y) dy - \int_{-r_2}^{r_2} \mu_{\mathcal{B}_i}(y) dy \right| \right\}$$

$$+ \frac{4r_2^3}{l\delta^2} \sum_{i=1}^{l} \|\alpha_{\mathcal{A}_i^k} - \alpha_{\mathcal{A}_i}\|_\infty.$$

Assume that $(\mathcal{A}^k, \mathcal{B}^k) \to (\mathcal{A}, \mathcal{B})$ in $\mathcal{F}_\delta$ and fix $(t, \zeta) \in [0, T] \times B_r$. Then it follows from the estimate above that

$$\lim_{k \to \infty} \sup_{x \in [-r_1, r_1]^n} |\rho_{\mathcal{A}^k \mathcal{B}^k}(x) - \rho_{\mathcal{AB}}(x)| = 0. \quad (18)$$

Hence, by (b) of Proposition 3,

$$\lim_{k \to \infty} \|x(t, \zeta, \rho_{\mathcal{A}^k \mathcal{B}^k}) - x(t, \zeta, \rho_{\mathcal{AB}})\| = 0. \quad (19)$$

Further, it follows from (18), (19) and (a) of Proposition 4 that

$$\lim_{k \to \infty} \rho_{\mathcal{A}^k \mathcal{B}^k}(x(t, \zeta, \rho_{\mathcal{A}^k \mathcal{B}^k}))$$

$$= \rho_{\mathcal{AB}}(x(t, \zeta, \rho_{\mathcal{AB}})). \quad (20)$$

Noting that $w : \mathbf{R}^n \times \mathbf{R} \to \mathbf{R}$ is positive and continuous, it follows from (19), (20) and the Lebesgue's dominated convergence theorem that the mapping

$$(\mathcal{A}, \mathcal{B}) \in \mathcal{F}_\delta \mapsto$$

$$\int_{B_r} \int_0^T w(x(t, \zeta, \rho_{\mathcal{AB}}), \rho_{\mathcal{AB}}(x(t, \zeta, \rho_{\mathcal{AB}}))) dt d\zeta$$

is continuous on the compact metric space $\mathcal{F}_\delta$. Thus it has a minimum (maximum) value on $\mathcal{F}_\delta$, and the proof is complete. □

5. CONCLUSION

The compactness of a membership function set $\mathcal{L}_\delta$ which is in L^2 space and its application to the existence of optimal control were discussed.

In this paper, because of the compactness of the set $\mathcal{L}_\delta$, the existence of an optimal fuzzy feedback control law was discussed. It is recognized that it could be a useful tool in analyzing the convergence of a fuzzy control rule modified recursively in various applications.

REFERENCES

Dunford, N. and J.T. Schwartz (1988). *Linear Operators Part I: General Theory*. John Wiley & Sons. New York.

Mamdani, E.H. (1974). Application of fuzzy algorithms for control of simple dynamic plant. *Proc. IEEE* **121, No. 12**, 1585–1588.

Miller, R.K. and A.N. Michel (1982). *Ordinary Differential Equations*. Academic Press. New York.

Shidama, Y., Y. Yang, M. Eguchi and H. Yamaura (1996). The compactness of a set of membership functions and its application to fuzzy optimal control. *The Japan Society for Industrial and Applied Mathematics* **6, No. 1**, 1–13.

Zadeh, L.A. (1968). Fuzzy algorithms. *Information and Control* **12**, 94–102.

NEURAL NETWORK MODELS USED FOR QUALITY PREDICTION AND CONTROL

Mika Järvensivu and Brian Seaworth

mika.jarvensivu@hut.fi
Helsinki University of Technology
Process Control Laboratory
02150 Espoo, Finland

gbs@gensym.com
Gensym Corporation
125 CambridgePark Drive
Cambridge, MA 02138

Abstract: The objective of this paper is to illustrate the applicability of neural networks (NNs) for predicting the behavior of an industrial nonlinear and multivariable process. This is demonstrated using industrial application of neural network predictor for burned lime residual $CaCO_3$ content. The neural network predictor is deployed as part of the real-time closed-loop control system for a lime mud reburning process. The presented work was carried out as a co-operative effort between academic and industrial representatives, and the developed system was implemented at the Pietarsaari pulp mill in western Finland. *Copyright ©1998 IFAC*

Keywords: Neural networks, Backpropagation, Optimal Control, Soft sensing, Expert Systems

1. INTRODUCTION

During the last decade, Advanced Process Control (APC) strategies and optimization techniques have been extensively applied in many industrial processes. Although there are many different ways of implementing APC, the common prerequisite for the success of each implementation principle is the availability of reliable, fast, accurate measurements and/or predictions (Deshpande, 1996; Friedmann, 1995; Anderson 1994; Boullard, 1992).

Obtaining an accurate and not too oversimplified phenomenological model to explain the underlying physics of an industrial process is a demanding and time-consuming task. In addition, there is often a lack of sufficient understanding of the exact physical phenomena involved in the process, and always a certain degree of uncertainty in the actual process measurements. It is therefore difficult to obtain a good phenomenological model for a complex nonlinear industrial process.

Another alternative is an empirical modeling approach, in which the models are obtained exclusively from the available process data. These may include measurements collected from normal operations and/or during process experiments, e.g. step tests. Empirical models can be based on models with definite structure or on models with an undefined structure, i.e. black-box models. However, knowing the required input-output structure of the model beforehand may be problematic, especially if the process involved is highly complex and there are significant nonlinearities. Therefore, the unstructured black-box approach, e.g. neural networks can be more appropriate (Sjöberg, 1995; Pradeep, 1997; Ramasamy, 1995; Brambilla, 1996; Willis, 1992; Song, 1993; Hunt, 1992).

2. DEVELOPMENT OF NNs QUALITY PREDICTOR

2.1 Aim of the presented work

Since 1996, the lime kiln process operation and environmental performance has been extensively studied at the Wisaforest pulp mill. The aim of the studies has been to gather new and more detailed knowledge about the kiln process and furthermore, based on the gathered knowledge to design and develop an intelligent control system for the lime kiln (Järvensivu, 1998a; Järvensivu, 1998b).

Product quality, which is determined by the residual $CaCO_3$ content of the lime, is not measure on-line. Produced lime is manually sampled with an 8-hour time interval. The residual $CaCO_3$ content of the sample is analyzed with a delay time of about 1 to 2 hours in the laboratory. The laboratory analysis is significantly delayed and therefore impracticable from the control point of view and useful only for the monitoring of the kiln operation and product quality.

2.2 NN model developed on the basis of historical data

Collection of the process data. The lime kiln process was studied by means of a 15-month field survey, which was comprised of both normal operations and process experiments. During the field survey, a large amount of the process data at 10-minute and one hour average intervals were collected and archived into the database. The collected data include manual samples analyzed in the laboratory, measurements of the on-line analyzers and all corresponding process measurements connected to the automation system.

Data preprocessing and input variable selection. First the large amount of the raw process data, i.e. about 11000 one hour average values of over 60 measured and calculated variables, were transferred in a more suitable form to be pre-processed and analyzed. In the pre-processing, the data were examined visually, i.e. by trend charts and histograms as well as by the statistical methods in order to reject statistical outliers, erroneous process measurements and the data that were gathered during the kiln shutdowns and startups.

After data pre-processing, comprehensive statistical analysis of the data were performed. The obtained results from statistical analysis were used together with visual observations and the prior theoretical and experimental knowledge of the lime kiln process behavior to determine the most relevant process measurement for the quality prediction. The past values of the selected process measurement as shown in the Table 1 and the results of the corresponding laboratory analysis were used for developing neural network predictor for the residual $CaCO_3$ content of the produced lime.

Table 1. Input variables used in the model for predicting the residual $CaCO_3$ content of the lime

Residual $CaCO_3$ content	
1	Cold end temperature with 2 hour time delay
2	Middle kiln temperature with 1 hour time delay
3	Hot end temperature with 1 hour time delay
4	Heavy fuel feed rate with 1 hour time delay
5	Production rate with 3 hour time delay

Data filtering and scaling. After data preprocessing and input variable selection so-called novelty filters were used to remove redundant data pairs and to make sure that the dataset spreads out and evenly covers the input space. In addition, all the input variables were normalized from 0 to 1. These normalized input variables were then used to generate the input patterns for a backpropagation network (BNP).

Network training and validation. Next, various types of network structure and the training algorithms were tested to select the appropriate network architecture for the model. Figure 1 shows an example of the application used to filter and scale the data, as well as to select appropriate network architecture, i.e. train and validate different networks.

The sequence of training with randomly selected sets of data, validation and re-randomization of weights was repeated 10 times for each architecture to achieve statistical representativeness of the results. Figure 2 shows the mean of the training and validation error with the different network architectures. This figure demonstrates clearly that the network with 5 hidden nodes is not sufficiently complex and the network with 15 hidden nodes is

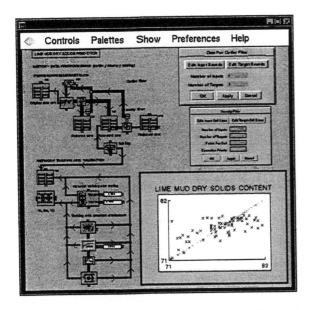

Figure 1. An example of data filtering, scaling and selecting an appropriate network architecture

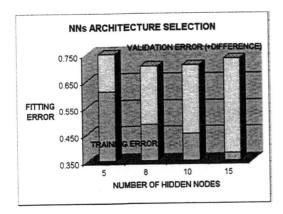

Figure 2. Fitting errors with different architectures

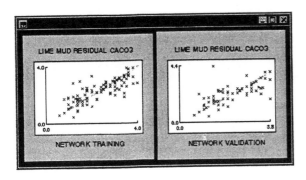

Figure 3. Fitting results with 8 hidden nodes

overly complex for the problem at hand. Finally, the best architecture with one hidden layer and 8 hidden nodes was trained using all the available data. A comparison of the laboratory analysis and the neural network outputs is shown in Figure 3.

Detection of the extrapolation. Neural networks have the ability to learn and adapt behavior of past

operational data, but neural networks are not capable of extrapolation. They will give unreliable results when presented with new data that are not represented by the training data. Hence, despite the good results shown above, the use of the quality prediction is used as a part of the closed-loop control system sets special requirements on the reliability. The system needed to be developed further to determine when the predictor starts to extrapolate.

Automatic detection of the extrapolation is based on another type of network, the radial basis function network (RBFN) (Moody, 1989; Leonard, 1992), trained with the same data as the BPN networks used in actual quality prediction. Cases that require extrapolation are automatically detected to avoid inaccurate predictions.

3. NNs IN QUALITY PREDICTION AND CONTROL

3.1 Intelligent system with embedded neural networks

Overall structure of the control system. The neural network predictor is one part of the intelligent system, which is connected to the process through the basic automation system and plant wide information system. The plant wide information system is used to collect process measurements from the basic automation system. This is then used to calculate the average values and to archive data into the real time database (RTDB). The intelligent system is connected to the RTDB, providing access to all data, both the current values and history values as well as write-access to the regulatory control loops in the basic automation system. The overall structure of the implemented system is shown in the Figure 4.

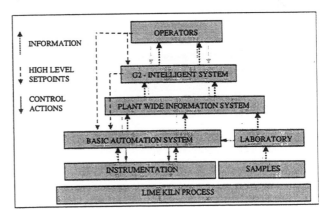

Figure 4. Overall structure of the intelligent system

Main functions of the intelligent system. The intelligent control system performs high-level control functions to stabilize and optimize the kiln operation. The application includes the following main functions as shown also in Figure 5: (Järvensivu and Ruunsunen, 1998)

- basic engineering calculations,
- neural network models based feedforward control of the kiln rotation, draft fan speed and fuel flow rate,
- neural network based calculation of the target value for the temperature and excess oxygen,
- neural network based prediction of the residual $CaCO_3$ content,
- high level feedback corrections based on the fuzzy logic principles and the natural language rules

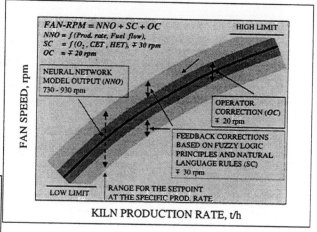

Figure 6. Principle of the feedforward control

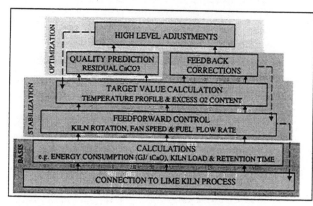

Figure 5. Main functions of the intelligent system

Feedforward control. Feedforward control of the kiln rotation, draft fan speed, and fuel flow rate are based on the neural network models which were developed using the same pre-processed data as the quality prediction. The models are used for changing the set point of the regulatory control loops during the production rate changes. Furthermore, the neural network models are used to determine bounds for the acceptable set point at the specific production rate. The principle of the feedforward control is show in Figure 6.

Target value calculation and feedback corrections. Neural network models are also used to calculate target values for the kiln temperature profile, and the excess oxygen content of the flue gas. The input variables used for the models are delayed production rate and heavy fuel oil flow.

Feedback corrections are based on the calculated target values, and the current process measurement and the change over the specified time period. The fuzzy logic principles and/or natural language type of rules determines required corrections. Figure 6 shows also how the feedback corrections are combined together with the feedforward part of the system. Some examples of the rules determining the required feedback correction are shown in Table 3.

Table 3. Examples of the rules used to determined feedback corrections

Rules for high level corrections	
Draft fan Speed	• if CET high and HET low and O2 low then decrease FAN speed
Kiln Rotation	• if KILN load high then increase KILN speed
Fuel flow rate	• if MKT (high+increas.) and HET (high+increas.) then decrease FUEL flow rapidly • if MKT (ok+increas.) and HET (low+increas.) then wait

CET = cold end temperature, MKT = middle kiln Temperature, HET = hot end temperature

High level adjustments. High level adjustments are functions designed to maximize the production capacity and energy efficiency of the kiln, and to optimize product quality and environmental performance.

The throughput maximization module determines the kiln production rate. The main function of the module is to determine the maximum production rate the kiln can sustain under current conditions. The maximum sustainable production rate is determined based on the prediction of the kiln load. The kiln load is predicted with neural networks by using the draft in the feed end of the kiln and the power required for rotating the kiln. In addition, the module determines the required kiln production rate based on the current value and trend of the lime mud storage level.

The minimization of the energy consumption and environmental protection are taken into account in the same module. The module adjusts the target value for excess oxygen content and wash water of the lime mud filters based on the total reduced sulfur (TRS) emission level.

The target value for excess oxygen is decreased with small steps to achieve improved energy efficiency (i.e. less cold air needed to be warmed up) if TRS emission level is low. In contrast, the target value for excess oxygen is increased to avoid environmental impact if the TRS emission level starts to increase. Similarly, wash water to the lime mud filters is reduced to achieve low moisture content of the lime mud fed into the kiln, if TRS emission is low. The amount of the wash water is increased to achieve better washing, if TRS emission level starts to increase.

Product quality optimization. The optimization module of the product quality (residual CaCO$_3$ content of the reburned lime), schematically represented in Figure 7, minimizes variations by maintaining the middle kiln and burning end temperature at the desired target values.

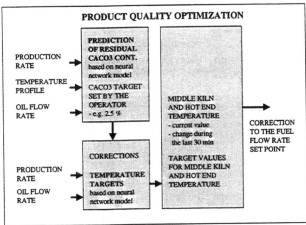

Figure 7. Product quality optimization principle

The product quality is maintained at the optimum level between 2 and 3 % by making adjustments for the middle kiln and hot end temperature targets. In addition to the laboratory analysis presented neural network prediction of the residual CaCO$_3$ content is used when the required adjustments are determined.

3.2 Results

The presented neural network predictor has been in use at the Pietarsaari pulp mill in Finland since March 1998 as part of the intelligent real-time control system, which has been in operation since April 1998.

The neural network prediction (BPN) and laboratory analysis of the residual CaCO$_3$ content of the reburned lime are shown in Figure 8. Maximum hidden node activation of the RBFN is also shown in the same Figure. Maximum hidden node activation indicates the confidence of the prediction and can therefore be used for extrapolation detection.

The upper chart shows all data collected during the first month in operation. The lower focuses on the highlighted portion of the data. As the top portion shows, extrapolation was detected six times during the examined period. Overall, the developed neural network model is predicting the product quality fairly well.

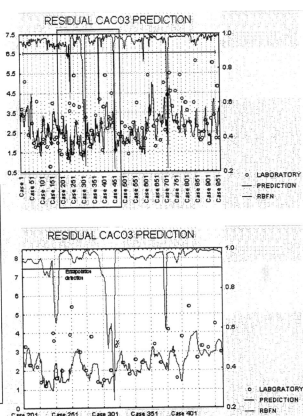

Figure 8. Quality predictions and laboratory analysis

Figure 9 shows that under manual control the mean value of the excess oxygen was about 4 % whereas automatic control averaged around 2.5% excess oxygen. Reduction of the excess oxygen makes the kiln more energy efficient and decreases draft in the kiln. This consequentially reduces dust circulation in

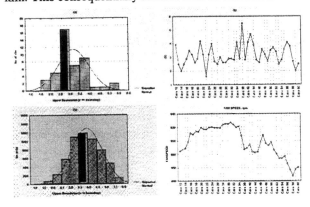

Figure 9. Excess oxygen variation. The excess oxygen content of the flue gas in automatic (upper) and in manual (lower) control.

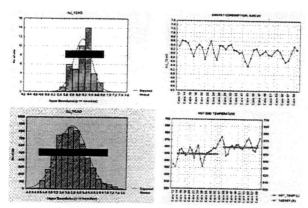

Figure 10. Variation of specific energy consumption in automatic (upper) and in manual (lower) control Also one-hour average values of the specific energy consumption and the actual and target value of the hot end temperature as a function of time.

the feed-end of the kiln, which in turn increase the attainable maximum production capacity. Figure 9 presents also one-hour average values of the excess oxygen content and draft fan speed as a function of the time. The excess oxygen was maintained above the critical environmental level of 1.5%.

As shown in Figure 10, manual control maintains the specific energy consumption between 4.6 and 7.0 GJ/t$_{CaO}$. On automatic control it was 5.4 and 6.8 GJ/t$_{CaO}$. The variation of the specific energy consumption when the intelligent system was in use was reduced over 30% compared to the manual mode. Reduction of the variation makes the kiln process more stabile and it will also significantly reduce lime quality variation.

4. CONCLUSIONS

This paper has illustrated the ability of neural network technologies to model a multivariate nonlinear process. The neural network model predicted the quality sufficiently accurately for control.

Furthermore the paper presents developed intelligent real-time control system with embedded neural networks, which has been successfully implemented at a Finnish pulp mill. Experience has shown that the experimental knowledge of the process behavior together with proper software tools can be utilized to improve process efficiency and produce considerable benefits both the economical and ecological respect.

5. ACKNOWLEDGEMENTS

This work was supported by UPM-Kymmene and the Academy of Finland. The authors wish to thank the staff of the Pietarsaari pulp mill for their valuable support and assistance during the field survey, as well as in the development and implementation stage of the intelligent control system

REFERENCES

Anderson J. (1994). Getting the most from advanced process control, *Chemical Engineering*, March 1994, p. 78-89.

Boullard, L. (1992). *Application of Artificial Intelligence in Process Control*, p. 455. Pergamon Press.

Brambilla, A. (1996). Estimate product quality with ANNs, *Hydrocarbon Processing*, September, 1996, p. 61-66.

Deshpande, P. (1996). Improve control with software monitoring technologies, *Hydrocarbon Processing*, September 1996, p. 81-88.

Friedmann, P. (1995). Economics of control improvement, *Instrument Society of America*, p. 162.

Hunt, K. (1992) Neural networks for control systems, *Automatica*, **28:6**, p.1083-1112.

Järvensivu M. (1998a). Evaluation of various alternatives to reduce TRS emission at the lime kiln, *TAPPI International Chemical recovery Conference*, Tampa, FL, USA.

Järvensivu M. (1998b). Empirical lime kiln process modeling with neural networks, EUROSIM' 98 Congress, April 14-15, 1998, Espoo, Finland.

Järvensivu M. and P. Ruusunen (1998) Intelligent lime-kiln control system, GUS'98 (Gensym Users Society Meeting), 13-15 May 1998, Newport, RI, USA.

Leonard, J. (1992) Using radial basis function to approximate a function and its error bounds, *IEEE Trans. Neural Network*, **3:4**, p. 624-627.

Moody, J. (1989). Fast learning in networks of locally-tuned processing units, Neural Computing, **1989:1**, p.281-294.

Pradeep, B. (1997) Predict difficult-to-measure properties with neural analyzer, *Control Engineering*, July 1997, p. 55-56.

Ramasamy, S. (1995). Consider neural networks for process identification, *Hydrocarbon processing*, June 1995, p.59-62.

Sjöberg, J. (1995) Nonlinear black-box modeling in system identification, *Automatica*, **31:12**, p. 1691-1724.

Song J. (1993). Neural model predictive control for nonlinear chemical processes, Journal of Chemical Engineering of Japan, 268:4, p. 347-354.

Willis, M. (1992). Artificial neural networks in process estimation and control, *Automatica*, **28:6**, p. 1181-1187.

NOVEL FUZZY ADAPTIVE COMBUSTION CONTROL

Igor Škrjanc* Juš Kocijan* Drago Matko*

* Faculty of Electrical Engineering
University of Ljubljana
Tržaška 25, 1000 Ljubljana
Slovenia

Abstract. In this paper a novel fuzzy adaptive control approach is presented and a case study of multivariable combustion control design is given. The fuzzy adaptive control in this case is of indirect type based on recursive fuzzy identification. The fuzzy model of the process is of Takagi-Sugeno type. On basis of this fuzzy model an adaptive controller is designed using pole placement design which gives a superior performance over the performance achieved by conventional linear controllers. *Copyright © 1998 IFAC*

Keywords. Combustion proces, nonlinear control, fuzzy control, adaptive control

1. INTRODUCTION

Combustion control is a control problem not infrequently met in process engineering as well as in some other areas (car industry, aircraft industry etc). The combustion process control problem is often approached as a univariable problem and, consequently, not all outputs are controlled. Outputs are usually various gas concentrations. The second very popular approach is the control of the air and fuel flow ratio (as input), which again represents a univariable approach. The most promising approach to combustion control, on the other hand is multivariable control. However, multivariable control methods based on linear models usually lead to a relatively poor closed-loop performance [3].

In the process industries, 90% of all control problems can be solved very satisfactorily by way of PID controllers tuned locally on a loop-by-loop basis, while the remaining 10% of control problems are multivariable, highly non-linear and deal with interactive processes which are not easily decoupled, nor is it advisable to do so. It is in many ways the very success of artificial intelligence methods in process control applications that has focused attention on research developments in that area. The combustion process with its inherited non-linearity as well with highly non-linear actu-

ators is a typical problem suited for an artificial intelligence method.

This paper presents a result, which is part of a research programme into combustion control design at the University of Ljubljana in collaboration with industry, with a view to increasing process efficiency and solving increasingly important ecological problems. The paper is organised as follows. An overview of the proposed control algorithm is given in Section 2. Section 3 provides description and a model of combustion process. Section 4 gives a brief description of proposed fuzzy identification algorithm. Section 5 describes the control strategy which is implied to the fuzzy adaptive scheme and Section 6 contains results and their assessment. Conclusions are stated in Section 7.

2. OVERVIEW OF THE ADAPTIVE FUZZY CONTROL ALGORITHM

Fuzzy model represents a non-linear mapping between input and output variables. Dynamic systems are usually modelled by feeding back delayed input and output signals. The common non-linear model structure is NARX (Non-linear AutoRegressive with eXogenous input) model, which gives the mapping between past input-output data and the predicted output. Each of the fuzzy model types has its own learning and fuzzy reasoning algorithm and set of free parameters. The structure identification in fuzzy model sense means the specification of operators for logical connectives, fuzzification, inference and defuzzification algorithms. Once the structure is determined the consequent parameters can be estimated using least squares method. Fuzzy modelling or identification aims at finding a set of fuzzy if-then rules with well defined parameters, that can describe the given I/O behaviour of the process. In the recent years many different approaches to fuzzy identification have been proposed in the literature. In our case the model is based on Sugeno fuzzy model type. In the case of combustion control the whole dynamics can be described by two different four input-one output fuzzy models.

Control of the system is based on adaptive controller which is designed by pole-placement design. The actual values of the controller parameters varies according to the operating point and should be calculated for each time instant. In the case of combustion control two adaptive PID controllers are needed because of the plant multivariable nature. Simulation results show a superior performance and great robustness of closed-loop control system. The proposed control policy represents a novel approach in the field of applied artificial intelligence. [4,6–9]

The proposed control method has been evaluated with a computer simulation of a non-linear model of the combustion in a steam boiler.

3. MODEL OF THE COMBUSTION PROCESS

The system under investigation is a process of combustion in a steam boiler PK 401 at Cinkarna Celje Company, Celje, Slovenia. It is controlled by the multiloop microcomputer controller MMC-90, which was developed at the Jožef Stefan Institute, Slovenia. This device enables the testing of numerous control algorithms and the verification of results obtained through simulation. The discussed multivariable control represents a possible and very promising control algorithm which can be applied to the combustion process.

The combustion process is relatively complex for modelling since it involves many not entirely deterministic reactions, as well as many possible inputs and especially outputs. Nevertheless, for control design purposes, a simplified, but still comprehensive model can be developed. The further described model has been successfully used for the design of the control system which has been implemented and commissioned [3].

According to [2] the combustion process model with two inputs (flow of air and fuel) and two outputs (O_2 and CO) can be described by the following equations:

$$\frac{dx_{O_2}}{dt} = \frac{1}{V_c}\{-x_{O_2}[\Phi_a + \Phi_f(V_g - V_0)]+$$
$$+21\Phi_a - 100V_0\Phi_f\} \quad (1)$$
$$\frac{dx_{CO}}{dt} = \frac{1}{V_c}\{-x_{CO}[\Phi_a + \Phi_f(V_g - V_0)]+$$
$$+(1-a)1.866C\Phi_f\} \quad (2)$$

where x_{O_2} is the percentage of O_2 (vol.%), x_{CO} is the percentage of CO (vol.%), V_c is the volume of the combustion chamber (m^3), Φ_f is the normalised total flow of fuel (kg/s), Φ_a is the normalised total flow of air (Nm^3/kg), V_0 is the theoretically required air volume for the combustion of one unit of fuel (Nm^3/kg), V_g is the theoretically obtained gas volume from one unit of fuel (Nm^3/kg), $(1-a)$ is the relative portion of carbon converted into CO (%/100), and C is the relative portion of carbon in the fuel (%/100). While V_c, V_g, C are constant values, $(1-a)$ or a alone is a function of Φ_f and Φ_a, and V_0 is a function of a and indirectly of Φ_f and Φ_a. This relations, although not directly shown in the given equations contribute to the complexity of process itself.

Equations (1) and (2) can be expressed in the form:

$$\dot{x}(t) = A_1 u(t)x(t) + A_2 v(t)x(t) +$$
$$+\mathbf{B}_1 u(t) + \mathbf{B}_2 v(t), \quad (3)$$
$$\mathbf{y}(t) = \mathbf{x}(t - T_d), \quad (4)$$

where

$$\mathbf{x}^T = [x_{O_2} \quad x_{CO}],$$
$$u = \Phi_a, \quad v = \Phi_f,$$
$$A_1 = -\frac{1}{V_c},$$
$$A_2 = \frac{V_0 - V_g}{V_c}$$
$$\mathbf{B}_1^T = \begin{bmatrix} \frac{21}{V_c} & 0 \end{bmatrix}$$
$$\mathbf{B}_2^T = \begin{bmatrix} \frac{-100V_0}{V_c} & \frac{(1-a)1.866C}{V_c} \end{bmatrix}$$

and T_d is the dead-time because of transport delay.

A more detailed scrutiny of equation (3) shows that it represents a bilinear (nonlinear) model with variable coefficients A_2 and $\mathbf{B}_2$. The behaviour of bilinear models is sometimes described as the behaviour of a linear system with a variant time constant whose value depends on the magnitude of the input signal.

The air flow is controlled through a damper. As it is part of the closed-loop, it has to be modelled and added to the combustion model. The damper is a nonlinear dynamical system. Its gain is described by the relations:

$$K_d = \frac{\Phi_m}{2} \exp\left(\frac{3(\phi - 45)}{45}\right), 0° \leq \phi \leq 45°, \quad (5)$$

$$K_d = \frac{\Phi_m}{2}\left(2 - \exp\left(\frac{-3(\phi - 45)}{45}\right)\right), 45° < \phi \leq 90°$$

where K_d is the damper gain, ϕ is the angle of the damper, and Φ_m is the maximum flow of air. Damper dynamics are described by the first-order transfer function:

$$G(s) = K_a \frac{0.5}{s + 0.5}. \quad (6)$$

It should be emphasised that the fuel flow rate is constrained by the load rate, which consequently means by the power of the plant. However, these constraints do not affect the fuel flow rate for the reasonably chosen operating conditions and selected set-point changes.

The developed multivariable model reveals that the gas concentrations are strongly interconnected. Therefore, to control the concentration setpoints independently, a sufficient number of manipulating variables is necessary for the system to be controllable. In our case two inputs are necessary for the independent control of two outputs. Nevertheless, the interconnections still severely affect the transient responses.

4. FUZZY IDENTIFICATION AND LINEARIZATION

In this section Takagi-Sugeno fuzzy model of the process is discused. Due to the multivariable nature of the combustion process it can be modelled using two different four input-one output fuzzy models. The proposed structure of the fuzzy model is described by the following fuzzy rule

$$\mathbf{R_i : if} \ y_1 \ is \ \mathbf{A_{i1}} \ and \ y_2 \ is \ \mathbf{A_{i2}}$$
$$and \ u_1 \ is \ \mathbf{B_{i1}}$$
$$and \ u_2 \ is \ \mathbf{B_{i2}}$$
$$\mathbf{then} \ y(k) = f_i(y_1, y_2, u_1, u_2) \quad (7)$$
$$i = 1, \ldots N \quad (8)$$

where $u_1 = u_1(k - 1)$ and $u_2 = u_2(k - 1)$ represents input variables of the combustion process and $y_1 = y(k-1)$ and $y_2 = y(k-2)$ where y stands for one of the output variables. A_{in} and B_{in} are fuzzy sets of each input variables. According to the previous notation input u_1 is equal to the angle of the damper ϕ, u_2 is equal to the flow of the fuel Φ_f. Multivariable structure of the process implies two fuzzy models with different outputs, the

first fuzzy model has output defined as x_{O_2} and the second one as x_{CO}.

The **if**-parts (antecedents) of the rules describe fuzzy regions in the space of input variables and the **then**-parts (consequent) are linear combinations of the inputs. In this case the linear combination is defined

$$f_i(y_1, y_2, u_1, u_2) = a_{i1}y_1 + a_{i2}y_2 + b_{i1}u_1 + b_{i2}u_2 \quad (9)$$

where a_{in} and b_{in} are the consequent parameters. Such a very simplified fuzzy model can be regarded as a collection of several linear models applied locally in the fuzzy regions defined by the rule antecedents. Smoth transition from one subspace to another is assured by the overlapping of the fuzzy regions. The consequent parameters of the fuzzy model are obtained by on-line recusive least square method. The method is described in paper [5].

The design of the control algotrithm is made on linearized model of the process which is obtained in each time instant by linearizing the obtained fuzzy model of the process. The method will be briefly discussed next. Estimated output value of the process can be defined by fuzzy model of the proces as

$$\hat{y}(k) = \frac{\sum_{i=1}^{K} \beta_i^l(\phi(k))(a_{i1}y_1 + a_{i2}y_2 b_{i1}u_1 + b_{i2}u_2 + c_i)}{\sum_{i=1}^{K} \beta_i^l(\phi(k))} \quad (10)$$

where K stands for the number of rules and

$$\beta_i^l(\phi(k)) = T(\mu_{A_{i1}}, \mu_{A_{i2}}, \mu_{B_{i1}}, \mu_{B_{i2}}) \quad (11)$$

where T represents the intersection operator, $\mu_{A_{ij}}$ and $\mu_{B_{ij}}$ are membership values of observed variables to the prescribed membership functions.

According to the previous notation the estimated output of the process is given with the following equation where $\beta_i(\phi(k))$ stands for normalized memebrship values

$$\hat{y}(k) = \sum_{i=1}^{K} \beta_i(\phi(k))(a_{i1}y_1 + a_{i2}y_2 + b_{i1}u_1 + b_{i2}u_2 + c_i) =$$
$$= \sum_{i=1}^{K} \beta_i(\phi(k))L_i(\phi(k)) = g(\phi(k))$$

Linear parameters of the process obtained by instantaneous linearization of Takagi-Sugeno fuzzy model around the operating point $\phi(k = \tau)$ are given by the following equations

$$\bar{a}_j = \frac{\partial g(\phi(k))}{\partial y_j} =$$
$$-\sum_{i=1}^{K}(a_{ij}\beta_i(\phi(k)) + L_i(\phi(k))\frac{\partial \beta_i(\phi(k))}{\partial y_j}) \quad (12)$$

The partial derivatives of fuzzy value β_i are given by following equation

$$\frac{\partial \beta_i(\phi(k))}{\partial y_j} = \frac{d\mu_{A_{ij}}(y_j)}{dy_j} \prod_{\substack{\lambda=1 \\ \lambda \neq j}}^{2} \mu_{A_{i\lambda}}(y_\lambda) \prod_{\lambda=1}^{2} \mu_{B_{i\lambda}}(u_\lambda)$$

The parameters $\tilde{b}_1$ and $\tilde{b}_2$ are calculated using the following equation

$$\tilde{b}_j = \frac{\partial g(\phi(k))}{\partial u_j} = -\sum_{i=1}^{K}(b_{ij}\beta_i(\phi(k)) + L_i(\phi(k))\frac{\partial \beta_i(\phi(k))}{\partial u_j})$$

where

$$\frac{\partial \beta_i(\phi(k))}{\partial u_j} = \frac{d\mu_{A_{ij}}(y_j)}{dy_j} \prod_{\lambda=1}^{2} \mu_{A_{i\lambda}}(y_\lambda) \prod_{\substack{\lambda=1 \\ \lambda \neq j}}^{2} \mu_{B_{i\lambda}}(u_\lambda)$$

The whole identification algorithem is composed of on-line recursive fuzzy identification which results in fuzzy model in the form of first order Takagi-Sugeno model and of instantaneous linearization of obtained fuzzy model. Obtained linearized process parameters which are functions of the operating point are than used to design the control.

5. ADAPTIVE CONTROL

Design of general adaptive linear controller in our case is based on pole placement [8]. Pole placement is a very attractive and often used design procedure for adaptive control. The purpose of the method is to design the controller so that all poles of the closed-loop system, or equivalently the coefficient of the closed-loop system characteristic polynomial

$$A(z^{-1}) = P(z^{-1})A(z^{-1}) + Q(z^{-1})B(z^{-1}) \quad (13)$$

where B and A are the nominator and denominator of the process transfer function, and Q and P are polynomial which defines the controller becomes equal to the prescribed polynomial. We assume prescribed values, i.e. that the polynomial $A(z^{-1})$ becomes equal to a prescribed polynomial $A_m(z^{-1})$. For a given process, that is, given a_i and b_i, $i = 1,...m$, and for the chosen $\alpha_i, i = 1,...,l$ which are the parameters of prescribed polynomial, equation 13 represents l equations. If, in addition, an integral-type controller should be designed in order to avoid the steady-state error, the polynomial $P(z^{-1})$ must fulfil the following condition: $P(1) = 0$ or equivalently

$$\sum_{i=1}^{m} P(i) = -1 \quad (14)$$

Unknown regulator parameters p_i and q_i are obtained solving the Diofantine equation, which has a solution if $A(z^{-1})$ and $B(z^{-1})$ do not have common factors, i.e. if the process is controllable. The

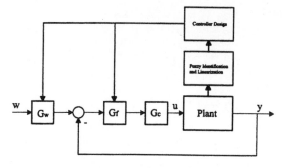

Fig. 1. The adaptive control scheme

poles of the disturbance transfer function are equal to the chosen poles of the closed-loop transfer function. It is also obvious that neither process zeros nor poles are cancelled by the controller.

The adaptive sheme which consists of on-line fuzzy identification and controller design based on pole-placement method is presented on Fig.1. The controller consists of two different part: first one is prefilter $G_w(z) = \frac{1}{Q(z)B(z)z^{-d}}$ and the second one is controller in direct loop $G_f(z) = \frac{Q(z)}{P(z)}$. According to the multivariable nature of the process with strong interactions between both channels, first the balance of the channels should be inproved. Inproved balance, which would facilitate the fulfilment of the control goal, can be achieved by separation of the channels bandwidth or by dynamical precompensator. In our case the balance of the channels is achived designing the precompensator $G_c(z)$ [3].

The main advantage of this approach in comparison to the conventional approach given in [3] is in fuzzy identification which can cope with the nonlinear dynamics of the described process.

6. SIMULATION RESULTS

The proposed adaptive algorithm has been tested on combustion process usin simulation method. Adaptive control has been applied only to control the concentration of O_2. So, the adaptive control was simulated only for one channel but it could be easily extended to both channels. The main goal of our approach was to design the control which would result in similar performance for the different operating points.

Figure 2 represents the time responses of the percentage of O_2 and CO in air flow in the case of adaptive control based on fuzzy identification. The time response of x_{O_2} exhibits similar performance for different operating points. Figure 3 represents the time responses of process input signals. The first input signal is the damper angle ϕ and the second one is flow of the fuel Φ_f. Figure 4 shows the time courses of the estimated process parameters. The first estimated parameters is time constant of the channel T_s which depends on

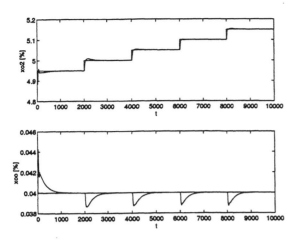

Fig. 2. Time responses of x_{O_2} and x_{CO} in the case of fuzzy adaptive control

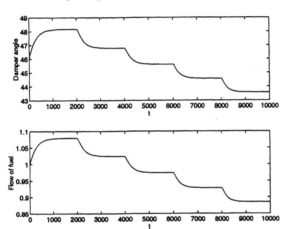

Fig. 3. Time responses of the damper angle ϕ and flow of the fuel Φ_f

ϕ and Φ_f, and the second estimated parameters is gain of the channel K_s which is also depending on damper angle ϕ and flow of the fuel Φ_f. In the case of purposed first order dynamics of the process the controller in the direct loop becomes a controller of PI-type. The third part of Figure 4 shows gain of the controller K_p in direct loop G_f, while the integral time constant T_i of the controller remain almost constant for the whole operating domain.

7. CONCLUSION

Combastion control continues to be an important problem due to the need to meet ever more stringent economic and environmental requirements. That is the reason to design novel algorithms to improve the efficiency of combustion control. In this paper a novel fuzzy adaptive control approach has been designed which gives promising results in case of combustion process control. The fuzzy adaptive control in this case is of indirect type based on recursive fuzzy identification. The type of identified fuzzy model is a Takagi-Sugeno structure model. On basis of fuzzy model an adaptive

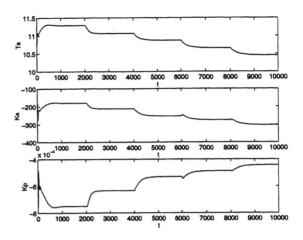

Fig. 4. Time courses of the estimated and controller parameters

controller is designed using pole placement design which gives a superior performance over the performance achieved by conventional controllers.

8. REFERENCES

[1] A. Bitenc, J. Čretnik, J. Petrovčič, Design and application of an industrial controller, Comp. and Contr. Eng. Journal, 1992, Vol.3, No. 1, pp.29-34

[2] J. Čretnik, S. Strmčnik, B. Zupančič, A model of combustion of fuel in the boiler, Proc. 3rd Symp. Simulationstechnik, 1985, D.P.F. Möller, Informatik-Fachberichte, pp.469-473, Bad Münster, Springer-Verlag

[3] J. Kocijan, An approach to multivariable combustion control design, J. Proc. Cont., 1997, Vol.7, No.4, pp.291-301

[4] T. Takagi and M. Sugeno, Fuzzy identification of systems and its applications to modelling and control, IEEE Trans. Systems, Man and Cybernetics, 1985, Vol.15, pp.116-132

[5] I.Škrjanc, K.Kavšek-Biasizze, D.Matko, Real-Time Fuzzy Adaptive Control, Engng.Applic.Artif.Intell., Vol. 10, No.1, pp.53-61,1997

[6] C.G. Moore and C.J. Harris, Indirect adaptive fuzzy control, Int. J. Control, 1992, Vol.56, No.2, pp.441-468

[7] K.J. Åström and B. Wittenmark, Computer-controlled systems, Theory and Design, Prentice Hall International, 1984

[8] R. Isermann, K.H. Lachmann, D. Matko, Adaptive control systems, Prentice Hall International, 1992, Prentice Hall International series in systems and control engineering, Hertfordshire, UK

[9] R. Babuška and H.B. Verbruggen, An overview of fuzzy modelling for control, Control Eng. Practice, 1996, Vol.4, No.11, pp.1593-1606

Solving Astrom Problem Based on Statistical Fuzzy PID Control

Hua Harry Li

Computer Engineering Department
College of Engineering
San Jose State University
San Jose, CA 95192
E-mail: huali@email.sjsu.edu

Astrom problem is referred to as a real time control of a nonlinear, time-variant beam-and-ball system. Previsously, the author of this paper has demonstrated using fuzzy logic and neural network technique, we can establish control objective in real time. However, in order to further improve the system performance, especially reduce the controller's over-correction under heavy random disturbances, there is a need to develop a more reliable control algorithm. In this paper, we discuss a technique combining the fuzzy logic PID controller with a conventional statistical process control (SPC.) The SPC technique is utilized as a supervisor to monitor the random disturbances and to identify the need of a control action against the randm noise. Simulation result of the SPC-fuzzy PID controller for the first order system is provided. *Copyright © 1998 IFAC*

Keywords: Fuzzy control, non-linear control, statistical process control.

I. Introduction

Increasing use of artificial neural networks and fuzzy logic to nonlinear and time varying system control becomes one of the focal points of the recent development in control community. Recently the author of this paper has reported work on combing statistical process control (SPC) and fuzzy PID control in semiconductor manufacturing (Li 1997).The purpose of this paper is to address the excessive control actions due to random noise which often causes undesirable system responses in "under-and/or-over" the set-point. In particular, we would like to consider applying this technique to solving the Astrom problem, a nonlinear real time control problem.

II. The Theoretical Foundation

Fuzzy Logic was introduced by L.A. Zadeh in 1965 [Zadeh 1965]. This technique is a tool to deal with uncertainty, to exploit the tolerance for imprecision and to deal with mathematically ill-defined problems. In the real world, there are many problems which can not be uniquely

2.1 Membership Functions

A Fuzzy Set A of an universe of discourse U can be denoted as follows,

Which consists of two critical components:

$$A = \int_U U_A(y)/y$$

1. A membership function $U_A: U \longrightarrow$ [0,1], which associates with each element x of U a membership in the interval [0,1], and
2. The support of the fuzzy set which is the set of elements x.

The value of the membership function represents the grade of x in A. Fuzzy membership function can be constructed based on heuristics of the human expert knowledge. This allows fuzzy logic to capture the human

expertise acquired through the engineering practice and to deal with the real world problems either too complex to be described in a close-form mathematical equation or lack of the precise knowledge of the parameters to construct a mathematical description. In particular,

1. Fuzzy membership function $u(x)$ describes the belongingness of a fuzzy variable x.
2. Many functions can be derived based on heuristics, and they can take the form of simple triangular functions, or piecewise linear functions.

2.2 Operations of Fuzzy Sets

Now we describe some commonly used fuzzy set operation. Among them, perhaps the most commonly used operations are "AND" and "OR" operations. Given both A and B are fuzzy sets, then

1. For "AND" operation, we have

$$A \cap B = \int_u \min(u_A(x), u_B(x))/x.$$

2. For "OR" operation, we have

$$A \cup B = \int_u Max(u_A(x), u_B(x))/x.$$

Note that the above integral sign is adopted in the Fuzzy Set theory, but the actual calculation is just simple *Min* or *Max* operation of the given collection of membership functions $u_A(x)$ and $u_B(x)$.

2.3 Fuzzy Inference Engine

In order to perform approximate reasoning, fuzzy inference engine plays an important role. The engine can be briefly described as follows,

1. The inference engine generally speaking consists of a set of IF-THEN rules which performs so called "Fuzzy Reasoning".
2. The design and creation of the fuzzy inference rules is based on the knowledge from either the human expert or from other empirical means, such as Neural Network based approach.

A simple example of such rules is given as follows,

Rule 1: If x_1 is A_{11} and x_2 is A_{12}, then y is B_1.
Rule 2: If x_1 is A_{21} and x_2 is A_{22}, then y is B_2.

Based on this example, we can construct more complex inference rules with connection of possible use of multiple "AND," or "OR."

2.4 Evaluating Fuzzy Reasoning Results

Now let's consider the evaluation of fuzzy reasoning. With lose of the generality, we will continue using the above simple example involving only two elements, x_1 and x_2 of a given fuzzy set for the matter of simplicity.

Suppose the true values of the premises are,

$$w_1 = \min(u_{A_{11}}(x_1), u_{A_{12}}(x_2))$$

And

$$w_2 = \min(u_{A_{21}}(x_1), u_{A_{22}}(x_2))$$

Then the actual result of this fuzzy reasoning is given by the following defuzzification formula. Note that there are many different defuzzification techniques, a good paper of the discussion on this subject can be found in Mizumoto's paper collected in Li and Gupta's edited book [Li, 1995a].

$$y = \frac{\sum_{k=1}^{2} w_k y_k}{\sum_{k=1}^{2} w_k}.$$

III. Fuzzy Controller for Solving Astrom Problem

An extensive discussion and design details of a fuzzy PID controller to solving Astrom problem is given in (Li, 1995b).

Astrom conceptualized a real time nonlinear system control problem: balancing a beam-and-ball system in real time [Astrom, 1984]. The control objective is to move the ball to the center of the beam with as little over shot as possible and as quickly as possible. This work is more challenging when no explicit parameters, such as mass, torque, friction distribution of the system are given. In our fuzzy logic controller design, we built two control systems:

1. The first is based on the human control through the joystick interface to drive the DC motor to tilt the beam in order to move the ball to the center. This human control system is used as a reference to allow the future comparison to the fuzzy logic controller.
2. The second system is fuzzy logic controller without human intervention. Based on the 5-step design practice, we have designed and implemented the controller. The algorithm was written in C and Assembly Language, and hardware was realized by using FPGA (Field Programmable Gate Array). The prototype system was tested and shipped to 1994 World Congress Neural Network Conference for 3 and a half day demo. The testing result demonstrated that the fuzzy logic controller outperforms most of the human operators.

Given in Figure 1 is the photo of the prototype system. Note this system only requires modest computation power, a 80286 computer which contrasts to otherwise a computation intensive modern control approach.

Generally speaking, SPC is a collection of techniuqes basd on the statistical modeling and analysis of a give process, to monito the operation of a given process or equipment, detect the occurrence of off-specification performance either due to the random disturbances, parameter drafting, or the

Figure 1. *The Prototype System of Real Time Fuzzy Logic Controller which was given "Industrial Neural Network Award" in 1994 World Congress Neural Network Conference, San Diego, California (Li, 1995b).*

equipment alfunctions, faults, and to locate, identify the cause of the problem, then to correct the problem. A brief review paper that summarizes some of the commonly used SPC technique can be found in (Brady 1992), (Dale 1991.) Inspired by Thomson's work and as the continuation of our work in SPC-Fuzzy PID controller design (Li 1997), we have the following SPC-PID algorithm:

1. Assume independent measurement, x, follow normal distribution $N(\mu, \sigma^2)$. Present to the random noise, 99.7% of the measurement data of a process will be in the range of $[\mu-3\sigma, \mu+3\sigma]$. data falling outside this range will lead to the warning which can be the indication of the process variation. In order to cope with the random disturbance and to have a better way to monitor the process. The x mean value is calculated and it follows normal distribution $N(\mu_\xi, \sigma_\xi)$, where x is the input and usually σ_ξ

$\le \sigma$ gives better sensitivity to the plant or process variation (σ_ξ is related to $\bar{x}$).

2. In order to detect small process deviation, a weighted moving average is employed,

$$x_t = \sigma x_t + (1-\sigma)z_{t-1},$$

where z_t is the system output which also follows normal distribution.

3. A pair of warning lines is defined at mean $\bar{x}$ plus and minus $2\sigma_\xi$ points (95.4%) to track the process shift against the mean chart.

4. A pair of action lines is defined at the mean $\bar{x}$ plus and minus $3\sigma_\xi$ points (99.7%). Any samples (system output) falls beyond these lines will trigge the controller to act.

5. In order to improve the performance, we also defined a second pair of waring lines which is defined in the similar form as in 3 except the mean and standard deviaton are all derived from x not $\bar{x}$. In addition, the second pair of action lines are also defined in the similar form as in 4 except the mean and standard deviation are all derived from x.

6. The control rules are defined as follows,

(1) Rule 1: when a data point, $\bar{x}$ or z_t is falling outside the predefined action lines, then the fuzzy-PID controller is turned on.

(2) Rule 2: When 2 consecutive data points, either from $\bar{x}$ or z_t are falling outside the action lines, then the fuzzy-PID controller is turned on.

(3) Rule 3: When a sequence of 7 consecutive data points, either from $\bar{x}$ or z_t are falling above or below μ or μ_ξ, then the integral control is on to correct the slow shifting change.

(4) Rule 4: When a sequence of 7 consecutive data points, either from $\bar{x}$ or z_t are either rise or fall, then the integral control is on.

(5) Rule 5: At any other situation, there will be no control action.

The SPC Fuzzy-PID controller is designed based on the system block diagram given in Figure 2. The details of the Fuzzy-PID controller is given in (Li, 1997.)

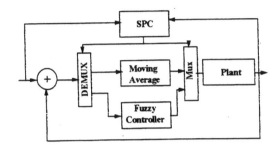

Figure 2. The system block diagram of the SPC-Fuzzy PID controller.

IV. Simulation Result and Summary

The simulation is conducted by using SIMULINK. The parameters of Fuzzy PID controller is fine tuned. The random noise follows normal distribution with zero mean and the standard deviation equal to 1. The noise level is scalled up to 6% of the full range of the position displacement. Given in Figure 3 is the fuzzy PID control result while Figure 4 gives the system response under SPC-Fuzzy PID control. As shown in the experiment, SPC-Fuzzy PID controller improves the system performance. However it has to be pointed out that due to the complexity of the beam-and-ball system, the close form plant model is very difficult to come by if it is not impossible. Therefore, for the purpose of proof-of-concept, the simulation is conducted for a 1st order system as a plant model.

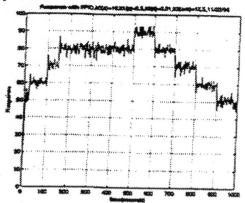

Figure 3. The system response under fuzzy PID control.

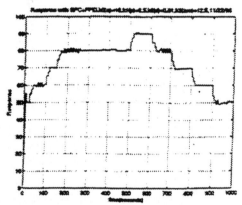

Figure 4. The system response under SPC-Fuzzy PID control.

V. References

1. K.J. Astrom and B. Wittenmark, Computer Controlled Systems, Theory and Design, Prentice Hall, Englewood Cliffs, NJ, 1984.
2. E.T. Brady, B.K. Hinds, F.G. Wilson, "Implementation of SPC - lessons to be learned," Measurement and Control, Vol. 28, pp. 74-77, April 1995.
3. B.B. Dale and P. Shaw, "Statistical Process Control: An Examination of Some Common Queries," International Journal of Production Economics, Elsevier, 22, pp. 33-41, 1991.
4. S. H. Huang, H.C. Zhang, S. Sun, and H. Li, "Function Approximation and Neural-Fuzzy Approach to Machining Process Selection," IEEE Transactions on Components, Packaging, and Manufacturing Technology, Part C, Vol. 19, No. 1, pp. 9, January 1996.
5. H. Li and N. Godfrey, "A Beam-and-Ball Controller Prototype" IEEE Macro, Vol. 15, No. 6, pp. 64, December 1995.
6. H. Li and M.M. Gupta, Fuzzy Logic and Intelligent Systems, an edited book, Kluwer Academic Publishers, 1995a.
7. H. Li, B. Vaidhyanathan, and S. Sun, "Statistical Fuzzy PID Controller Design," Proceedings of the 6th IEEE International Conference on Fuzzy Systems, IEEE catalog number: 97CH36032, Vol. III, pp. 1499-1504, 1997.
8. M. Parekh, M. Desai, H. Li and R.R. Rhinehart, "In-Line Control of Nonlinear pH Neutralization Based on Fuzzy Logic," IEEE Transactions on Components, Packaging, and Manufacturing Technology, Part A, Vol. 17, No. 2, pp. 192-201, June, 1994.
9. L.A. Zadeh, "Fuzzy Set," Information Control, Vol. 8, pp. 338-353, Academic Press, 1965

MIMO PREDICTIVE CONTROL BY MULTIPLE-STEP LINEARIZATION OF TAKAGI-SUGENO FUZZY MODELS

S. Mollov   R. Babuška   J. A. Roubos   H.B. Verbruggen

Delft University of Technology, Faculty of Information Technology and Systems
Control Engineering Laboratory, Mekelweg 4, P.O. Box 5031, 2600 GA Delft
The Netherlands, fax: +31 15 2786679, e-mail: s.mollov@its.tudelft.nl

Abstract: The main bottleneck for the real-time implementation of a predictive controller based on a nonlinear process model is the complexity of the associated optimization problem. The computational costs can be lowered by linearizing the model at a current operating point and using linear predictive control. For long-range predictive control, however, this leads to the accumulation of linearization errors. In this paper, multi-step linearization around the working points within the prediction horizon is investigated. Takagi-Sugeno (TS) fuzzy models are chosen, as the model structure as local linear models can be derived from the linear rule consequents in a straightforward way. *Copyright ©1998 IFAC.*

Keywords: Model-based predictive control, nonlinear systems, fuzzy modeling, linearization, multivariable (MIMO) systems.

1. INTRODUCTION

Model-based predictive control (MBPC) is a powerful method for the control of multivariable systems. It has become a major research topic during the last few decades and, unlike many other advanced techniques, it has also been successfully applied in industry (Richalet, 1993). The main reason for this success is the ability of MBPC to control multivariable systems under various constraints in an optimal way. However, two major issues limit the possible application of MBPC to nonlinear systems:

1. A model must be available, that can predict the process variables over the specified prediction horizon with sufficient accuracy.

2. Given a nonlinear process model, a nonlinear (and usually non-convex) optimization problem must be solved at each sampling period.

The first factor hampers the application of MBPC to complex or partially known systems for which reliable models cannot be obtained. The second factor hampers the application to fast systems where iterative optimization techniques cannot be properly used, due to short sampling times.

Fuzzy modeling can be efficiently applied to address the first issue. It has large advantages when when different pieces of knowledge (e.g., measurements, information based on first principles, qualitative knowledge) must be integrated, in order to obtain a good process model (Yager and Filev, 1994; Babuška and Verbruggen, 1996). If little prior knowledge is available, fuzzy models can be extracted from the process measurements, using various techniques such as fuzzy clustering, neuro-fuzzy learning, orthogonal least squares, etc. In this paper, a fuzzy model of the Takagi–Sugeno (TS) type (Takagi and Sugeno, 1985) is employed.

To avoid non-convex optimization, it is proposed in this paper to use local linearization of the nonlinear fuzzy model along the calculated trajectory. A standard state-space formulation of the MBPC algorithm is modified to accomodate the set of local linear models. A constrained convex optimization problem results, which can effectively be solved by quadratic programming.

2. FUZZY MODEL STRUCTURE

Takagi-Sugeno (TS) fuzzy models are suitable to model a large class of nonlinear systems (Babuška, 1998; Takagi and Sugeno, 1985; Babuška, et al., 1998). Consider a MIMO system with n_i inputs: $\mathbf{u} \in U \subset \mathbb{R}^{n_i}$, and n_o outputs: $\mathbf{y} \in Y \subset \mathbb{R}^{n_o}$. This system will be approximated by a collection of coupled MISO discrete-time fuzzy models. Denote q^{-1} the backward shift operator: $q^{-1}y(k) \overset{\text{def}}{=} y(k-1)$, where y is a signal sampled at discrete time instants k. Given two integers, $i \leq j$, define an ordered sequence of delayed samples of the signal y as:

$$\{y(k)\}_i^j \overset{\text{def}}{=} [y(k-i), \; y(k-i-1), \; \ldots, \; y(k-i-j+1)].$$

The MISO models are of the input–output NARX type:

$$y_l(k+1) = \mathcal{F}_l\big(\xi_l(k)\big), \quad l = 1, 2, \ldots, n_o, \quad (1)$$

where the regression vector $\xi_l(k)$ is given by:

$$\xi_l(k) = \Big[\{y_1(k)\}_0^{n_{yl1}}, \ldots, \{y_{n_o}(k)\}_0^{n_{yln_o}}, \\ \{u_1(k+1)\}_{n_{dl1}}^{n_{ul1}}, \ldots, \{u_{n_i}(k+1)\}_{n_{dln_i}}^{n_{uln_i}}\Big].$$

Here n_y and n_u are the number of delayed outputs and inputs, respectively, and n_d is the number of pure (transport) delays from the input to the output. n_y is an $n_o \times n_o$ matrix, and n_u, n_d are $n_o \times n_i$ matrices. $\mathcal{F}_l$ are rule-based fuzzy models of the Takagi–Sugeno (TS) type (Takagi and Sugeno, 1985):

$$R_{li}: \quad \textbf{If } \xi_{l1}(k) \text{ is } \Omega_{li1} \textbf{ and } \ldots \textbf{ and } \xi_{lp}(k) \text{ is } \Omega_{lip}$$
$$\textbf{then } y_{li}(k+1) = \zeta_{li}\mathbf{y}(k) + \eta_{li}\mathbf{u}(k) + \theta_{li}$$
$$i = 1, 2, \ldots, K_l. \quad (2)$$

Here Ω_{li} are the antecedent fuzzy sets of the ith rule, ζ and η are vectors of polynomials in q^{-1}, e.g., $\zeta = \zeta_0 + \zeta_1 q^{-1} + \zeta_2 q^{-2} + \ldots$, and θ is the offset. K_l is the number of rules in the lth model. The MIMO TS rules are estimated from input-output system data (Yager and Filev, 1994; Babuška, et al., 1998).

3. LOCAL LINEARIZATION

Local linearization of TS fuzzy models in a state space form is investigated in order to extend the operation range of linear model based predictive controllers. The local linear model is presented in a state-space form for which well-developed MBPC techniques exist. At each sample time k, given the operating point $\mathbf{u}(k-1)$ and $\mathbf{y}(k)$ (Fig. 1), the local state-space model is calculated as follows:

Calculate the degrees of fulfillment $\omega_i\big(\mathbf{x}(k)\big)$ of the antecedents, using product as the fuzzy logic **and** operator. The rule inference gives:

$$y_l(k+1) = \frac{\sum_{i=1}^{K_l} \omega_{li}(\mathbf{x}_l(k)) \cdot y_{li}(k+1)}{\sum_{i=1}^{K_l} \omega_{li}(\mathbf{x}_l(k))}, \quad (3)$$

$$y_{li}(k+1) = (\zeta_{li}\mathbf{y}(k) + \eta_{li}\mathbf{u}(k) + \theta_{li}). \quad (4)$$

Define ζ_l^* and η_l^* as:

$$\zeta_l^* = \frac{\sum_{i=1}^{K_l} \omega_{li}(\mathbf{x}_l(k)) \cdot \zeta_{li}}{\sum_{i=1}^{K_l} \omega_{li}(\mathbf{x}_l(k))}, \quad (5)$$

$$\eta_l^* = \frac{\sum_{i=1}^{K_l} \omega_{li}(\mathbf{x}_l(k)) \cdot \eta_{li}}{\sum_{i=1}^{K_l} \omega_{li}(\mathbf{x}_l(k))}. \quad (6)$$

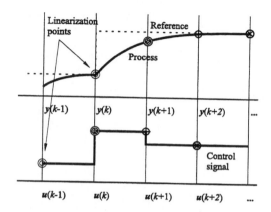

Fig. 1. Operating points for local linearization.

Define $\mathbf{x}$, $\mathbf{u}$ and $\mathbf{y}$ for state-space description as:

$$\mathbf{x}(k) = [x_1(k), x_1(k-1), \ldots, x_1(k-n_{y1}), \ldots, \\ x_{n_o}(k), x_{n_o}(k-1), \ldots, x_{n_o}(k-n_{yn_o})]^T,$$
$$\mathbf{u}(k) = [u_1(k-n_{d1}+1), u_1(k-n_{d1}), \ldots, \\ u_1(k-n_{d1}-n_{u1}+1), \ldots, u_{n_i}(k-n_{d_{n_i}}+1), \\ u_{n_i}(k-n_{d_{n_i}}), \ldots, u_{n_i}(k-n_{d_{n_i}}-n_{un_i}+1)]^T,$$
$$\mathbf{y}(k) = [x_1(k), x_2(k), \ldots, x_{n_o}(k)]^T.$$

The local linear system matrices are now derived as follows: $\mathbf{A}^*$ is an $\alpha_1 \times \alpha_1$ matrix, where α_1 stands for $\sum_{j=1}^{n_o} n_{yj}$:

$$\mathbf{A}^* = \begin{bmatrix} \zeta_{1,1}^* & \zeta_{1,2}^* & \cdots & \cdots & \cdots & \cdots & \zeta_{1,\alpha_1}^* \\ 1 & 0 & 0 & & \cdots & & 0 \\ 0 & 1 & \vdots & & \ddots & & 0 \\ 0 & \cdots & \ddots & & \ddots & & \vdots \\ \zeta_{2,1}^* & \zeta_{2,2}^* & \cdots & \cdots & \cdots & \cdots & \zeta_{2,\alpha_1}^* \\ 0 & \vdots & \ddots & & \ddots & & \vdots \\ \zeta_{n_o,1}^* & \zeta_{n_o,2}^* & \cdots & \cdots & \cdots & \cdots & \zeta_{n_o,\alpha_1}^* \\ 0 & \cdots & 0 & 1 & \cdots & 0 & 0 \\ \vdots & & \vdots & \vdots & \ddots & \vdots & \vdots \\ 0 & \cdots & 0 & 0 & \cdots & 1 & 0 \end{bmatrix}, \quad (7)$$

$\mathbf{B}^*$ is an $\alpha_2 \times \alpha_1$ matrix, where $\alpha_2 = \sum_{j=1}^{n_i} n_{uj}$:

$$\mathbf{B}^* = \begin{bmatrix} \eta_{1,1}^* & \eta_{1,2}^* & \cdots & \eta_{1,\alpha_2}^* \\ 0 & \cdots & \cdots & 0 \\ \vdots & & & \vdots \\ 0 & \cdots & \cdots & 0 \\ \eta_{2,1}^* & \eta_{2,2}^* & \cdots & \eta_{2,\alpha_2}^* \\ \vdots & \ddots & \ddots & \vdots \\ \eta_{n_o,1}^* & \eta_{n_o,2}^* & \cdots & \eta_{n_o,\alpha_2}^* \\ \vdots & \ddots & \ddots & \vdots \end{bmatrix}, \quad (8)$$

and $\mathbf{C}$ is a $n_o \times \alpha_1$ matrix:

$$\mathbf{C} = \begin{bmatrix} 1 & 0 & \ldots & \ldots & \ldots & \ldots & 0 \\ \vdots & & \ddots & & \ddots & & \vdots \\ 0 & \ldots & \ldots & 1 & 0 & \ldots & 0 \end{bmatrix}. \quad (9)$$

The ones in $\mathbf{C}$ are positioned such that $y_l(k) = x_l(k)$.

4. MODEL BASED PREDICTIVE CONTROL

Two types local linearization of Takagi-Sugeno fuzzy models can be exploited in combination with linear model based predictive controllers: single-step and multi-step linearization.

4.1. Linear State-Space MBPC

In linear MBPC (García, et al., 1989), a linear model is used to predict the output $\hat{\mathbf{y}}$ as a function of the control signal $\Delta\mathbf{u}(k, \ldots, k + H_p)$, with H_p the prediction horizon. The objective function

$$J = \sum_{i=1}^{H_p} \|(\mathbf{r}_{k+i} - \hat{\mathbf{y}}_{k+i})\|_{\mathbf{P}_i}^2 + \sum_{i=1}^{H_c} \|(\Delta\mathbf{u}_{k+i-1})\|_{\mathbf{Q}_i}^2, \quad (10)$$

is minimized for a given reference trajectory. The signal $\Delta\mathbf{u}$ may change over the control horizon H_c ($H_c \leq H_p$) and remains constant between H_c and H_p. In single-step linearization, the linearized model, obtained at the current time instant k

$$M(k) = \{\mathbf{A}_k, \mathbf{B}_k, \mathbf{C}_k\}, \quad (11)$$

is used over the entire prediction horizon (Roubos, et al., 1998). The optimal control signal can be found by means of quadratic programming. For multiple-step-ahead control, however, the influence of the linearization error may significantly deteriorate the performance. This can be remedied by linearizing the model along the reference trajectory within the prediction horizon, i.e., a sequence of models $M(i)$ is obtained over $i = k, \ldots, k + H_p$. The procedure, called multi-step linearization is as follows:

1. Using the already obtained linear model $M(k)$, compute the control signal $\mathbf{u}(k)$ over the entire prediction horizon.

2. Taking $\mathbf{u}(k)$ compute $\mathbf{y}_m(k + 1)$.

3. By linearizing the TS model locally around $\big(\mathbf{y}_m(k + 1), \mathbf{u}(k)\big)$, $M(k + 1)$ is obtained.

4. Using $M(k)$ and $M(k+1)$, compute new control sequence $\mathbf{u}$ over the entire prediction horizon.

5. Now take $\mathbf{u}(k)$ and $\mathbf{u}(k+1)$ and compute $\mathbf{y}_m(k+2)$.

6. By linearizing around $\big(\mathbf{y}_m(k + 2), \mathbf{u}(k + 1)\big)$, $M(k + 2)$ is obtained.

7. Using $M(k)$, $M(k + 1)$ and $M(k + 2)$ compute new control sequence $\mathbf{u}$ over the entire prediction horizon.

Steps 5 through 7 are repeated for $i = k + 1, \ldots, k + H_p$. Then, based on $M(k), M(k+1), \ldots, M(k+H_p)$, the final control $\mathbf{u}$ is computed. There are two ways to calculate the control sequence $\mathbf{u}$. At step 1, when only a model $M(k)$ is available, $\mathbf{u}$ is obtained as in the single-step case. Hereafter a set linear models $\{M(k + i)\}_{i=1}^{H_p}$ is used. The quadratic programming (QP) problem is modified as follows:

$$\min_{\Delta\mathbf{u}} \left\{ \frac{1}{2}\Delta\mathbf{u}^T\mathbf{H}\Delta\mathbf{u} + \mathbf{c}^T\Delta\mathbf{u} \right\}, \quad (12)$$

with:

$$\begin{cases} \mathbf{H} = 2(\mathbf{R}_u^T\mathbf{P}\mathbf{R}_u + \mathbf{Q}) \\ \mathbf{c} = 2[\mathbf{R}_u^T\mathbf{P}^T(\mathbf{R}_x\mathbf{A}_k\mathbf{x}(k) - \mathbf{r})]^T, \end{cases} \quad (13)$$

and the constraints on $\mathbf{u}$, $\Delta\mathbf{u}$, and $\mathbf{y}$:

$$\Lambda\Delta\mathbf{u} \leq \omega, \quad (14)$$

with

$$\Lambda = \begin{bmatrix} \mathbf{I}_{\Delta u} \\ -\mathbf{I}_{\Delta u} \\ \mathbf{I}^{H_p m} \\ -\mathbf{I}^{H_p m} \\ \mathbf{R}_u \\ -\mathbf{R}_u \\ \mathbf{R}_{u1} \\ d\mathbf{R}_u \\ -\mathbf{R}_{u1} \\ -d\mathbf{R}_u \end{bmatrix} \quad \omega = \begin{bmatrix} \mathbf{u}^{max} - \mathbf{I}_u u_{k-1} \\ -\mathbf{u}^{min} - \mathbf{I}_u u_{k-1} \\ \Delta\mathbf{u}^{max} \\ -\Delta\mathbf{u}^{min} \\ \mathbf{y}^{max} - \mathbf{R}_x\mathbf{A}_k x_k \\ -\mathbf{y}^{min} - \mathbf{R}_x\mathbf{A}_k x_k \\ \mathbf{dy}^{max} - \mathbf{R}_{x1}\mathbf{A}_k x_k + y_k \\ \mathbf{dy}^{max1} - d\mathbf{R}_x\mathbf{A}_k x_k \\ -\mathbf{dy}^{min} + \mathbf{R}_{x1}\mathbf{A}_k x_k - y_k \\ -\mathbf{dy}^{min1} + d\mathbf{R}_x\mathbf{A}_k x_k \end{bmatrix}$$
$$(15)$$

where $\mathbf{I}^{H_p m}$ is a $(H_p m \times H_p m)$ unity matrix. The matrices $\mathbf{I}_u, \mathbf{I}_{\Delta u}, \mathbf{R}_x, \mathbf{R}_u, d\mathbf{R}_x$ and $d\mathbf{R}_u$ are defined by:

$$\begin{bmatrix} u_k \\ \tilde{u}_{k+1} \\ \vdots \\ \tilde{u}_{k+H_c-1} \end{bmatrix} = \underbrace{\begin{bmatrix} I \\ I \\ \vdots \\ I \end{bmatrix}}_{\mathbf{I}_u} u_{k-1} + \underbrace{\begin{bmatrix} I0\ldots0 \\ II\ldots0 \\ \vdots \ddots \ddots \\ II\ldots I \end{bmatrix}}_{\mathbf{I}_{\Delta u}} \begin{bmatrix} \Delta u_k \\ \Delta\tilde{u}_{k+1} \\ \vdots \\ \Delta\tilde{u}_{k+H_c-1} \end{bmatrix}$$
$$(16)$$

$$\mathbf{R}_x = \begin{bmatrix} \mathbf{C}_k \\ \mathbf{C}_{k+1}\mathbf{A}_k \\ \vdots \\ \mathbf{C}_{k+H_p}\mathbf{A}_{k+H_p-1}\ldots\mathbf{A}_k \end{bmatrix}, \quad (17)$$

$$\mathbf{R}_u = \begin{bmatrix} \mathbf{C}_{k+1}\mathbf{B}_k & \ldots \\ \mathbf{C}_{k+2}\mathbf{A}_{k+1}\mathbf{B}_k & \ldots \\ \vdots & \ddots \\ \mathbf{C}_{k+H_p}\mathbf{A}_{k+H_p-1}\ldots\mathbf{A}_{k+1}\mathbf{B}_k & \ldots \\ & \\ \mathbf{0} \\ \mathbf{0} \\ \vdots \\ \mathbf{C}_{k+H_p}\mathbf{A}_{k+H_p-H_c}\ldots\mathbf{A}_{k+1}\mathbf{B}_{k+H_c} \end{bmatrix},$$
$$(18)$$

$$dR_x = \begin{bmatrix} C_{k+1}A_k - C_k \\ C_{k+2}A_{k+1}A_k - C_{k+1}A_k \\ \vdots \\ C_{k+H_p}A_{k+H_p-1}\cdots A_k - \\ C_{k+H_p-1}A_{k+H_p-2}\cdots A_k \end{bmatrix}, \quad (19)$$

$$dR_u = \begin{bmatrix} C_{k+2}A_{k+1}B_k - \\ C_{k+1}B_k \\ C_{k+3}A_{k+2}A_{k+1}B_k - & \cdots \\ C_{k+2}A_{k+1}B_k & \cdots \\ \vdots & \ddots \\ C_{k+H_p}A_{k+H_p-1}\cdots A_{k+1}B_k - \\ C_{k+H_p-1}A_{k+H_p-2}\cdots A_{k+1}B_k & \cdots \end{bmatrix}$$

$$\begin{bmatrix} 0 \\ 0 \\ \vdots \\ C_{k+H_p}A_{k+H_p-1}\cdots A_{k+H_c+1}B_{k+H_c} - \\ C_{k+H_p-1}A_{k+H_p-2}\cdots A_{k+H_c1}B_{k+H_c} \end{bmatrix}, \quad (20)$$

The matrices dR_{x1} and dR_{u1} are defined as the the hypothetical first rows in dR_x and dR_u: $dR_{x1} = C_k$ and $dR_{u1} = C_{k+1}B_k - C_k$.

4.2. IMC Scheme

The linear MBPC algorithm is used within the internal model control (IMC) scheme, which is used to compensate for process disturbances, measurement noise and modeling errors (Babuška, 1998; Economou, et al., 1986). In general, the IMC scheme consists of three parts (Garcia and Morari, 1982): 1) an internal model to predict the effect of control actions on the process output, 2) a feedback filter to achieve robustness, and 3) a controller to optimize the process. Fig. 2 gives the local linearized model in the LMBPC scheme with an internal model and a feedback to compensate for disturbances and modeling errors. y_m is the output of the at every sample step locally multi-step linearized TS model. This block has its own state and is integrated internally. The difference between the process output y and the linearized model output y_m is fed back through a filter.

5. EXAMPLES

In this section, two examples with water vessel systems are presented.

5.1. SISO Liquid Level Process

The proposed controller was simulated for a SISO liquid level process. The process involves a tank with a constant horizontal cross-section, with a restricted liquid outlet in its base. The control problem is to follow level setpoint changes by adjusting the flow rate of the liquid entering the tank. This example is taken from

(Postlethwaite, 1996), where the process is described by the equation:

$$\dot{h} = 1/S(F_i - \beta\sqrt{h})), \quad (21)$$

where S is the horizontal cross-sectional area (10^{-3} m^2), h is the level of liquid in the tank, β is a flow coefficient (equal to 1), and F_i is the inlet liquid flow rate. This model was simulated in SIMULINK in order to obtain input-output data sequences for identification. The control signal is limited to $[0.02, 0.1]$ (l/s) and has a sample time of 10 (s). The fuzzy model contains three rules with a first order ARX structure in the consequents. To compare the performance achieved with both linearization methods, this model is used in the IMC scheme (Fig. 2). The prediction and control horizons are $Hp = 5$ and $Hc = 3$, respectively. The criterion weights P and Q are 1 and 0, and the feedback gain was set at 0.5. The control and prediction horizons are selected such that to take into consideration the possible aggregated error coming along with the single-step linearization. The tracking performance for a reference trajectories with step changes of 0.20 and 0.35 (m) is shown in Fig. 3. The multi-step linearization method performs better, as there is no overshot and the response is smoother. The difference measure is 0.4099 and 1.2025 for multi-step and single-step case respectively. The controller based on the multi-step linearization anticipates better the future reference.

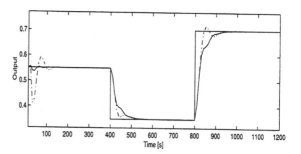

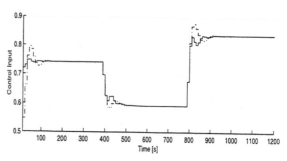

Fig. 3. System performance and control $H_p = 5$ and $H_c = 3$; - - - single-step linearization, — multi-step linearization.

5.2. MIMO Liquid Level Process

The same controller is used for a liquid level process in a dual cascaded configuration (Fig. 4). The four

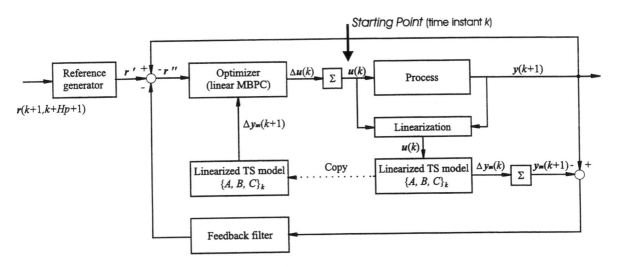

Fig. 2. Locally linearized TS model in the MBPC scheme with an internal model and a feedback to compensate for disturbances and modeling errors.

levels are denoted by $\mathbf{y} = [h_1, h_2, h_3, h_4]^T$,

$$
\dot{h} = \begin{bmatrix}
-\frac{S_{2,1}}{S_{1,1}} & 0 & r_{3,1}\frac{S_{2,3}}{S_{1,1}} & r_{4,1}\frac{S_{2,4}}{A_{1,1}} \\
0 & -\frac{S_{2,2}}{S_{1,2}} & r_{3,2}\frac{S_{2,3}}{S_{1,2}} & r_{4,2}\frac{S_{2,4}}{A_{1,2}} \\
0 & 0 & -\frac{S_{2,3}}{S_{1,3}} & 0 \\
0 & 0 & 0 & -\frac{S_{2,4}}{S_{1,4}}
\end{bmatrix} \sqrt{2gh}
$$
$$
+ \begin{bmatrix}
0 & 0 \\
0 & 0 \\
\frac{1}{S_{1,3}} & 0 \\
0 & \frac{1}{S_{1,4}}
\end{bmatrix} F_{in} , \tag{22}
$$

where $S_{1,j}$ and $S_{2,j}$ are the inlet and the outlet area of tank j, respectfully g is the gravity constant (equal to 9.81), $r_{i,j}$ is the restriction parameters from vessel i to vessel j and $f_{in,j}$ is the water flow into vessel j. In this case, the variables are chosen as:

$$
\begin{bmatrix} r_{3,1} & r_{3,2} \\ r_{4,1} & r_{4,2} \end{bmatrix} = \begin{bmatrix} 0.8 & 0.2 \\ 0.2 & 0.8 \end{bmatrix} , \tag{23}
$$

$$
S = \begin{bmatrix} 10^{-3} & 10^{-3} & 10^{-3} & 10^{-3} \\ 10^{-6} & 10^{-6} & 10^{-6} & 10^{-6} \end{bmatrix} . \tag{24}
$$

The structure of the MIMO model is selected by using insight in the physical structure of the system as follows:

$$
n_y = \begin{bmatrix} 1 & 0 & 1 & 1 \\ 0 & 1 & 1 & 1 \\ 0 & 0 & 1 & 0 \\ 0 & 0 & 0 & 1 \end{bmatrix} , \quad n_u = \begin{bmatrix} 0 & 0 \\ 0 & 0 \\ 1 & 0 \\ 0 & 1 \end{bmatrix} , \quad n_d = \begin{bmatrix} 0 & 0 \\ 0 & 0 \\ 1 & 0 \\ 0 & 1 \end{bmatrix} . \tag{25}
$$

Vessels 1 and 2 cannot influence vessels 3 and 4 because the water cannot flow back. Also vessel 1 cannot influence vessel 2, and vessel 3 cannot influence vessel 4.

Both linearization methods were tested in the proposed LMBPC structure. The controller is simulated for various setpoint changes using $H_p = 5$ and $H_c = 3$. First, a set of steady states is determined for the system using constant inputs. The reference trajectories with setpoint changes are designed using these steady states. The feedback gain is set at 0.5. Simulation results for a reference trajectory is given in Fig. 5a. The weight matrices in Eq.(10) are $\mathbf{P} = diag([1, 1, 1, 1])$ and $\mathbf{Q} = diag([0, 0])$. The calculated control actions

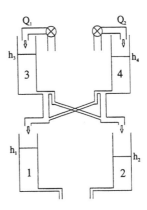

Fig. 4. Liquid level process with four cascaded vessels.

are given in Fig. 5b. It is evident that the controller using multi-step linearization performs better for all four outputs. The control signals are smoother as well. However, the computational load is H_p times higher, compared with the single-step case, due to the linearization performed around each point within the prediction horizon.

6. CONCLUSIONS

Model based predictive control of nonlinear processes by using locally linearized TS models shows good results for the given examples. Moreover, the presented approach using multi-step linearization of the TS fuzzy model performs better than the one with single-step linearization. Some of its advantages are: fast computation of the resulting QP problems and efficient dealing with system and control constraints. The proposed controller with a locally linearized TS model results in a convex optimization problem, which is a main advantage in comparison with the use of relational models (Linkens and Kadiah, 1996; Postlethwaite, 1996; de Oliveira and Lemos, 1995), where nonlinear optimization methods are necessary. When relatively

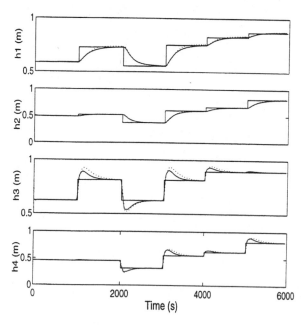

(a) Outputs for four vessel liquid level process.

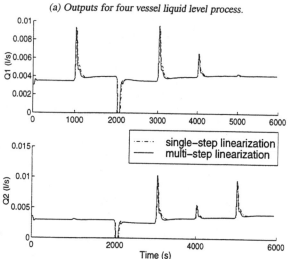

single-step linearization
multi-step linearization

(b) Control inputs four vessel liquid level process.

Fig. 5. Controller performance for $H_p = 5$ and $H_c = 3$.

long prediction and control horizons are necessary, the proposed method performs better than the single-step linearization. The computational load of the multi-step linearization method is significantly higher than in the single-step method. In the latter case, at each iteration there is only one linearization around the current working point and the control signal is calculated directly. When multi-step linearization is applied, the fuzzy model is simulated at every point within the prediction horizon, starting each time from the current state and a set linear models is used to compute the control signal. The computational load is approximately H_p times the load at the single-step case.

Acknowledgement. This research is supported by the FAMIMO Esprit project LTR 219 11. More information is available at http://iridia.ulb.ac.be/FAMIMO

REFERENCES

Babuška, R. (1998). *Fuzzy Modeling for Control.* Kluwer Academic Publishers, Boston.

Babuška, R., J.A. Roubos and H.B. Verbruggen (1998). Identification of MIMO systems by input-output TS fuzzy models. In *Proceesings FUZZ-IEEE'98*, Anchorage, Alaska.

Babuška, R. and H.B. Verbruggen (1996). An overview of fuzzy modeling for control. *Control Engineering Practice 4(11)*, 1593–1606.

Economou, C.G., M. Morari and B.O. Palsson (1986). Internal model control. 5. Extension to nonlinear systems. *Ind. Eng. Chem. Process Des. Dev. 25*, 403–411.

Garcia, C.E. and M. Morari (1982). Internal model control: 1. a unifying review and some new results. *Ind. Eng. Chem. Process Des. Dev. 21*(308-323).

García, C.E., D.M. Prett and M. Morari (1989). Model predictive control: Theory and practice – a survey. *Automatica 25*, 335–348.

Linkens, D.A. and S. Kadiah (1996). Long-range predictive control using fuzzy process models. *IChemE 74*, 77–88.

Oliveira, de, J. Valente and J.M. Lemos (1995). Long-range predictive adaptive fuzzy relational control. *Fuzzy Sets and Systems 70*, 337–357.

Postlethwaite, B.E. (1996). Building a model-based fuzzy controller. *Fuzzy Sets and Systems 79*, 3–13.

Richalet, J. (1993). Industrial applications of model based predictive control. *Automatica 29*, 1251–1274.

Roubos, J.A., R. Babuška and H.B. Verbruggen (1998). Predictive control by local linearization of a Takagi-Sugeno fuzzy model. In *FUZZ-IEEE*, Anchorage, Alaska, pp. 37–42.

Takagi, T. and M. Sugeno (1985). Fuzzy identification of systems and its application to modeling and control. *IEEE Transactions on Systems, Man and Cybernetics 15*, 116–132.

Yager, R.R. and D.P. Filev (1994). *Essentials of Fuzzy Modeling and Control.* Johnn Wiley, New York.

FRAMEWORK FOR APPROXIMATE TIME ROUGH CONTROL SYSTEMS: A ROUGH-FUZZY APPROACH

J.F. Peters[1], S. Ramanna[2], K. Ziaei[3]

[1] Computational Intelligence Laboratory, Department of Electrical and Computer
Engineering, University of Manitoba, Winnipeg, Manitoba R3T 2N2 Canada
[2] Department of Business Computing, University of Winnipeg,
Winnipeg, Manitoba R3B 2E9, Canada
[3] Department of Mechanical Engineering, University of Manitoba,
Winnipeg, Manitoba R3T 2N2 Canada

Abstract: This paper presents an approach to the design of a framework for an adaptive approximate time rough controller. The class of rough control systems described in this paper utilizes a new form of clock called an approximate time window (atw). Each atw is wired to a single-layer fuzzy neural network constructed with logic-processing neurons. Granulated measurements of the speed and quality of controller behavior provide input to a fuzzy neural network, which is calibratable. Controller rules are derived from a decision table filled with network values. A rough fuzzy Petri net model of such a controller is included. An example application of the this form of adaptive rough controller is given. Copyright © 1998 IFAC.

Keywords: adaptive control, approximate, decision-making, fuzzy neuron, information granule, real-time, rough control, rough sets.

1. INTRODUCTION

Considerable work on rough controllers relative to a rapidly growing range of problems has been reported (Czogala et al., 1995; Munakata and Pawlak, 1996; Pawlak, 1995; Pawlak, 1997; Peters and Ziaei, 1998; Peters, Ziaei, and Ramanna, 1998; Polkowski and Skowron, 1995; Polkowski and Skowron, 1998a; Polkowski and Skowron, 1998b; Szladow and Ziarko, 1996). Recent work on adaptive rough controllers can be found in (Peters and Ziaei, 1998; Peters, Ziaei, and Ramanna, 1998). The form of adaptive rough controller reported in this paper has resulted from a search for a controller capable of fast response to a variety of disturbances, and which achieves an optimum level of performance under varying inputs and varying conditions of operation. Experiments with this form of rough control have been carried out in the design of an attitude control system for a small, scientific satellite. Controller gains are estimated on-line relative to step response analysis of the speed and quality of system response. Estimates of gains are guided by a decision-making process rooted in rough set and granulation theory. An adaptive fuzzy rough controller is guided by rules derived from an approximate time decision-making system. Rough sets theory is used to derive controller rules (Pawlak, 1982; Pawlak, 1991;

Skowron, 1995). Granulation theory is used to model decision system sensors as fuzzy implications.

In this paper, the clocks used to measure durations required to achieve controller objectives are modeled as approximate time windows, which were introduced in (Peters, 1998), which are an extension of the notion of time window (Petri, 1996), fuzzy clock (Peters and Sohi, 1996), and the theory of approximate real-time decision-making (Peters, Skowron, Suraj, and Ramanna, 1998). An approximate time window (atw) partitions time relative to granules (clumps of similar timing measurements) such as early, ontime, late. An atw determines the degree of membership of each observed duration in each of its temporal partitions. Based on observations of the degree of overshoot, rise time, and settling time during the operation of a control system, the architecture of a fuzzy rough control system is established. Controller gains are estimated on-line relative to step response analysis of the speed and quality of system response.

The paper is organized as follows. The basic idea of an approximate time window is given in Section 2. An approach to constructing an adaptive approximate time rough controller is given in Section 3. A description of an application of the new form of rough controller is given in Section 4.

2. APPROXIMATE TIME WINDOWS

Based on observations concerning time and its measurement, approximate time windows (atws) are introduced relative to vaguely known partitions of temporal intervals. The basic approach in designing an atw is to correlate accuracy of time measurements with intervals. In related work, this leads to the introduction of rough (approximate) functions which are mappings resulting from measurements rather than abstract definitions (Pawlak, 1987; Pawlak, 1995). Time is commonly viewed as an interval between successive events or acts. In this article, time is identified with the duration between firings of (not necessarily successive) transitions within a process modeled with a Petri net. Knowledge concerning the duration between transitions in a process tends to be vague, imprecise. Time windows are defined relative to durations between firings of transitions in Petri nets.

Def. A _time window_ is a process which measures durations between transitions in a net.

The design of a time window hinges on the articulation of a time interval and the position of a pointer marking the current elapsed time inside a time interval (see Fig. 1).

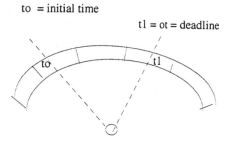

Fig. 1 Time Window

An approximate time window results from attempts to establish a interval over which elapsed time is measured. Time window intervals are vague and found to be approximate (i.e.,a boundary of an interval represents a time limit which is considered _near_ rather than exactly the required duration).

Def. An _approximate time window_ is a process which measures the degree-of-membership of each observed duration in one or more time granules.

An approximate time window (atw) is partitioned into subintervals named with linguistic labels such as early and late (see Fig. 2.). The penumbra covering each boundary of the time intervals is intended to represent the approximate character of interval boundary estimates. The vagueness of the partitions of a time window stems from a lack of crisp knowledge of deadlines imposed on the activities (tasks) performed by agents. Approximate time windows measure the speed of response of systems relative to the degree that a response fits within a time window partition.

Axiom of approximate time: approximate time results from uncertainty about time measurement relative to a timing constraint placed on a transition in a net (i.e., uncertainty in the articulation of an upper bound in the duration measured by a time window).

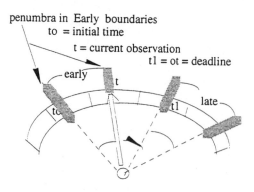

Fig. 2 Approximate Time Window

Axiom of approximate time measurement: approximate time measurements are a set of atomic judgments = $\{m_i(t) \mid m_i(t) \in [0,1], t \in [t_o, t_1], i \in \{early, ontime, late\}\}$.

In measuring the time-varying behavior of controllers, clocks are designed relative to deadlines. In designing such clocks, a distinction is made between a time-originating mechanism (a pulse generator) and a clock, which consists of a time monitor, time interface (connections between monitor and monitor), and clock interface (connections supplying settings, deadlines, and display) as in Fig. 3.

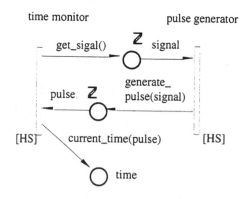

Fig. 3 _Simplified Clock_

The pulse generator and time monitor transitions in Figures 3 are hierarchical. The details of how pulses are generated, how pulse-frequency is controlled, and how time is monitored are hidden. The basic monitoring mechanism in Fig. 3 is a sum-of-pulses variable (a local variable hidden inside the time monitor). This mechanism is represented with the following pseudocode:

```
sum-of-pulses := sum-of-pulses + 1;
current_time(pulse) := sum-of-pulses;
```

The current_time(pulse) outputs the value of sum_of_pulses each time a pulse is received from the pulse generator. The clock in Fig. 3 is "wired" to a calibratable net which generates rules which guide a rough controller in responding to the varying speed-of-response of a controller. The result is a clock information system shown in Fig. 4.

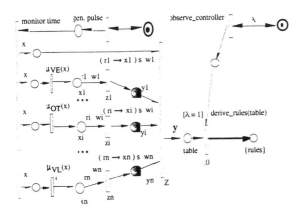

Fig. 4 *ATW Decision System*

In Fig. 4, let $\mathbf{x}(k)$, $\mathbf{r}$, $\mathbf{w}$ denote input places, vector $\mathbf{r}$ (modulators or cut-offs), and vector $\mathbf{w}$ (connections), respectively. Usually the objective function (performance index) Q is defined as a sum of squared errors (see (1)).

$$Q = \sum_{k=1}^{N} [target(k) - y(\mathbf{x}(k), \mathbf{r}, \mathbf{w})]^2 \qquad (1)$$

The gradient - based optimization method is driven by the increments in the parameter space (2).

$$-\frac{\partial Q}{\partial \mathbf{param}} \qquad (2)$$

where the gradient is taken over the connections (here denoted as **param**) of the neurons forming the Petri net. Thus the formula reads as given in (3).

$$\mathbf{param}(new) = \mathbf{param} - \alpha \frac{\partial Q}{\partial \mathbf{param}} \qquad (3)$$

Coefficient $\alpha > 0$ in (3) is the learning rate. The detailed expressions can be easily derived once we confine ourselves to some specific forms of the triangular norms. Let $z(k) = y(\mathbf{x}(k); \mathbf{r}, \mathbf{w})$. Omitting the intermediate steps, we subsequently derive (4) and (5).

$$\frac{\partial z(k)}{\partial w_i} = \frac{\partial}{\partial w_i}\left[\mathop{T}_{j=1}^{n} [(r_j \to x_j)\, s\, w_j]\right] \qquad (4)$$

$$\frac{\partial z(k)}{\partial r_i} = \frac{\partial}{\partial r_i}\left[\mathop{T}_{j=1}^{n} [(r_j \to x_j)\, s\, w_j]\right] \qquad (5)$$

Let $r_j \to w_j$ denote the implication operator defined in (6).

$$r_j \to x_j = \min\left(1, \frac{x_j}{r_j}\right),\ x_j \in [0,1],\ r_j \in (0,1] \qquad (6)$$

In addition, let A be defined in (7).

$$A = \mathop{T}_{j \neq i}^{n}\left[(r_j \to x_j)\, s\, w_j\right] \qquad (7)$$

and assume that the triangular norm s is given as the probabilistic sum (i.e., B s C is rewritten as B + C - B*C). Then from (4) we derive the formula for the connections:

$$\frac{\partial z(k)}{\partial w_i} = \frac{\partial}{\partial w_i}\left[A\left[\left(r_i \to x_i\right) s\, w_i\right]\right] = A\left[1 - \left(r_i \to x_i\right)\right]$$

A similar formula to compute r-values can also derived from (5).

3. DESIGNING ROUGH CONTROLLER

For conciseness, the adaptive rough-fuzzy controller design methodology is presented relative to (a) the analysis of system response based on observations of overshoot as well as durations t_s and t_r and (b) the instrumentation of an approximate time window with attributes identified with sensors used to evaluate durations as well as sensors to evaluate overshoot (see Figures 5a, 5b, and 5c).

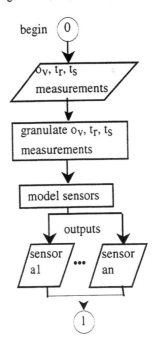

Fig. 5a. Granulation

Overshoot (ov) is the biggest deviation of step response from a particular steady state after the step response has reached a tolerance band for the first time. Rise time t_r is the time when a step response reaches 90% of its steady-state value for the first time, and settling time t_s is measured relative to rise time (i.e., the clock for t_s is reset at $t = t_r$). Time itself is computed relative to a clock which measures durations in the context of information granules named early, ontime, and late. An information granule is collection of similar or related values (Zadeh, 1996). In other words, based on observed rise time and settling time, approximate time windows are designed relative to fuzzy partitions of temporal intervals. Overshoot is conceptualized in terms of fuzzy sets relative to linguistic labels acceptable, big, and very big. Each granule has its own membership function. Membership function values computed relative to the speed of response and quality of response of a system provide input to sensors. A sensor maps each input to a value in a decision system table. This part of the controller design methodology is represented in Fig. 5a. Each firing of sensors $a_1,...,a_n$ in Fig. 5a leads to the creation of a new row in a decision table. Rules are then derived from a completed decision table (see Fig. 5b).

Rules r_1, ..., r_k in Fig. 5b indirectly define a reference model of a system where each rule defines the ideal selection of gains based on measurement of speed-of-response and quality-of-response of the system. Based on the conclusion of a selected rule, controller gains are modified to achieve an optimum system response. In the method shown in Fig. 5c, two approximate time windows concurrently compute rise time and settling time while the quality of response (overshoot) is analyzed relative to the degree of membership in ov granules. A rule is selected which most closely approximates actual overshoot, rise time and settling time. The conclusion of a selected rule is used to adjust controller gains.

4. APPLICATION

As an application of the approximate time rough control methodology, we consider the attitude control of a satellite. A high-level block diagram showing the principal components of the rough control system are shown in Fig. 6.

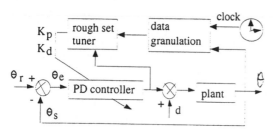

Fig. 6 Rough Control System

For the sake of simplicity, we consider only one degree of freedom. The control methodology is illustrated relative to the control of a satellite pitch angle (see Fig. 2), which is decoupled from satellite yaw and roll angle control. Using a PD pitch controller, the close-loop system response is given by:

$$\theta = \frac{K_p + K_d s}{Js^2 + K_d s + K_p} r + \frac{1}{Js^2 + K_d s + K_p} d$$

where θ is the pitch angular position; r, setpoint; d, disturbance; J, moment of inertia of the plant; and K_d and K_p are the controller differential and proportional gain parameters.

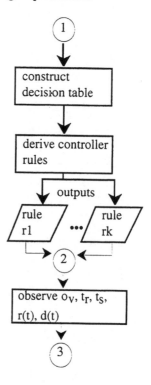

Fig. 5b. Rules

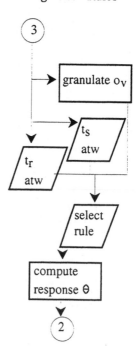

Fig. 5c. Select rule

The distribution of degree of membership values in a granule associated with a sensor a_i, $1 \leq i \leq 9$, is assumed to be approximately normal in a Gaussian distribution with mean (modal point) m and standard deviation s (spread). Let g, x be the name of a granule associated with sensor a_i and measurement x (e.g. overshoot at given instant in time), respectively. Hence, the membership function used in modeling sensor is given by

$$g(x) = e^{\left(\frac{-(x-m)^2}{s^2}\right)}$$

In modeling sensors a_4, a_5, a_6 (rise time sensors) and a_7, a_8, a_9 (settling time sensors), we also introduce a modulator r and strength-of-connection w. Taken collectively, each trio of sensors constitutes an atw (approximate time window). A modulator imposes a threshhold on stimuli, and a strength-of-connection raises or lowers the impact of an input in an atw. Then atw sensor a_i is modeled as an aggregation of a fuzzy implication value and strength-of-connection w:

$$a_i(x) = (r \rightarrow g(x)) \; s \; w, \text{ where } (r \rightarrow g(x)) = \min\left(1, \frac{g(x)}{r}\right)$$

In this research, the operator s (s-norm operator) computes a probabilistic sum. It has been shown that the modulator and strength-of-connection parameters in approximate time windows can be calibrated (Peters, 1998). It should also be noted that each application of a rule relative to an observed step response of the control system results in changes in both K_p and K_d. An algorithm for finding the rule which most closely approximates observed speed of response and quality of response is given in (Peters, Ziaei, and Ramanna, 1998). Sample rules for changing proportional and differential gains is given in (8) and (9), respectively.

[a_3(0.0) AND a_5(1.00) AND a_6(1.00) AND a_7(0.81)] OR [a_3(0.0) AND a_5(1.00) AND a_6(0.93) AND a_7(0.62)] => kp_fac(2.00)   (8)

a_2(0.1) AND a_5(1.00) AND a_7(0.62) => kd_fac(1.80)   (9)

Such rules are derived from a real-time decision system table based on a sufficient number of prototypical experimental measurements of controller performance and the granulation of these measurements. A Petri net model of the adaptive rough control system described in this section is given in Fig. 7.

Attached to the block diagram for a closed loop, adaptive controller is a model of the rule derivation process (done off-line) in the form of what is known as a rough fuzzy Petri net. It is assumed that the reader is familiar with classical Petri nets (Petri, 1962) as well as the basic structure of coloured Petri nets found in (Jensen, 1992) and fuzzy Petri nets in (Pedrycz and Gomide, 1994). Briefly, rough Petri nets are derived from coloured and hierarchical Petri nets as well as from rough set theory. The complete process of deriving rules for a decision system can be modeled with a rough Petri net at a sufficiently high level to facilitate an understanding of a particular rule-derivation process. In addition, the strengths of connections (weightings of attribute computations) in a rough Petri net can be calibrated. Rough fuzzy Petri nets are derived from rough Petri

nets and fuzzy Petri nets (Peters, Skowron, Suraj, Ramanna, 1998). This form of Petri net makes it possible to add learning capability to a decision-making rule-derivation process, which combines the use of fuzzy sets and rough sets in rule formulation.

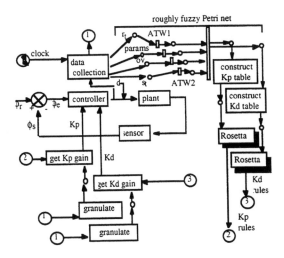

Fig. 7 Petri net model

K_d and K_p correction rules for a PD controller have been derived using a combination of a roughly fuzzy Petri net and Rosetta. Tuning information from a number of controller simulations was used to build an information system and to generate some tuning decisions. These rules can be used to tune a satellite pitch controller on-line. The new information collected after each tuning were added to the rough control system and dynamic reducts were employed to modify decision rules periodically. A comparison of the performance of rough control with a PD control of the pitch angle in the presence of a rectangular (square wave form of) disturbance is shown in Figs. 8 and 9. Fine-pointing is achieved rapidly. Each step response of the rough controller is due to the firing of a pair of rules (see Fig. 7) used to select appropriate changes in proportional and differential gains of the PD controller. This approach differs from classical PD control, since the gains are changing dynamically depending on the degree of disturbance. Other forms of disturbance have also be investigated with similar fine-pointing results. Further, good results have been achieved in cases where the system is underdamped or overdamped.

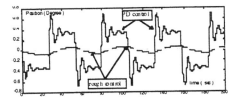

Fig. 8 Closed-loop response to rectangular disturbance

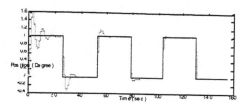

Fig. 9 Closed-loop response during self-tuning

5. CONCLUDING REMARKS

A framework for designing adaptive rough controllers has been given. This framework provides an application of approximate real-time decision-making. The approach given in this paper includes the use of approximate time windows which measure the speed of response of a system approximately relative to the degree of membership of observed speeds in temporal information granules. In addition, the application of real-time decision-making in selecting gains for a satellite attitude control system has been presented. The approach described in this paper has been applied in a variety of control systems (hydraulic servo system, flood water diversion control system, temperature control system for a vertical mixer for preparing solid propellant for rocket engines, and air traffic control system), software deployability control system and in control functions for interacting situated robots in a computer zoo.

ACKNOWLEDGMENT

Research supported by a Canadian Natural Sciences and Engineering Research Council (NSERC) operating grant, NSERC Industrial-Research Partnership grant, Bristol Aerospace, Ltd. grant, and Canadian Space Agency grant.

REFERENCES

Czogala, E., Mrozek, A., Pawlak, Z. (1995). The idea of a rough fuzzy controller and its application to the stabilization of a pendulum-car system. Fuzzy Sets and Systems 72, 61-73.

Jensen, K. (1992). Coloured Petri Nets--Basic Concepts, Analysis Methods and Practical Use 1. Berlin, Springer-Verlag, 1992.

Munakata, T., Pawlak, Z. (1996). Rough control: Application of rough set theory to control. In Proc. EUFIT'96, 209-218.

Pawlak, Z. (1987). Rough functions. Bull. PAS, Tech. Ser. 35 (5-6), 249-251.

Pawlak, Z. (1982). Rough sets, Int. J. of Computer and Information Sciences 11, 341-356.

Pawlak, Z. (1991). Rough Sets: Theoretical Aspects of Reasoning About Data. Boston, MA, Kluwer Academic Publishers.

Pawlak, Z. (1995). Rough real functions and rough controllers. ICS Research Report 1/95, Warsaw.

Pawlak, Z. (1997). Rough real functions and rough controllers. In: T.Y. Lin, N. Cercone (Eds.), Rough Sets and Data Mining: Analysis of Imprecise Data. Boston, MA, Kluwer Academic Publishers (1997) 139-147

Pedrycz, W., Gomide, F. (1994). A generalized fuzzy Petri net model". IEEE Trans. on Fuzzy Systems 2/4, 295-301.

Peters, J.F., Ziaei, K., Ramanna, S. (1998). Approximate Time Rough Control, Concepts and Application to Satellite Attitude Control. In: L. Polkowski and A. Skowron (Eds.), Rough Sets and Current Trends in Computing, Lecture Notes in Artificial Intelligence 1424. Berlin, Springer-Verlag, 1998, 491-498.

Peters, J.F., Ziaei, K. (1998). Generating rules in selecting controller gains, "A combined rough sets/fuzzy sets approach". In: Proc. Canadian Conference on Electrical and Computer Engineering (CCECE'98), Waterloo, Ontario, 25-28 May 1998, pp. 233-236.

Peters, J.F. (1998). Time and clock information systems: Concepts and roughly fuzzy Petri net models. In: Polkowsk, L. and Skowron, A. (Eds.), Rough Sets and Knowledge Discovery 2: Applications, Case Studies and Software Systems. Berlin, Physica Verlag, a division of Springer Verlag [in press].

Peters, J.F., Sohi, N. (1996). Coordination of multiagent systems with fuzzy clocks. Concurrent Engineering: Research and Applications 4, 73-88.

Peters, J.F., Skowron, A., Suraj, Z., Ramanna, S. (1998). Approximate Real-Time Decision Making: Concepts and Rough Fuzzy Petri Net Models. International Journal of Intelligent Systems [in press].

Petri, C.A. (1962). Kommunikation mit Automaten. Schriften des IIM Nr. 3, Institut für Instrumentelle Mathematik, Bonn, West Germany, 1962. See, also, Communication with Automata (in English). Griffiss Air Force Base, New York Technical Report RADC-Tr-65-377, 1, Suppl. 1.

Petri, C.A. (1996). Nets, time and space. Theoretical Computer Science 153, 3-48.

Polkowski, L., Skowron, A. (1998a). Rough mereological approach--A survey. Bulletin of Int. Rough Set Society, 2(1), 1-14.

Polkowski, L., Skowron, A. (1998b). Rough mereological foundations for design, analysis, synthesis, and control in distributed systems. Information Sciences 104 (1-2), 129-156.

Polkowski, L., Skowron, A. (1995). Introducing rough mereological controllers: Rough quality control. In: T.Y.Lin, A.M. Wildberger (Eds.), Soft Computing, 240-243.

Skowron, A. (1995). Extracting laws from decision tables: a rough set approach". Computational Intelligence 11/2, 371-388.

Szladow, A.J., Ziarko, W. (1993). Adaptive process control using rough sets. Paper 93-384, Instrument Society of America, 1421-1430.

Zadeh, L.A. (1996). Fuzzy logic = computing with words". IEEE Trans. on Fuzzy Systems 4/2, 103-111

USING FUZZY SYSTEM IDENTIFICATION TO EXTRACT A DETECTOR OF DISCRETE CYCLIC EVENTS

Margaret M. Skelly and Howard Jay Chizeck

*Motion Study Laboratory, Cleveland V.A. Medical Center;
Department of Electrical, Systems, Computer Engineering and Science,
Case Western Reserve University, Cleveland, Ohio, USA*

Abstract: This paper describes the use of fuzzy systems identification for the real time detection of discrete events of a cyclic process. The process is the locomotion of a paraplegic individual, generated using electrical stimulation of paralyzed muscles. Locomotion was represented as a finite state model consisting of five states with five discrete events at the transitions between the states. Event detection was performed in two-part procedure. First, the sensor signals were classified into one of the five states using fuzzy logic. Second, a supervisory algorithm monitored the state classification at each time sample to determine if a state transition occurred. This second supervisory portion forced the "forward" progression through the finite state model and eliminated the occurrence of certain errors. *Copyright © 1998 IFAC*

Keywords: Finite state machines, Discrete-event systems, Fuzzy modelling, Electrical stimulation

1. INTRODUCTION

Bipedal locomotion is achieved by the repetition of coordinated motion of two legs. For paralyzed individuals, this motion can be generated through the use of Functional Electrical Stimulation (FES). One area of research is the design of feedback controllers to improve the quality of this artificial gait. One such controller requires the knowledge of when the legs were in certain *functional phases* of the cyclic walking motion. Since the motion of both legs during walking is the same, but out of phase with each other, the motion of the two legs can be analyzed separately. The motion of a single leg can be represented in a finite state model (Fig. 1). The *functional phases* are the states in the model and the transitions between the states become discrete events that must be detected in real time for the feedback controller.

Functional Electrical Stimulation (FES) uses electrodes, either on the surface of the skin or implanted in the muscles, and pulses of electrical current to generate muscle force (Stein, *et al.*, 1992).

The relationship between the stimulus pulse width and pulse frequency to the resulting muscle force is nonlinear and time varying. It is not readily described in equation form. FES activated muscles are usually weaker and fatigue faster than normal muscles. Electrical stimulation reduces the ability to accurately modulate muscle force usually resulting in a few quantized levels of force available to drive the nonlinear biomechanical system. In addition, there is a significant and variable time delay between electrical stimulation and muscle force production. This time delay may vary between 50 ms and 300 ms depending upon the muscle and its potentiation or fatigue state (Kobetic, *et al.*, 1994).

The goal of most FES walking systems is to stimulate the muscles to produce limb motions similar to those observed in normal walking. The length of the time delay between electrical stimulus and force production makes infeasible the use of traditional trajectory-following feedback control (such as those used in robotics). For this reason cycle-to-cycle control has been suggested, where the input (electrical stimulation) is modified for the next step

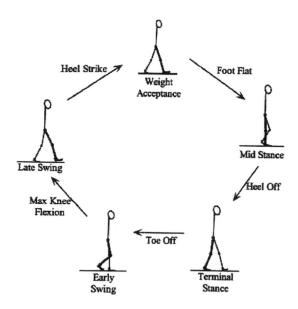

Fig. 1. Finite State Model of the Gait Cycle. Functionally, the gait cycle for a single leg (shown as bold) is divided into five phases depicted by the stick figures. The transitions between the phases, shown as arrows, are discrete gait events.

(gait cycle) after assessment of the performance achieved in the previous cycle.

Evaluation or control of gait on a cycle-to-cycle basis requires the knowledge of how and when the system progressed through the cycle. The gait cycle of a single leg is divided into five phases (Fig. 1). The transitions between the phases are discrete events. The work presented here is method for detecting the discrete events in this cyclic process.

Two general approaches have been taken to this problem in the past. One approach has been to use a finite state model to describe the states of the legs and to detect the transitions between the states by detecting when specific sensors exceed predefined thresholds (Franken, *et al.*, 1993). Another approach is to use classification methods to estimate the phase of gait from the combination of the sensor signals (Ng and Chizeck, 1997; Popovic, 1993). Because of sensor inaccuracies (see below) and the fact that the sensed quantities are only indirectly related to gait phase transitions, the method described here uses a combination of these two approaches.

The information available to the event detector is limited to sensors that can be worn by the FES subject. Minimizing the number of sensors will increase the ease of use and cosmesis of the complete walking system. The difficulty in securely mounting sensors to humans means that the sensors will be susceptible to movement noise. The sensors may also be sensitive to the step-to-step variability present in normal gait and amplified in FES gait.

The best sensors (in terms of mounting, signal strength and reliability) for each subject may be different. For this reason a classifier with an *identified* or empirically determined relationship between the inputs and output (state) was chosen.

2. METHODS

The application of machine intelligence methods to this real time gait phase detection problem is motivated by the lack of underlying mathematical models of the phenomena, signal corruption due to sensor and sensor mounting problems, and the subject-specific nature of gait.

A fuzzy logic model was chosen to perform the initial phase classification because it has been shown to be able to deal with the sensor noise and cycle-to-cycle variability present in this application. The identification of the fuzzy model requires relatively few training samples compared to neural net training, an important factor in this application. Because it must be obtained from experiments where subjects walk with FES, limited training data is available.

The event detector is composed of two parts. First, an identified, fuzzy-based model is used to classify the output into one of the five phases. Then, a finite state model is superimposed upon the phase estimates to detect the discrete events.

2.1 First Level: Identified Fuzzy Rules

The fuzzy logic classifier is an identified model. The fuzzy rule base is identified with known input-output data pairs. The identified fuzzy model is then used to estimate output when presented with new input. Before identification, the structure of the fuzzy model must be defined. The inputs, fuzzy sets, inference method, and defuzzification method must be chosen.

Fuzzy Model Structure. The basic components of this fuzzy model are the (multiple) inputs and (single) output, their reference sets, and the fuzzy inference method.

Several inputs are derived from the available sensors and the phase of gait is the output. The specific inputs to the fuzzy classifier are the low pass filtered voltages from force sensitive resistors (FSRs) and stimulator breakpoint information. FSRs are small pressure-sensitive sensors worn in the subject's shoe insole. The voltage from an FSR is nonlinearly related to the amount of pressure applied to the sensor. The derivative, computed as a 2-point difference of each filtered FSR voltage, was also used as an input to the fuzzy model. The stimulation breakpoint information is an integer {0, 1, 2, 3, 4} indicating the progression of the stimulator through the predefined pattern of electrical stimulus pulses. Fig. 2 shows typical traces of the inputs.

Each of the inputs and the output is divided into reference sets. The number of reference sets for the inputs varied, but the number of sets for the output was always five, one corresponding to each phase of gait. Although the definition of the reference sets can influence the performance of the model, there is no known "best" method to define sets.

In this work, the reference sets are defined by an expert through examination of histograms of the input data for the five output phases. The sets are always defined as trapezoids, and the overlap between adjacent sets is defined so that for any input value, the sum of its memberships in all sets is 1. By defining unevenly sized sets, the expert attempts to minimize the number of different output phases that will be represented in each input reference set.

The fuzzy inference method consists of two components, the relationship *combination* and the *composition* methods. *Product combination* (input memberships for a given rule are multiplied together with the rule's weight to form the degree to which the rule is activated) combined with *product composition* (the given rule's output set is scaled by the resulant degree) was chosen.

Fuzzy Identification. Once the model structure is established, the fuzzy rule base is identified from the input-output data of a single walking trial consisting of between four and nine gait cycles. The following two steps are performed for each data pair.

(1) Fuzzify the processed inputs and output. Fuzzification is the assignment of degrees of membership for each input and output into its prescribed reference sets. Values outside the defined input range are assigned a membership of 1 in one of the end sets. Since the sets are defined so that no more than two sets overlap for any input value, each input will have non-zero membership in at least one set and at most two sets. The integer representation

of the output state will always fuzzify into a single set with a membership of 1.

(2) Generate the fuzzy rules. Each combination of reference sets, one set per input and output, is used to generate a rule. Since an individual input value may fuzzify into more than one set, a particular data pair may generate several rules. Because *product combination* is used, the product of the memberships of the input and output sets are recorded as the rule's degree. Each rule is recorded as a list consisting of the reference set corresponding to each input, the output reference set, and the rule's degree.

Additional rules are generated as subsequent data samples are fuzzified and encoded into rules. Frequently, multiple data samples will generate the same rule, i.e. same sets specified for each input and output, but to different degrees. In this case, the *maximum* degree that any data sample generated the rule is saved.

After all the samples have been processed, the resulting list of rules is weighted by the number of samples that generated each rule. This additional weighting places confidence in rules generated by many data samples. The influence of rules generated by very few samples is reduced, since they were most likely generated by "atypical" data. The motivation for this additional weighting was provided by Wang and Mendel (1992) who multiplied each rule by an expert's confidence (assigned a value over the range of 0 to 1) for each data pair.

Fuzzy Estimation. Once the rule base has been identified, it can be used to estimate the phase of gait. Output estimation involves three steps: (1) fuzzification of the processed inputs, (2) composition with the rule base, and

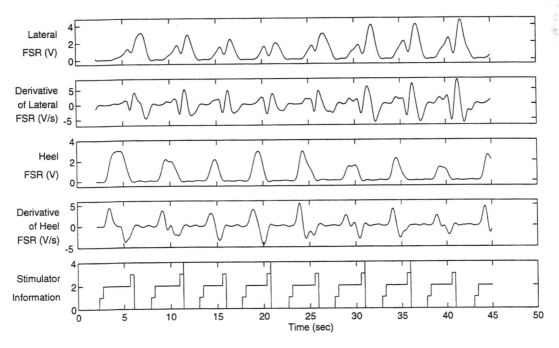

Fig. 2. Inputs to the Gait Event Detector. The sensors for one leg of a walking trial are shown. The eight peaks in the first and third inputs roughly correspond with the eight gait cycles. The variability of the subject's gait is evident by the varying peak heights and slight differences in peak shape and duration.

211

(3) defuzzification of the output. The same filtering and fuzzification used during identification is applied to the new input data pair. The fuzzy inputs are composed with the rule base to produce a fuzzy output estimate. For each rule, the product of the memberships of the reference sets specified and the rule's degree is used to scale the output set specified in the rule.

Although each rule must be applied to the inputs, most rules produce a membership of zero in the specified output set. Exclusion of these zero-valued memberships has no effect on the output, and reduces the computational burden (Postlethwaite, 1991). Only fuzzified inputs that have non-zero memberships in the all the input sets defined in a rule will produce non-zero output. Thus, computation time was reduced by eliminating many calculations whose product is zero by searching for only those rules whose input sets matched the non-zero reference sets of the fuzzified inputs.

The summation of the output sets produced by each rule forms the fuzzy output. The final step is to defuzzify the output. Centroid defuzzification was chosen because it involves a relatively simple calculation that captures the contribution of all the rules within the fuzzy output. In the case of product composition and trapezoid-shaped reference sets, the calculation becomes one dependent upon the output sets' areas and centroids and the degree to which each rule is activated (Kosko, 1992). Therefore, the defuzzified output, y, is the centroid computed from:

$$y = \frac{\sum_{i=1}^{r} w_i c_i I_i}{\sum_{i=1}^{r} w_i I_i} \qquad (1)$$

where I_i and c_i are defined as the respective area and centroid of the ith rule's consequent, or output set. There are r rules, and w_i is the degree to which the rule is implemented. Because product composition is used w_i is the product of the rule's antecedent set memberships.

$$w_i = U_1^i \cdot U_2^i \cdot \ldots \cdot U_k^i \cdot d_i \qquad (2)$$

where

w_i = the weight to which rule i affects the output

$U_1^i, \ldots, U_k^i$ = the memberships of the k inputs in the sets defined by rule i

d_i = the degree of rule i

In this particular application, the sizes of the output reference sets corresponding to the phases of gait are all equal. So, the centroid expression simplifies to one including only the rule weight, w_i, and the output set centroid, c_i:

$$y = \frac{\sum_{i=1}^{r} w_i c_i}{\sum_{i=1}^{r} w_i} \qquad (3)$$

Because of the cyclic nature of the output sets, they are mapped onto a circle and the centroid is computed in polar coordinates (Ross, 1995). If a linear coordinate system were used, the estimate whose fuzzy output is restricted to the sets at the ends of the defined range would inappropriately (for cyclic outputs) defuzzify to a set in the center of the range. The calculation carried out in polar coordinates correctly defuzzifies into one of the end sets. The crisp output value is then converted into one of the five phases of gait by choosing the set with the center.

Testing of the fuzzy rule base with data samples that had not been used for training revealed that the output state could be estimated to an accuracy of approximately 85%. Fig. 3 is an example of the estimated and actual phases of gait. The incorrect state estimations were usually brief, i.e. the estimated state returned to the correct value within a few data samples, and often occurred near the actual transition between states. The presence of these errors in the estimated state leads to the use of supervision. The primary purpose of the supervisory rules is to smooth or reduce these short-interval errors. In fulfilling this objective, the transitions between states, i.e. the discrete events, are detected. In addition, the discrete events are a more convenient and compact format for other parts of a controller to access the phases of gait.

2.2 Second Level: Supervisory Rules

The supervisor tracks the estimated state. By comparing the current and previous samples' estimates, the discrete events are detected. Several internal counters and flags keep track of past states, so when questionable transitions between states are encountered, intelligent decisions can be made.

There are three basic supervisory rules. The form of the supervisory rules differs from the identified fuzzy rules. The fuzzy rules can be expressed as *if...then* statements with common antecedents and consequents, and all rules are executed in parallel. The supervisory rules, however, are imbedded in a series of interdependent decisions.

Rule 1 verifies that the cycle's duration is "reasonable." Once each time through the finite state model, the just-detected series of discrete events are verified to have spanned a reasonable duration in time. If the duration of the gait cycle is less than 25% of the average, then the erroneously detected events are removed. If the gait cycle is within 25% to 150%

of the existing average, then the gait cycle is accepted and the average gait cycle duration is updated.

Rule 2. The second supervisory rule removes brief intervals of incorrectly estimated states. The time spent in each state, as it is estimated by the fuzzy model, is monitored. Transitions to other than the next anticipated state that last for less than the previous state's duration are rejected. Thus, the supervisor forces "forward" progression through the finite state model.

Rule 3. The third rule allows for acceptance of the complete gait cycle even if two specific consecutive states are not estimated by the fuzzy model. Specifically, *early swing* and *late swing* can be incorrectly estimated as *terminal stance* or *weight acceptance* or a combination of the two. When these states are not estimated but the duration of the completed cycle has been verified (Rule 1), then an artificial event (from *terminal stance* to *late swing*) is generated at the time sample before the detected *heel strike* event.

3.   RESULTS AND DISCUSSION

This cyclic event detection was tested with three accomplished walking FES subjects. A single trial consisted of recording sensor data and video as the subject walked for approximately 16 steps (8 gait cycles) with the aid of a walker or crutches. Fig. 2 shows typical traces of the inputs for a single trial. The sensor signals along with the phases of gait that were determined from the video of a single walking trial were used to train a fuzzy rule base. Two rule bases were generated for each subject—one for each leg. The same supervisory rules, however, were used in all cases.

The ability of the identified fuzzy model to properly classify the output state was analyzed. Then, the effect of the supervisory to enhance the detector's performance was quantified.

As part of the fuzzy model's structure, the number of sets for each input must be defined. It was found that between three and five sets yielded better performance. The use of more than five sets degraded the model's phase estimation accuracy. The additional sets reduced the "fuzziness" of the model. When between 4 and 8 reference sets were used, the identified rules could accurately estimate the phase using data with which it was trained. But when tested with data from a different trial, the accuracy degraded. The use of too many reference sets identifies a rule base very specific to the training data, and the capacity of the rule base to generalize the relationship between inputs and output is diminished.

The effect of the weighting based upon the number of samples that generated each rule was assessed. The additional weighting enabled the model to estimate the phase better than when no weighting was used. For three days in which trials were conducted for one subject, the accuracy of estimating the phase of gait was compared when the rule bases were formed with the weighting and without the weighting. The comparison listed for each day in Table 1Table 1 consisted of training two rule bases (one for each leg) including the weighting for each of eight walking trials. The output (phase of gait) was then estimated for each of the eight trials using each of the rule bases (8 rule bases x 8 data trials = 64 estimated trials). The accuracy, defined as the percent of correctly estimated data samples when compared to the video, is listed in the first column. The second column contains the results of the same analysis excluding the weighting during the rule base formation. The average accuracy across all trials in a day was higher

Table 1. Accuracy With and Without Weighting of the Fuzzy Rules

	Accuracy		Paired t-test
	with weighting	without weighting	
Day 1	80.5%	78.6%	p<0.001
Day 2	86.5%	83.8%	p<0.001
Day 3	84.6%	81.8%	p<0.001

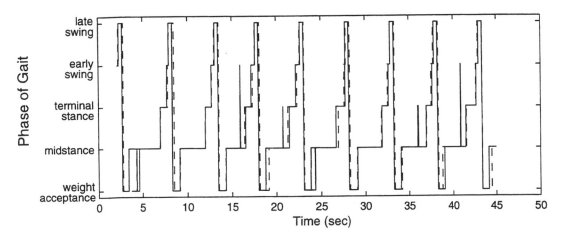

Fig. 3. Phase of Gait Estimated by the Fuzzy Rules. The solid line is the output as classified by the fuzzy model. The dashed lines are the phases extracted from a videotape of the experimental trial.

Table 2. Improved Accuracy in Estimated Phase Due to Supervisory Rules

Subject-Leg	Median Accuracy Before	After	Significant change?
DN - Left	85.9%	87.0%	yes, p<0.001 (n=56)
DN - Right	75.3%	77.1%	yes, p<0.001 (n=56)
DC - Left	86.1%	86.4%	yes, p<0.001 (n=42)
DC - Right	80.8%	80.8%	no, p=0.232 (n=42)

with weighting than without it (paired t-test, $p < 0.001$, n=128 = 8 rule bases x 8 trials x 2 legs).

The supervisory rules detected 99% of the gait cycles correctly. The discrete events that the supervisory rules detected were the same as the ones a human expert examining the trace of estimated phases gait would have chosen. The few errors were due to very poor state estimates by the fuzzy model.

To quantify the effect of the supervisory rules, the accuracy of the estimated phases of gait were compared before and after supervisory rules detected the gait events. The median percent accuracy of phase of gait estimation for two subjects is listed in Table 2. The non-parametric, signed-rank test was used since the several of the populations tested were not normally distributed (Kolmogarov-Smirnov test, p<0.02). For three of the four legs, the supervisory rules improved the accuracy of the estimated phase, and had no effect on the fourth leg.

In this paper it has been shown that is possible to detect the discrete events associated with gait from multiple sensors containing various degrees of noise, including cycle-to-cycle variability of the underlying system. This particular two-level approach combined two methods that had previously been applied to this same application with varying degrees of success. An identified fuzzy model initially classified the output state, then heuristic *if...then* rules constrained the output to the cyclic finite state model. The resulting cyclic discrete event detector has also been implemented and tested in real time making it appropriate for other applications where it is desired to track a cyclic phenomena in the presence of significant variability from cycle to cycle.

REFERENCES

Franken, H.M., W. de Vries, P.H. Veltink, G. Baardman and H.B.K. Boom (1993). *State detection during paraplegic gait as part of a finite state based controller*. IEEE Eng Med Biol Soc 15th Ann Int Conf, San Diego, CA, 1322-1323.

Kobetic, R., E.B. Marsolais, P. Samame and G. Borges (1994). The Next Step: Artificial Walking. In: *Human Walking* (Rose, J. and J.G. Gamble. (Ed)). Williams & Wilkins, Baltimore, MD.

Kosko, B. (1992). *Neural Networks and Fuzzy Systems, A Dynamical Systems Approach to Machine Intelligence*. Prentice Hall, Inc., Englewood Cliffs, NJ.

Ng, S.K. and H.J. Chizeck (1997). Fuzzy model identification for classification of gait events in paraplegics. *IEEE Trans Fuzzy Sys* **5**(4): 536-544.

Popovic, D.B. (1993). Finite state model of locomotion for functional electrical stimulation systems. *Prog Brain Res* **97**: 397-407.

Postlethwaite (1991). Empirical Comparison of Methods of Fuzzy Relational Identification. *IEE Proceedings-D* **138**(3): 199-206.

Ross, T.J. (1995). *Fuzzy Logic with Engineering Applications*. McGraw-Hill, Inc, New York.

Stein, R.B., P.H. Peckham and D.P. Popovic, Eds. (1992). *Neural Prostheses, Replacing Motor Function After Disease of Disability*. New York, Oxford University Press.

Wang, L.-X. and J.M. Mendel (1992). Generating fuzzy rules by learning from examples. *IEEE Trans Sys Man Cybern* **22**(6): 1414-1427.

CONTINUOUS CASTER SCHEDULING SYSTEM
ANALYSIS WITH FUZZY TECHNOLOGY

I. B. Türksen, and Michael S. Dudzic

*Department of Mechanical and Industrial Engineering,
University of Toronto, Toronto, Ontario, Canada, M5S 3G8;
Dofasco Inc., Hamilton, Ontario, Canada, L8N 3J5*

Abstract: A "proof of concept" project was initiated to investigate if Fuzzy technology could help Dofasco, a fully integrated steel company, better understand complex scheduling situations at its steelmaking facility. The approach and methodology taken to investigate these opportunities involved the development of a "Fuzzy Expert System" model. The objective of the study was to investigate and analyse complex fuzzy interactions of variables affecting a scheduling system and its performance measure under current operating conditions. Copyright © 1998 IFAC

Keywords: Artificial Intelligence, Fuzzy Expert Systems, Fuzzy Modelling, Scheduling Algorithms, and Steel Industry.

1. INTRODUCTION

In the North American steel industry to date, the use of Fuzzy Logic technologies in supervisory control and business applications has been limited. Some of the key issues involved in this are:

1) The lack of understanding of Fuzzy technologies as they can apply to solving industrial problems.
2) A need for a robust methodology for the use of Fuzzy technologies in industry.
3) The need for appropriate tools and training for applying Fuzzy Logic methodologies.

These were issues that made applying Fuzzy technologies at Dofasco a difficult task. To help overcome this and to prove that Fuzzy technologies can play a role in the solution to a class of complex problems in the steel industry, a "proof of concept" project was initiated. The problem presented was to investigate if Fuzzy technology could help Dofasco better understand two complex Continuous Caster scheduling situations; the relationship of customer delivery due dates and mixed grade steel processing to the schedules used. The objective was to use Fuzzy modeling to provide an aggregate analysis and hence an overall view of the operating conditions for Caster scheduling and its behaviour patterns under the current structure of scheduling procedures.

2. MOTIVATION FOR FUZZY TECHNOLOGIES

The motivation for using Fuzzy technologies versus classical statistical methods such as multivariate regression is based on the perceived subjective nature of the data interactions in this application. The advantage of Fuzzy system modeling to be proposed here are that:

1) Fuzzy knowledge representation techniques allow the extraction of highly complex interactions of input variables and their effect on the output variables.
2) Fuzzy inference techniques allow reasoning with Fuzzy knowledge representation for both crisp or fuzzy input patterns and hence provide an assessment and prediction of system performance measures for new schedules.
3) The model structure is not predefined and is optimally generated by the data.

That is, Fuzzy system models would depict and demonstrate non-linear and complex interactions in a flexible and robust manner amongst system variables and represent their trade-offs more clearly as well as forecast performance measure indicators under the current operating conditions.

3. FUZZY SYSTEM MODELING "PARADIGM"

The modeling methodology used for this project involved the development of a "Fuzzy Expert System" model. The objective was to develop a collection of Fuzzy membership functions and rules for representation of the interaction of variables and reasoning, and to be able to provide prediction using new scheduling data. In general, the methodology consisted of the following high-level activities:

1) Identification of "Fuzzy" patterns in the input-output data ("Fuzzy" data mining).
2) Establishment of a model of system behaviour. This involved:
   a) Discovering critical rules.
   b) Discovering critical variables.
3) Proving the Fuzzy model against traditional industrial technologies.

A key feature of Fuzzy modeling is the identification of fuzzy information "granules". As depicted in Figure 1, given a scatter diagram, there is the option of determining fuzzy information granules with fuzzy clustering techniques or fitting a non-linear curve with multivariate regression techniques. In this manner, fuzzy information granules are identified for each variable that affects the performance measure in combination with all input variables and their information granules.

There are two key phases involved in the development of the Fuzzy Expert System model; Unsupervised and supervised learning. Unsupervised learning is the knowledge acquisition and representation phase. This is where learning the features of the input-output data set occurs without requiring the input of human expert knowledge. It provides the structure of system knowledge via the determination of the "IF... THEN" fuzzy rules, that expose the "hidden" behaviour of the system. Key features of this phase are:

1) Fuzzy cluster analysis of the output.
2) Determination of the output membership functions.
3) Projection of the output clusters onto the input variables spaces for significant input variable selection.
4) Determination of the optimal input cluster and membership functions.

Supervised learning is the model determination and tuning phase. This is where the training data set is used to minimize the error of the model with respect to the output measure. Key features of this phase are:

1) Determination of the "Approximate Reasoning" (i.e. Fuzzy model) parameters.
2) Membership function tuning.

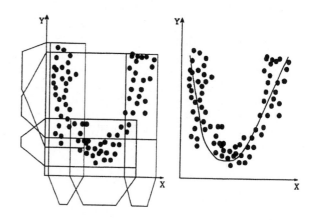

Fig.1. Formation of Fuzzy Information Granules Versus Multivariate Regression Curve Fitting.

3) Validation of the model via a minimum error calculation.

The following are brief highlights of the theoretical aspects involved in the paradigm used in this application. Technical details can be found in the references listed.

3.1 Unsupervised Learning

Knowledge representation is based on a modified and improved Fuzzy C-Means (FCM) algorithm which is an extension of classical FCM algorithms. It is formulated with a structure identification technique that first elicits output patterns in terms of fuzzy clusters, then relates them to the input variables' fuzzy clusters and hence leads to the formation of fuzzy rules in "IF...THEN" patterns (Sugeno, and Yasukawa, 1993; Türksen (1986, 1995a,b, 1996a,b)). The fuzzy clustering algorithm has three parameters:

1) Order or "level" of fuzziness, $m \in [1, 8]$ that identifies the optimal level of fuzziness between the crisp and infinitely fuzzy representation.
2) The number of fuzzy clusters, $c \in [2, ..., 20]$.
3) The location of fuzzy cluster centres.

As a result, the structure identification consists of the rule generation activity that has three sub-modules:

1) Output and input clustering.
2) Input and output membership function assignments.
3) Input selection.

Input and Output Clustering; There are three sub-modules in the input and output clustering module: (a) Initial location of output clusters, (b) Number of fuzzy output clusters and the level of fuzziness, and (c) Formation of input membership functions.

a) Determination of the initial location of the output cluster centers. Initially, cluster centers should be determined, for example, as hard cluster centers with Agglomerative Hierarchical Cluster Algorithm (Türksen, 1996) as a procedure to find a suitable guess for the initial locations of cluster prototypes or by some other clustering technique (Kaufman, and Rousseeuw, 1990) for the FCM algorithm (Bezdek, et al., 1980; Bezdek, 1981).

b) Determination of a suitable number of fuzzy output clusters, c, in interaction with a suitable level of fuzziness, m, by implementing FCM (Bezdek, et al., 1980; Bezdek, 1981; Emami, et al., 1996; Fukuyama, and Sugeno, 1989; Kandel, 1982; Türksen, 1996).

c) Formation of input membership functions. Firstly, one projects the output membership values into the space of input variables with the projection method (Sugeno, and Yasukawa, 1993). Secondly, one classifies the data points for each variable, for example, with the application of the K-Nearest Neighbor Method in order to extend the assigned fuzzy clusters to the entire output space (Keller, et al., 1985).

Input and Output Membership Function Assignments; For the given output and input variables, a curve fitting algorithm is used to generate membership functions for each cluster of each variable, such as triangular, trapezoidal, or other type with the use of scatter diagram information. This method was improved by a line balancing approach that assigns a membership value of one to those scatter points that are close to one. A closeness threshold must be chosen for this purpose.

Input Selection; A significance measure test (Türksen, 1996) is used to select the significant input variables from the data set. Ineffective candidate input variables should then be removed from the model.

3.2 Supervised Learning

There are four parameters that are subjected to supervised learning for a given input-output data set and error minimization criterion. They are: (1) Two parameters p and q are associated with (a) the "AND" connective of the antecedents, i.e., input variable clusters, and (b) the "IMPLICATION" connective, depending on the type of implication that is to be chosen, that dictates the impact of inputs on the output variable, i.e., the consequent, respectively; and (2) Two other parameters β and α are associated with: (a) the combination of alternative reasoning methods, and (b) the power of the generalized defuzzification method, respectively (Sugeno, and Yasukawa, 1993; Türksen (1986, 1995a,b, 1996a,b)). The calculation is accomplished via a non-linear optimization program minimizing a performance index.

A general defuzzification method is utilized, based on the probabilistic nature of the selection process among the values of a fuzzy set, called the Basic Defuzzification Distribution (BADD) method (Yager, and Filev, 1994). The BADD method is essentially a family of defuzzification methods parameterized by α. By varying α continuously in the real interval, it is possible to have more appropriate mappings from the fuzzy set to the crisp value depending on the system behaviour.

Finally, the model is validated with a totally new set of input-output data in order to increase the confidence on the "goodness" of the model developed. In this manner, it is further assured that the fuzzy system model responds effectively with a better performance with respect to minimization of model based error in comparison to actual system behaviour for a given performance index. The "goodness" of the model may be assessed with ME, mean error, or RMSE, root mean square of error, for a given performance index.

4. STEEL APPLICATION – CONTINUOUS CASTING SCHEDULING

The purpose of the Continuous Caster is to solidify newly made molton steel into a desired shape such that it can be handled for further processing. Figure 2 illustrates the main components of a Caster. Many different factors influence the operation thus productivity of the Caster including steel grade, width of the resulting "slab" of steel, desired quality, etc... These factors are greatly influenced by the scheduling of the steel products to be made at the Caster. The ability to optimize the scheduling of the Caster provides additional flexibility in meeting customer delivery requirements.

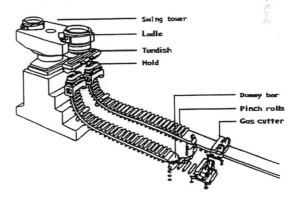

Fig. 2.: Schematic Of A Continuous Caster

In scheduling the Caster, several objectives must be considered such as:

1) The maximization of the productivity of the system.
2) The generation of high quality products.

3) The minimization of mixed grade steel products.
4) Meeting customer delivery due dates.

These objectives are often in conflict. For example, scheduling for just-in-time delivery, i.e., zero tardiness, may result in many different set-ups with grade changes that lead to increases in mixed grade production which reduces the value of the product. Scheduling for high productivity needs long casting sequences of the same product without considering the due date requirements that produce tardy deliveries and a high level of inventory. Moreover, the change of steel grades from one heat to the next is an important factor in scheduling with respect to quality considerations.

In general, there are many factors that can be considered in modeling and analyzing Continuous Caster scheduling. For the "proof of concept" study, two specific scheduling situations were proposed as follows:

1) To understand the relationship between customer delivery due dates and the schedule inputs and how the schedules can be optimized to best meet these delivery due dates
2) To understand the relationship between mixed steel grade processing at the Caster and the schedule inputs and how the schedules can be optimized to minimize mixed grade processing

These two issues were felt to be good candidates for Fuzzy modeling due to the subjective nature of some of the input data involved. For example, priority of the customer order is subjectively measurable and has an indirect but definite influence on the scheduling procedures, which ultimately impacts customer delivery. The ability of the Caster to make steel continuously over many different chemical grades produces transition pieces or "mixed" grade steel. Quantifying the different grade transitions of steel and its financial impact is also subjective. Table 1 illustrates the linguistic classification for mixed steel grades used in the analysis. To date, analytical analyses of these type of scheduling situations have not provided the optimal solutions required. The challenge was to see if Fuzzy technologies could provide new and unique perspectives on these situations.

Table 1: Grade Transition Classification

Categories	Grade Change Between Adjacent Heats
Class 1	No change (same grade)
Class 2	Like grade (compatible)
Class 3	Similar grade (slight grade difference)
Class 4	Unlike grade (large grade difference)
Class 5	Very unlike grade (severe grade difference)

In support of the above, a similar Continuous Caster scheduling problem is discussed by Slany (1996) where he states; "Real-world scheduling is decision making under vague constraints of different importance, often using uncertain data, where compromises between antagonistic criteria are allowed".

4.1 Data Acquisition

Based on discussions with Caster scheduling and production experts about the key variables related to the performance of customer delivery and the mixed grade steel production, two different sets of data have been defined. The variables that relate to customer delivery were selected to be:

(1) Priorities of customer orders; (2) grade transition penalties; (3) customer order widths; (4) minimum width; (5) maximum width; (6) the quality of steel; (7) weights of customer orders; (8) number of the slabs needed for each customer order; (9) weights of the slabs allocated to the customer orders; (10) measure of delivery performance which is the actual delivery date minus the promised delivery due date (output performance measure).

The variables that relate to mixed grade steel production were selected to be:

(1)-(10) Various chemical elements in the steel; (11) tundish start weight; (12) tundish end weight; (13) average casting speed; (14) maximum casting speed; (15) minimum casting speed; (16) mixed grade steel production (output performance measure).

In the study, 800 data vectors were collected for the customer delivery analysis, and 200 data vectors for the mixed grade analysis.

5. FUZZY MODEL ANALYSIS RESULTS

Following the Fuzzy Expert System paradigm, models were generated for both scheduling cases. Four major "hidden" rules were discovered for the customer delivery case with two being designated critical. For example, the rule depicted in Figure 3 shows how a good delivery performance measure may be obtained when the indicated information granules of the input variables interact in a non-linear manner. The analysis of this rule, for example, indicates that when "low to moderate" priority and "very low to moderate" grade transition penalty and "somewhat high" customer order widths and "somewhat high" minimum slab widths and "somewhat average" maximum slab widths and "more or less average" qualities and "very low and low" customer order widths and "very low and low" number of slabs that are required per order and "average to large" slab weights interact together, then missing customer delivery due dates will be "low".

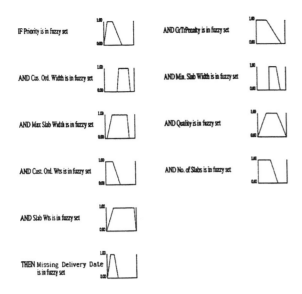

Figure 3: Fuzzy Rule That Shows Interactions That Result In Good Customer Delivery Performance

Further analysis showed that all the input variables were deemed significant with the critical (i.e. most important) variables being the following:

1) Grade transition penalties.
2) Maximum width.
3) The quality of steel.
4) Weights of customer orders.
5) The number of the slabs needed for each customer order.

In a similar manner, five major rules were discovered for the mixed grade production case with two being designated critical. As illustrated in Figure 4, "Low" mixed zones tonnage is produced when input variables have values in the clusters shown in the rule.

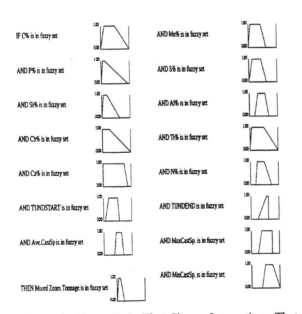

Figure 4: Fuzzy Rule That Shows Interactions That Result in Low Mixed Grade Steel Production

In this analysis, all variables are concluded to be significant with the critical variables being the following:

(1) % Carbon
(2) % Manganese
(3) % Phosphorous
(4) % Sulfur
(5) % Silicon
(6) % Titanium
(7) Tundish start weight
(8) Average casting speed

6. COMPARISON WITH REGRESSION MODELS

In order to compare the resulting Fuzzy system models, the same data sets were modeled using multivariate regression techniques. The results of the multivariate regression analyses are shown in equations 1 and 2 for the customer delivery performance and mixed grade production situations respectively:

Figure 4: Fuzzy Rule That Shows Interactions That Result In Low Mixed Grade Steel Production

$$Y = 63.398 - 0.108X_1 - 0.00189X_2 + 0.02555X_5 - 0.319X_6 + 0.371X_8 - 0.00164X_9 \quad (1)$$

(Where X_n is the input variable number n)
Note: the input variables: (a) Customer ordered width, (b) Minimum slab width and (c) Customer ordered weights are excluded due to zero coefficients. Input variables determined as critical were:

1) Priorities of customer orders.
2) Grade transition penalties.
3) Maximum width.
4) Weights of the slabs allocated to the customer orders.

$$Y = 73.09 - 22.144X_1 + 16.145X_2 + 7.684X_3 + 172.705X_4 + 5.941X_5 - 265.857X_6 - 155.857X_7 + 153.163X_8 + 2247.507X_9 - 1090.946X_{10} - 0.0000167X_{11} - 0.00022X_{12} + 0.007583X_{13} + 0.07117X_{14} - 0.0106X_{15} \quad (2)$$

For this analysis, no variables were selected as critical.

Further analyses of the results reveal that the standard deviation of residuals (errors) for the multiple regression model is 14.69 and 3.9 for the Fuzzy model in the customer delivery performance analysis. In the mixed grade production analysis, the standard deviation of residuals (errors) for the multiple regression model is 14.86 and 4.33 for the Fuzzy model.

7. CONCLUSIONS

To date, initial Fuzzy Expert System models have been developed and demonstrated at Dofasco. Feedback from key scheduling and production people involved in the project were that they understood how the Fuzzy technology was applied and that the models provided some insight into their problems. A better fit of the data was observed with the Fuzzy models when compared to multivariate regression models.

The ability to extract knowledge from complex, subjective data (Fuzzy data mining) is an attractive feature that can benefit industry such as steel.

8. ACKNOWLEDGEMENTS

We gratefully acknowledge J. Aultman, K. Churm, J. Borysewicz, B. Gojmerac, K. Jawahir, M. Moore, B. Prowse, P. Rasmussen and A. Young for their assistance and co-operation in providing valuable operation and scheduling information related to the Continuous Caster. We also acknowledge R. Dibbley for his support of this project and Fazel Zarandi for all his efforts in the model development.

REFERENCES

Bezdek, J.C., Windham, M.P. and Ehrlich, R., (1980). Statistical Parameters of Cluster Validity Functional, *International Journal of Comput. Inf. Sci.*, **9 (4)**, pp. 324-336.

Bezdek, J.C., (1981). *Pattern Recognition with Fuzzy Objective Function Algorithms*, Plenum Press, New York.

Emami, M.R., Türksen, I.B. and Goldenberg, A.A., (1996). An Improved Fuzzy Modeling Algorithm, Part II: System Identification, *Proceedings of 1996 Biennial Conference of NAFIPS*, pp. 294-298.

Fukuyama, Y. and Sugeno, M., (1989). A New Method of Choosing the Number of Clusters for Fuzzy C-Means Method, *Proceedings of 5th Fuzzy Systems Symposium* (in Japanese), pp. 247-250.

Kandel, A., (1982). *Fuzzy Techniques in Pattern Recognition*, Wiley, New York.

Kaufman, L. and Rousseeuw, P.J., (1990) *Finding Groups in Data*, Wiley, New York.

Keller, J.M., Gray, M.R. and Givens, J.A., (1985). A Fuzzy K-Nearest Algorithm, *IEEE Trans. Systems, Man. and Cybernetics*, **SMC-15 (4)**, pp. 580-585.

Slany, W., (1992). Scheduling as a Fuzzy Multiple Criteria Optimization Problem, *Fuzzy Sets and Systems*, **78**, pp. 197-222.

Sugeno, M. and Yasukawa, T., (1993). A Fuzzy-Logic Based Approach to Qualitative Modeling, *IEEE Trans. Fuzzy Systems*, **1**, No.1, pp.7-31.

Türksen, I.B., (1986). Interval-valued Fuzzy Sets Based on Normal Forms, *Fuzzy Sets and Systems*, **20**, No. 2, pp. 191-210.

Türksen, I.B., (1995). Fuzzy Normal Forms, *Fuzzy Sets and Systems*, **69**, pp. 319-346.

Türksen, I.B., (1995). Type 1 and Interval-valued Type II Fuzzy Sets and Logics, in: *P.P. Wang, Advances in Fuzzy Theory and Technology*, Bookright Press Raleight, NC., **3**, pp. 31-82.

Türksen, I.B., (1996). Fuzzy Truth Tables and Normal Forms, *Proceedings of BOFL'96*, TIT, Nagatsuta, Yokohama, Japan, pp. 7-12.

Türksen, I.B., (1996). Type I and Type II Fuzzy System Models, Special Issue, *Fuzzy Set and Systems*.

Yager, R.R. and Filev, D.P., (1994). *Essentials of Fuzzy Modeling and Control*, Wiley, NewYork.

FUZZY EQUATION BASED SIMULATION OF A LIQUID LEVEL CONTROL SYSTEM

N.A. Hickey, U. Keller, D.S. Reay, R.R. Leitch,

Intelligent Systems Laboratory, Heriot-Watt University, Riccarton Edinburgh, EH14 4AS, UK

Abstract-- Fuzzy equation based modelling is a technique that can be used to reason explicitly with both imprecision and uncertainty, both of which must be accounted for if accurate predications of real world behaviour are to be made. This paper details the application of fuzzy equation based models in the form of Fuzzy Differential Equations (FDEs), to the modelling of a tank liquid level control scheme. *Copyright © 1998 IFAC*

Keywords: Precision, Accuracy, Fuzzy Modelling, Dynamic Systems, Control

1   INTRODUCTION

This paper introduces a methodology for dealing explicitly with the related, but separate issues of uncertainty and imprecision which must be considered when modelling real world dynamic systems. The specific meaning of these two terms, as used when discussing the modelling of dynamic systems, is defined and explained before explaining the mechanisms by which they occur.   It is then shown how treating these two terms as distinct and separate properties, allows measurements of their magnitude to be obtained from experimental results. Experimental results are obtained from a laboratory tank liquid level control scheme, and are used to produce a FDE model of the system. The predictions of the system behaviour produced by the model are then compared to the experimental results obtained.

2   PRECISION, ACCURRACY AND UNCERTAINTY

2.1   Measuring Dynamic Systems

In order to acquire analytic knowledge of a dynamic system it is necessary to measure its required output behaviour for known system inputs.   If an ideal measuring process was applied to an ideal system in a steady state, then the measurement system would return the same value i.e. the *correct* or *true* value each time. However in reality such an ideal dynamic system or an ideal measuring process does not exist. In the non-ideal real world uncertainty and imprecision exist and must be accounted for.

When commencing to model a dynamic system it must decided what source of modelling knowledge will be used to build the model. This choice can be broadly broken down into two general areas of *subjective knowledge* or *analytic knowledge* or indeed a combination of both. Subjective knowledge would be obtained from people who have experience of the system such as the modeller or operator of the system.   This knowledge may be precise and or vague and may cover a wide range of operating conditions of the system. Analytic knowledge of the system would be obtained from measurements of the system carried out during its normal operation or specific tests designed for the purposes of building a model. This knowledge gained from measurements of the system may be precise but may not cover all of its operating conditions.   Obviously the choice of knowledge source would determine the structure of the model to be designed and vice versa.

In this paper the system will be modelled using only analytic knowledge, which is available in the form of data gained during pertinent test and measurement of the system.

2.2 Precision and Accuracy

The technical differences between precision and accuracy has been previously discussed (Leitch 1994), however it is worthwhile re-affirming these differences before going on to describe the modelling of a dynamic system, since they are often used ambiguously in everyday language.

The precision of a measured value refers to its resolution with respect to the quantity being measured. For example a measurement process that can measure distances with the resolution of one centimetre in one kilometre would be said to have a high precision, whereas a process that could only measure one centimetre in three centimetres would be said to have a low precision.

For a single measurement of a quantity to be accurate it must equal the true value of the measured quantity, otherwise it is inaccurate. A population of measurements could be termed as accurate if some indicator of their *best* value such as their mean corresponded to the true value. Also if a measured value was stated as a range and the true value lay within this range then that measurement can be termed as accurate.

As an illustrative example consider a measurement of the present time of day as 12.05 p.m., answers to the question "what is the time now?" could take such forms as:

"12.05 p.m." precise and accurate

"12.06 p.m." precise but inaccurate

"around noon" imprecise but accurate

"between 11.45 a.m. and 12.15 p.m." imprecise but accurate

"between 11.45a.m. and 12.00 p.m." imprecise and inaccurate

Since precision relates to how closely the measured values are spread out, it is more appropriate to discuss how far apart measurements are spread and talk instead of the imprecision of the measurements.

2.3 Uncertainty

The lack of the two prerequisites for obtaining correct measurements i.e. an ideal dynamic system and an ideal measurement process, introduce the sources of uncertainty in measurements of the system behaviour. Consider first, as an example of a non-ideal dynamic system, measuring the surface level of a tank of liquid, in actuality even at a supposed steady state it is never at a constant level. At different points on the surface the liquid will be at different heights, this can be due to meniscus effects, external vibrations causing ripples or Brownian motion of air particles causing surface pressure variations. All of these are what would be considered to be normal aspects of the system behaviour, and will cause the true actual level of the tank to vary. These differing factors affecting the system will also vary from day to day, and between similar tanks in different locations, causing variability or uncertainty in the measurements of the system. This source of uncertainty in the measurements of the system is termed *between* group effects (Mandel 1991). Between group effects, is also termed *reproducibility*, and refers to the uncertainty found between sets of results taken from separate measuring instances.

2.4 Imprecision

There are at least two sources of imprecision in modelling a dynamic system, namely in the measuring process and the construction of the model.

Measuring

Referring to the proposed ideal measuring process discussed above, in reality what does happen is that over a time period a statistical population of measurements is obtained. The narrower the range or spread of these values then the more precise the measuring process is, and conversely the wider the spread the more imprecise the measuring process. Every measurement process has a finite level of resolution of measurement, thus all changes that occur in the dynamic system with a higher resolution will not be detected by the measurement process, thus imprecision is introduced. The term *within* group effects refer to the imprecision found in results taken during one measuring instance, where instance could refer to one time period, one system of a set of similar systems or one location. Within group effects are also termed *repeatability* (Mandel 1991).

Modelling

Consider again the example of the tank, if it was a large tank and it was required to model the tank for large changes in level, then a model of the system would ignore those aspects mentioned above which it is assumed would only affect the level to a small negligible extent. Although these aspects are ignored in the model of the system they will affect measurements of the system. Thus it is not the dynamic system that is non-ideal, rather it is the model chosen to represent the behaviour of the system that is non-ideal. Since the model ignores certain behaviours of the real system up to a set resolution, the model is imprecise. In modelling a system if only those aspects of the system which caused small effects on the level were ignored then the imprecision would be low. However if aspects of the system which had larger effects on the level of the tank were ignored then the imprecision of the model would increase (Leitch 1998). The model simplification is necessary due to imprecise human knowledge about the dynamic system, and the need to avoid over complexity in the model.

The terms *repeatability* and *reproducibility* recognise the fact that there will be imprecision associated in a single set of measurements, and uncertainty between sets of measurements.

3 FUZZY DIFFERENTIAL EQUATIONS (FDES)

3.1 Simulation of FDEs

In order to provide a solution of the FDEs over time a simulator has been developed, which will now be described. The initial uncertainty and imprecision of both the initial conditions and the parameters of the FDE is represented by a fuzzy region. Bonarini & Bontempi (1994) proved that the evolution of the trajectories of the external surface of this fuzzy region will determine the evolution in time of the whole fuzzy region area itself. However the number of trajectories from these surfaces will be infinite. The problem can be reduced by use of Nguyen's identity Nguyen (1978) to split the fuzzy region into different regions defined by alpha cuts, which can then be evolved separately. However the number of trajectories from these alpha cuts will still be infinite and therefore some sampling procedure must be used so as to only project those trajectories that are relevant to the evolution of the fuzzy region. These relevant trajectories are the maximum and minimum values of the non fuzzy function evaluated as the initial conditions range over the initial region of uncertainty. Bonarini & Bontempi (1994) utilised a multivariable optimisation algorithm to find these extremes. In this paper the initial external surface is represented by corner points and the vertices are represented by cubic splines. Although the fuzzy region can be mapped back to an individual fuzzy set, for example to monitor the output behaviour, all simulations proceed on the combined fuzzy region and not an individual fuzzy set. Thus any interaction between the system variables is maintained, and no spurious behaviours are produced. A FDE solved on the simulator will produce a set of trajectories corresponding to the minimum and maximum of the alpha cuts, originally defined in the model, as they evolve over time. These trajectories are calculated from all combinations of trajectories within the

defined level and do not represent actual system trajectories.

4 LIQUID LEVEL CONTROL SCHEME

In order to assess the ability of the modelling method to predict the behaviours of real world dynamic systems, data was gathered from the closed loop behaviour of a tank level control system shown in Fig. 1.. The tank level control scheme comprises a simple centrifugal type pump which is used to fill a header tank, which has a constant drain, to a desired level. The pump is operated by means of a computer implemented PI control algorithm which measures the actual level in the tank h, calculates the desired control input u and issues this to the pump refer Fig. 2.. To exhibit various examples of dynamic response for a given tank drain the level control system is set-up to respond to different demanded tank levels r.

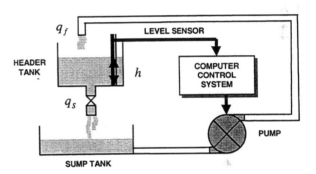

Fig. 1. Pump tank system

4.1 The Tank

The header tank comprises of a uniform upright cylinder with some extraneous heating and stirring apparatus located in the bottom. By examining Fig. 1. and Fig. 2. it can be determined that the increase in tank height with respect to time, will equal the flow rate into the tank q_f minus the flow rate from the tank q_s, multiplied by the gain constant K_R (1). Where K_R is the gain relating the volume of liquid in the tank to the height of the liquid in the tank.

$$\frac{dh}{dt} = K_R * \left(q_f - q_s \right) \tag{1}$$

However the drain from the tank is known and assumed to be constant with time. Thus (1) can be re-written as (2).

$$\frac{dh}{dt} = K_R q_f - D_s \qquad (2)$$

Where D_s is effect of the drain on the height of the tank for the sample time period. D_s was constant with time and was measured directly for each experiment.

4.2   The Pump

The centrifugal pump actually comprises an electric DC motor directly coupled to an impeller mounted in the pump housing. Although it would be possible to model this in depth using complex models of high order, an abstraction is made and the pump is modelled as a simple gain K_V which relates applied pump voltage u to change in tank liquid level (3). This abstraction is made to account for the lack of precise knowledge about the dynamics of pump, and to produce a model that is simple to construct and use.

$$q_f = u * K_V \qquad (3)$$

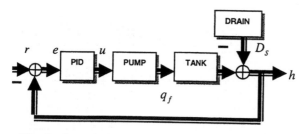

Fig. 2. Control system set-up

5   INCORPORATING UNCERTAINTY AND IMPRECISION IN A MODEL

The complete system comprises of three main components: the PI controller; the centrifugal pump and the header tank, all of which require to be modelled in the dynamic simulation of the system. The PI controller is left as it is implemented as a real valued differential equation, however the header tank and the pump are modelled as a FDE. Combining (2) and (3) produces (4).

$$\frac{dh}{dt} = K_{RV} * u - D_s \qquad (4)$$

Where the gain K_{RV} is a combination of K_R and K_V which therefore converts an applied pump voltage to a specific change in tank level. The complete pump and tank system has therefore been modelled as a single gain times the pump control signal, minus the tank drain rate. This simplifying abstraction is a typical modelling approach where a single gain value is chosen to represent a system over an operating region. However the penalty for this simplification is that the actual pump response over its usable operating range varies widely from any single gain chosen to represent it. This is verified in Fig. 3. where the pump response over its entire operating region is shown. Here it can clearly be seen that the system cannot be correctly represented by a single gain value. Also in the real system application the liquid height value is subject to a degree of imprecision and uncertainty. This is due to the limited accuracy of the level sensor and the aberrations in assuming one single gain value K_R will always accurately relate net liquid input to liquid level for the entire height of the tank.

It is in modelling the operation of the system that the ability to incorporate imprecision and uncertainty explicitly in the FDE is of most benefit in simulation. Thus the differential equation in (4) can be converted to a fuzzy differential equation by using a fuzzy interval for the gain K_{RV} (5).

$$\dot{\overline{h}}(t) = \overline{K_{RV}} * u(t) - D_s \qquad (5)$$

Where the gain parameter and initial conditions are represented by a trapezoidal fuzzy set (6) and (7), as defined in Fig. 4.

$$\overline{h}(0) = (f_1, f_2, f_3, f_4, f_5, f_6) \qquad (6)$$
$$\overline{K_{RV}} = (f_1, f_2, f_3, f_4, f_5, f_6) \qquad (7)$$

The values of imprecision and uncertainty incorporated in K_{RV} are directly obtained via measurements and test filling of the tank. The procedure to do this is first outlined and then discussed in detail :

1.   Define and specify a set of desired nominal conditions for the system.
2.   Run one initial experiment with those conditions.
3.   Derive minimum (f_5) and maximum (f_6) values for the model parameters from this experiment, to set the imprecision of the fuzzy set.

4. Run various other experiments under these (similar) nominal conditions.

5. Derive min and max model parameter values for each of these separate experiments.

6. Select the max of the max (f_2) and the min of the min (f_1) of these set of measurements, these are then used to define the borders for the uncertainty of the fuzzy set.

1. The nominal conditions represent the conditions in which the system operates for most of the time. This set includes such details as time of day and year, typical instrumentation used, location of system and operating levels and settings.

2.,3. Running one initial experiment with these nominal conditions will produce a set of model data for the system which is assumed to represent the within group variability in the dynamic system. This data represents the imprecision encountered when the dynamic system is measured and modelled for the assumed nominal operating conditions.

4., 5., 6. Running other experiments under these similar conditions will produce a set of measurement populations which represent the between group uncertainty of the system. Although each time the experiment is run it is assumed the nominal conditions will remain constant, in reality there will always be changes in these conditions that will affect the dynamic system. The spread of measurement values produced represent the uncertainty that has been encountered in measuring and modelling the dynamic system.

The top graph in Fig. 5. shows the filling of the tank from 10% to 80% of its height. This is achieved by applying a stepped input to the pump which increases from just over 70% to just over 82% of its input signal, in steps of 2%. The bottom graph in Fig. 5. shows the gradients produced from each of these steps. The gradients represent the % change in tank height due to the % change in pump input, i.e. the gain value K_{RV}. The range of gradient values obtained in the experiments define the limits of the fuzzy interval as described in the steps *1.* to *6.* above.

6   EXPERIMENTAL RESULTS

The pumps total operating range is divided into three smaller ranges, namely from 20% to 50%, 20% to 70% and 20% to 80%, as marked in Fig. 3. The variation in the gain values found in each of these

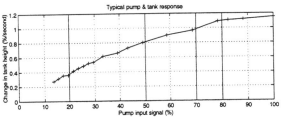

Fig. 3. Pump & tank response

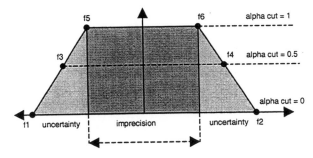

Fig. 4. Definition of fuzzy interval

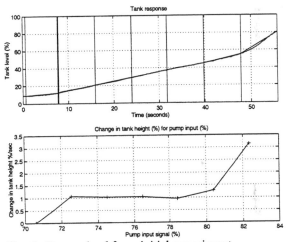

Fig. 5. Data gained from initial experiment

ranges, increases with the width of the range. This variation in the gain represents the imprecision and uncertainty in the model parameter K_{RV}. The fuzzy intervals defined for each range are shown in the top left plot of Fig. 6. to Fig. 8. .Whilst operating the pump in each of these ranges, three experiments are carried out. The three separate experiments involve setting different tank level demands *r* namely, 10%, 20% and 30% of full tank level. The experimental results obtained are then compared to the FDE dynamic simulation results obtained using the relevant fuzzy interval value of K_{RV}, in Fig. 6., Fig. 7. and Fig. 8. It can be seen that the actual behaviours measured are contained within the alpha cut trajectories produced by the model. The bounded behaviours predicted by the model are accurate, in that they contain the actual behaviours measured. However due to the range of imprecision and uncertainty inherent in the actual system and incorporated in the FDE dynamic model they are imprecise. Although the imprecision and uncertainty in the parameter K_{RV} is greatest when the pump is operated over the widest 20% to 80% range, the FDE simulation produced does not vary as greatly compared to the smaller 20% to 50% and 20% to 70% ranges. This suggests that the main causes of imprecision and uncertainty in modelling the system are present even when the pump is operated over its apparently most linear range.

7   CONCLUSIONS

It is shown that using FDEs in a fuzzy equation based simulation of a real system can accurately predict all behaviours of the system over the operating range. However the levels of imprecision in the predictions are equal to the levels of imprecision and uncertainty evident in the system and explicitly incorporated in the fuzzy equation based model. Thus although the predictions are imprecise they are accurate. This is in contrast to a single valued numeric model where a single gain is selected to represent the operation of the system, in which case the prediction is precise but inaccurate.

REFERENCES

Leitch, R.R., (1994) Qualitativeness Does Not Imply Fuzziness. *Proc. 3rd IEEE Int. Conf. On Fuzzy Systems* **Vol. 2** pp. 1257-1262 Florida

Leitch (1998) A qualitative approach to model reference adaptive control (QMRAC). *Eng. App. Of Artificial .Intelligence.* 11 269-278

Mandel, John, (1991) *Evaluation and Control of Measurements*. Marcel Dekker Inc. New York

Zadeh, L.A. (1975) The Concept of Linguistic Variable and it's Application to Approximate Reasoning. *Information Science* 8,9:199-249, 301-357, 43-80,

Bonarini, Andrea, Bontempi, Gianluca, (1994) A Qualitative Simulation Approach for Fuzzy Dynamical Models. *ACM Transactions on Modelling and Computer Simulation*, **Vol. 4**, October 1994, pages 285-313.

Nguyen, H.T. (1978) A Note on the Extension Principle for Fuzzy Sets. *J. Math. Anal. Appl.* **64,2**, 369-380

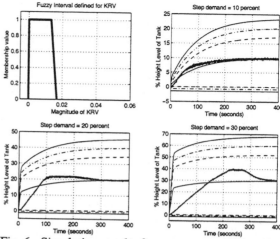

Fig. 6. Simulation results for pump range 20% - 50%

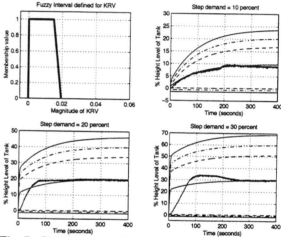

Fig. 7. Simulation results for pump range 20% - 70%

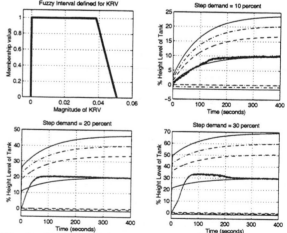

Fig. 8. Simulation results for pump range 20% - 80%

A Fuzzy Controller for Multiple Hyper-Redundant Cooperative Robots

Mircea Ivanescu, Viorel Stoian, Nicu Bizdoaca

Automatic and Computer Dept., University of Craiova, 1100, Craiova, Romania
e-mail: rector@alpha.comp-craiova.ro

Abstract. *A control system is proposed to solve the local control for a multi-chain robotic system formed by tentacle manipulators grasping a common object with hard point contacts. The two-level hierarchical control is adopted. The upper level coordinator gathers all the necessary information to resolve the force distribution. Then, the lower-level local control problem is treated as a open-chain redundant manipulator control problem. A closed loop control based on the inverse dynamic model and the DSMC procedure are proposed. The fuzzy rules are established. Simulation results are presented and discussed. Copyright © 1998 IFAC*

Keywords: Fuzzy Control, Hyper-Redundant Robots

1. INTRODUCTION

In the past few years, the research in coordinating robotic systems with multiple chains has received considerable attention. A tentacle manipulator is a manipulator with a great flexibility, with a distributed mass and torque that can take any arbitrary shape. Technologically, such systems can be obtained by using a cellular structure for each element of the arm. The control can be produced using an electrohydraulic or pneumatic action that determines the contraction or dilatation of the peripheral cells. The first problem is the global coordination problem that involves coordination of several tentacles in order to assure a desired trajectory of a load. The second problem is the local control problem, which involves the control of the individual elements of the tentacle arm to achieve the desired position. To resolve this large-scale control problem, a two-level hierarchical control scheme (Cheng,1995) is used. The upper-level system collects all the necessary information and solves the inter-chain coordination problem. Then, the problem is decoupled into j lower-level subsystems, for every arm. The local fuzzy controllers are assigned to solve the local control. In order to obtain the fuzzy control an approximate model of the integral-differential equation that describe the tentacle arm is used. A closed loop control that is based on the inverse dynamic model is used. A decoupling controller is introduced and the conditions that assure the stability of the motion are determined. The control strategy is based on the Direct Sliding Mode Control which controls the trajectory towards the switching line and then the motion is forced directly to the origin, on the switching line. A fuzzy controller is proposed and the fuzzy rules are established by using the DSMC procedures.

Efficiently considerations of the method are discussed. Numerical simulations for several control problems are presented.

2. MODEL FOR COOPERATIVE TENTACLE ARMS

A multiple-chain tentacle robotics system is presented in Fig. 1. With the chains of the system forming closed-kinematics loops, the responses of individual chains are tightly coupled with one another through the reference member (object or load). The complexity of the problem is considerable increased by the presence of the tentacle robots,

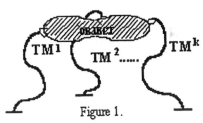

Figure 1.

(TM^j , j = 1..k), the systems with, theoretically, a great mobility, which can take any position and orientation in space. In the Figure 2 is presented a plane model, a simplified structure which can take any shape in the X0Z plane. The dynamic equations for each chain of the system are (Ivanescu,1984):

$$TM^j : \quad \rho_j A^j \int_0^s \left[\sin\left(q^j - q'^j\right) \dot{q}'^{j2} + \cos\left(q^j - q'^j\right) \ddot{q}'^j \right] ds' +$$

$$+ \rho A g \int_0^s \cos q^j \, ds' + \tau^j = T^j \, , j = 1,2 \qquad (2.1)$$

$$\int_0^{L^j} \tau^j \, ds = F_x^j \int_0^{L^j} \left(-\sin q^j \right) ds + F_x^j \int_0^{L^j} \cos q^j \, ds \qquad (2.2)$$

where we assume that each manipulator (TM^j) has a uniform distributed mass, with a linear density ρ' and a section A^j. We denote by s the spatial variable upon the length of the arm, $s \in [0, L^j]$. We also use the

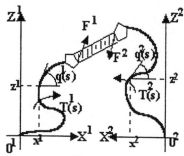

Figure 2.

notations: q^j – Lagrange generalised coordinate for TM^j (the absolute angle), $q^j = q^j(s,t), s \in [0, L^j]$, $t \in [0, t_f], q'^j = q^j(s',t), s' \in [0, L^j], t \in [0, t_f]$;

$T^j = T^j(s,t)$ - the distributed torque over the arm; $\tau^j = \tau^j(s,t)$ - the distributed moment to give the desired motion specified on the reference member. All these sizes are expressed in the coordinate frame of the arm TM^j. The k integral equations are tightly coupled through the terms τ^j, F_x^j, F_z^j where all of these terms determine the desired motion. We propose a two-level hierarchical control scheme for this multiple-chain robotic system. The control strategy is to decouple the system into k lower-level subsystems that are coordinated at the upper level. The function of the upper-level coordinator is to gather all the necessary information so as to formulate the corresponding force distribution problem and then to solve this constrained, optimization problem such that optimal solutions for the contact forces F^j are generated. These optimal contact forces are then the set-points for the lower-level subsystems. We consider the hard point contact with friction and the force balance equations on the object may be written as: $F^0 = \Sigma\, ^0D_j F^j$ (2.3)

where: F^0- the resultant force vector applied to object expressed in the inertial coordinate frame (0), ^0D_j- the partial spatial transform from the coordinate frame for the arm TM^j to the inertial coordinate frame (0). The object dynamic equations are obtined by the form $M_0 \ddot{r} = GF^0$ (2.4)

where M_0 is inertial matrix of the object and r defines the object coordinate vector $r = (x, z, \varphi)^T$ (2.5) and r(t) represents the desired trajectory of the motion. The inequality constraints which include the friction constraints and the maximum force constraints may be associated to (2.3), $\Sigma A^j F^j \le B$ (2.6)

where A^j is a coefficient matrix of inequality constraints and B is a boundary-value vector of inequality constraints. The problem of the contact forces F^j can be treated as an optimal control

problem if we associate to the relations (2.3) - (2.6) an optimal index $\Psi = \Sigma C^j F^j$ (2.7) This problem is solved in several papers by the general methods of the optimization or by the specific procedures (Cheng,1991).After all of the contact forces F^j are determinated, the dynamics of each arm TM^j are decoupled. Now, the equations (2.1), (2.2) can be interpreted as same decoupled equations with a given $\tau(s)$, $s \in [0, L^j]$ acting on the tip of the arm. A discrete and simplified model of (2.1), (2.2) can be obtained by using a spatial discretization.

$s_1, s_2, \dots s_N$; $\quad s_i - s_{i-1} = \Delta$; $q^j(s_i) - q^j(s_k)| < \varepsilon$; i, k = 1, 2, ... n^j (2.8)
where ε, Δ are constants and ε is sufficiently small. We denote $s_i = i\Delta$, $L^j = n^j \Delta$, $q^j(s_i) = q_i^j$,

$$T^j(s_i) = T_i^j, \tau^j(s_i) = \tau_i^j \quad (2.9)$$

and considering the arm as a light weight arm, from (2.1), (2.2) it results (Ivanescu,1986):

$$\ddot{q}^j + f(q^j) + F_x^j a(q^j) + F_z^j c(q^j) = B\, T^j \quad (2.10)$$

where $q^j = col(q_1^j, q_2^j, \dots q_{n^j}^j)$;

$$T^j = col(T_1^j, T_2^j \dots T_{n^j}^j) \quad (2.11)$$

$$a(q^j) = \begin{bmatrix} -b\sin q_1^j \\ -b\sin q_2^j + b\sin q_1^j \\ \vdots \\ -b\sin q_{n^j}^j + b\sin q_{n^j-1}^j \end{bmatrix} \quad (2.12)$$

$$c(q^j) = \begin{bmatrix} b\cos q_1^j \\ b\cos q_2^j - b\cos q_1^j \\ \vdots \\ b\cos q_{n^j}^j - b\cos q_{n^j-1}^j \end{bmatrix} \quad (2.13)$$

$$B = b\begin{bmatrix} 1 & 0 & \cdots\cdots & 0 \\ -1 & 1 & 0 & \cdots\cdots & 0 \\ 0 & -1 & 1 & 0 & \cdots & 0 \\ \cdots & \cdots\cdots\cdots\cdots \\ 0 & 0 & \cdots & 0 & -1 & 1 \end{bmatrix}; f(q^j) = bg\begin{bmatrix} \cos q_1 \\ \cos q_2 \\ \cdots \\ \cos q_{n^j} \end{bmatrix} \quad (2.14)$$

3. CONTROL SYSTEM

The control problem asks for determining the manipulatable torques (control variable) T_i^j such that the trajectory of the overall system (object and manipulators) will correspond as closely as possible to the behaviour. In order to obtain the control law for a prescribed motion, we shall use the inverse model. A closed-loop control system is used (Fig. 3).

Let $q_d^j, \dot{q}_d^j, \ddot{q}_d^j$ be the desired parameters of the trajectory, F_d^j the desired force applied at the j - contact point of the object, and , $q^j, \dot{q}^j, \ddot{q}^j, F^j$ the

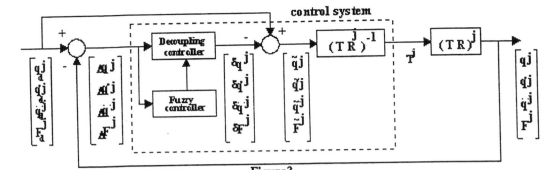

Figure3.

same sizes measured on the real system, the error of the feedback system is given by: $\Delta q^j = q_d^j - q^j$;

$\Delta \dot{q}^j = \dot{q}_d^j - \dot{q}^j; \Delta \ddot{q}^j = \ddot{q}_d^j - \ddot{q}^j ; \Delta F^j = F_d^j - F^j;$ (3.1)

The trajectory controller serves as the trajectory perturbation controller which generates the new variations $\delta q^j, \delta \dot{q}^j, \delta \ddot{q}^j, \delta F^j$ in order to assure the performances of the motion for the overall system. The control law is proposed as,

$\delta q^j = K_{11}^j \Delta q^j + K_{12}^j \Delta \dot{q}^j + K_{13}^j \Delta \ddot{q}^j ;$

$\delta \dot{q}^j = K_{21}^j \Delta q^j + K_{22}^j \Delta \dot{q}^j + K_{23}^j \Delta \ddot{q}^j$

$\delta \ddot{q}^j = K_{31}^j \Delta q^j + K_{32}^j \Delta \dot{q}^j + K_{33}^j \Delta \ddot{q}^j ;$

$\delta F_x^j = K_{f_{1x}}^j \Delta F_x^j + K_{f_{2x}}^j \Delta \dot{F}_x^j + K_{f_{3x}}^j \Delta \ddot{F}_x^j$

$\delta F_z^j = K_{f_{1z}}^j \Delta F_z^j + K_{f_{2z}}^j \Delta \dot{F}_z^j + K_{f_{3z}}^j \Delta \ddot{F}_z^j$ (3.2)

where the relations (3.2) define a complex control for the motion parameters and for the force. From (3.1), (3.2) results that the evolutions of direct and inverse system take place on the trajectories, $q^j = q_d^j - \Delta q_j; \dot{q}^j = \dot{q}_d^j - \Delta \dot{q}_j; \ddot{q}^j = \ddot{q}_d^j - \Delta \ddot{q}_j$ and

$\tilde{q}^j = q_d^j - \delta q_j; \tilde{\dot{q}}^j = \dot{q}_d^j - \delta \dot{q}_j; \tilde{\ddot{q}}^j = \ddot{q}_d^j - \delta \ddot{q}_j.$ (3.3)

$\tilde{\ddot{q}}^j + f(\tilde{q}^j) + F_x^j a(\tilde{q}^j) + F_z^j c(\tilde{q}^j) = BT^j$ (3.4)

Assuming that $\delta q^j << q_d^j; \delta \dot{q}^j << \dot{q}_d^j; \delta \ddot{q}^j << \ddot{q}_d^j,$

$\Delta q^j << q_d^j; \Delta \dot{q}^j << \dot{q}_d^j; \Delta \ddot{q}^j << \ddot{q}_d^j$ then using the Taylor-series expansion, neglecting the high-order terms, from (2.11), (3.3) we have

$\left(\Delta \ddot{q}^j - \delta \ddot{q}^j\right) + d\left(q_d^j, F_d^j\right)\left(\Delta q^j - \delta q^j\right) + d\left(q_d^j\right)\left(\Delta F_x^j - \delta F_x^j\right) + d\left(q_d^j\right)\left(\Delta F_z^j - \delta F_z^j\right) = 0$

$d\left(q_d^j, F_d^j\right) = \left(\frac{\partial f}{\partial q}\right)_{q=q_d^j} + F_{xd}^j \left(\frac{\partial a}{\partial q}\right)_{q=q_d^j} + F_{zd}^j \left(\frac{\partial c}{\partial q}\right)_{q=q_d^j}$ (3.5)

where d is a $n^j \times n^j$ matrix. Using (3.2) in (3.4) we obtain the following control equations:

$\left(I - K_{33}^j - dK_{13}^j\right)\Delta \ddot{q}^j - \left(K_{32}^j + dK_{12}^j\right)\Delta \dot{q}^j + \left(d\left(I - K_{11}^j\right) - K_{31}^j\right)\Delta q^j = 0$

$K_{f_{3x}}^j \Delta \ddot{F}_x^j + K_{f_{2x}}^j \Delta \dot{F}_x^j + \left(1 - K_{f_{1x}}^j\right)\Delta F_x^j = 0$

$K_{f_{3z}}^j \Delta \ddot{F}_z^j + K_{f_{2z}}^j \Delta \dot{F}_z^j + \left(1 - K_{f_{1z}}^j\right)\Delta F_z^j = 0$

For the nesingular matrix $(I - K_{33}^j - d K_{13}^j)$ these equations of the motion can be written as,

$\Delta \ddot{q}^j - \left(P^j\right)^{-1} Q^j \Delta \dot{q}^j - \left(P^j\right)^{-1} R^j \Delta q^j = 0$ (3.6)

and for the force

$K_{f_3}^j \Delta \ddot{F}^j + K_{f_2}^j \Delta \dot{F}^j + \left(1 - K_{f_1}^j\right)\Delta F^j = 0$ (3.7)

where we used the notations,

$P^j = (I - K_{33}^j - d K_{13}^j); Q^j = (K_{32}^j + d K_{12}^j);$

$R^j = d (I - K_{11}^j) - K_{31}^j$ (3.8)

and we considered $K_{f_{1x}}^j = K_{f_{1z}}^j = K_{f_1}^j;$

$K_{f_{2x}}^j = K_{f_{2z}}^j = K_{f_2}^j; K_{f_{3x}}^j = K_{f_{3z}}^j = K_{f_3}^j$ (3.9)

From (3.6), (3.7) we obtain immediately,

<u>Proposition 1.</u> The control law for the motion and force control requires

$\begin{bmatrix} 0 & I \\ -P^{-1}R & -P^{-1}Q \end{bmatrix}$ to be stable, and

$K_{f_2}^{j2} \leq 4K_{f_3}^j \left(1 - K_{f_1}^j\right)$ (3.10)

The relations (3.10) define the main conditions imposed to the controller in order to assure the global stability for the motion of the arm and for the force F_d^j at the terminal point of the arm. If the the inequality in (3.10) is easy to apply, the stability of the matrix (3.10) is more difficult to use. We can obtain a simplified procedure if we choose suitable matrices $K_{m,n}^j$ (m,n =1,2,3) in the control law (3.2):

$I - K_{33}^j - d K_{13}^j = \alpha I, \alpha$ - integer number ;

$K_{32}^j + d K_{12}^j = 2\Xi^j; d (I - K_{11}^j) - K_{31}^j = \Omega^j$ (3.11)

Where $\Xi^j = diag\left(\xi_1^j .. \xi_n^j\right); \Omega^j = diag\left(\omega_1^j .. \omega_n^j\right)$ (3.12)

The equations (3.6) become

$\alpha \cdot \Delta \ddot{q}_i^j - 2\xi_i^j \Delta \dot{q}_i^j + \omega_i^{j2} \Delta q_i^j = 0$ (3.13)

The equations (3.6) for the control of the arm parameters and (3.7) for the control of the force offer a simple control for a Direct Sliding Mod Control (DSMC), (Ivanescu,1995).

The DSMC is a control method which operates in two steps. First step assures the motion towards the switching line S_q (or S_F), $\Delta \dot{q}_i^j + p_i^j \Delta q_i^j = 0$

$$\Delta \dot{F}^j + p_{iF}^j \Delta F^j = 0 \qquad (3.14)$$

by the general stability conditions (Fig. 4).

$$\xi_i^j < \min_s \left[\alpha \omega_i^{j2}(s) \right]^{\frac{1}{2}}$$

$$K_{f2}^j \leq 2 \left(K_{f3}^j \left(1 - K_{f1}^j \right) \right)^{\frac{1}{2}}; \ j = 1, 2 \qquad (3.15)$$

When the trajectory penetrates S_q (or S_F), the damping coefficients ξ_i^j, K_{f2}^j are increased,

$$\xi_i^j > \max_s \left[\alpha \omega_i^{j2}(s) \right]^{\frac{1}{2}}; \ K_{f2}^j > 2 \left(K_{f3}^j \left(1 - K_{f1}^j \right) \right)^{\frac{1}{2}}$$
$$j = 1, 2 \qquad (3.16)$$

The system is moving towards the origin, directly, on the switching line S_q (or S_F).

4. FUZZY CONTROLLER

A fuzzy control is proposed by using the control of the damping coefficient ξ_i^j, K_{f2}^j in (3.15) - (3.16). We consider a DSMC strategy with the switching of a control variable on the switching line (3.14), (Figure 4). We shall let the errors $\Delta q_i^j, \Delta F^j$ and the

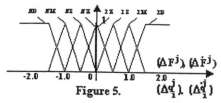

Figure 4.

error rates $\Delta \dot{q}_i^j, \Delta \dot{F}^j$ be defined by eight linguistic variables: NB,NM,NS,NZ,PZ,PS,PM,PB partitioned on the error spaces $[-\Delta q_m, \Delta q_m]$, $[-\Delta F_m, \Delta F_m]$ and the error rate spaces $\left[-\Delta \dot{q}_m, \Delta \dot{q}_m \right], \left[-\Delta \dot{F}_m, \Delta \dot{F}_m \right]$ where all these quantities are normalized at the same interval. The membership functions for these quantities are shown in Figure 5. The fuzzy output

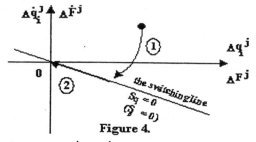

Figure 5.

variables,the control coefficients ξ_i^j, K_{f2}^j, will use four fuzzy variables on the normalized universe. $\xi_i^{*j} = F_i^{*j} = \{ 0, 0.5, 1.0, 1.5, 2.0, 2.5 \}$ where the range of the values is chosen such that

$$1 = \frac{\xi_i^{*j}}{\xi_{i\,max}^j} \left\{ \frac{1}{2} \left[\min_s \left(\left(\alpha \omega_i^{j2}(s) \right)^{\frac{1}{2}} \right) + \max_s \left(\left(\alpha \omega_i^{j2}(s) \right)^{\frac{1}{2}} \right) \right] \right\}$$

for the damping coefficient ξ_i^j, and for the force:

$$1 = \frac{F^{*j}}{K_{f2\,max}^j} 2 \left(K_{f3}^j \left(1 - K_{f1}^j \right) \right)^{\frac{1}{2}} \qquad (4.1)$$

The memberships of the output variables are represented in Figure 6, where ST1, BT1 define

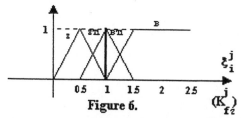

Figure 6.

linguistic variable: SMALLER THAN1 and BIGGER THAN 1,respectively. According to the theoretical results obtained in the previous part of the paper, we can generate the control rules which establish a fuzzy control for a DSMC control (Table 1). The main idea is to assure the normal control

Table 1.

ΔF^j Δq_i^j \ ΔF^j Δq_i^j	NB	NM	NS	NZ	PZ	PS	PM	PB
PB	B	BT1	BT1	BT1	S	S	S	S
PM	BT1	B	BT1	BT1	S	S	S	S
PS	BT1	BT1	B	BT1	S	S	S	S
PZ	BT1	BT1	BT1	B	ST1	ST1	ST1	ST1
NZ	ST1	ST1	ST1	ST1	B	BT1	BT1	BT1
NS	S	S	S	S	BT1	B	BT1	BT1
NM	S	S	S	S	BT1	BT1	B	BT1
NB	S	S	S	S	BT1	BT1	BT1	B

towards the switching line and direct control when the trajectory penetrates this line. A standard defuzzification procedure based on the centroid method is then used.

5. NUMERICAL RESULTS

The purpose of this section is to demonstrate the effectiveness of the method. This is illustrated by solving a fuzzy control problem for a two tentacle manipulator system which operates in X0Z plane (Fig. 7). These two manipulators form a closed-chain

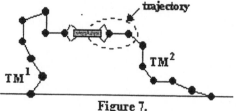

Figure 7.

robotic system by a common object which is manipulated. An approximate model (2.10) with $\Delta = 0.06$ m and $n^j = 7$ is used. Also, the length and the mass of the object are 0.2 m and 1 kg, respectively.

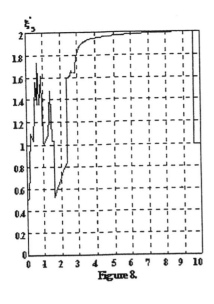

Figure 8.

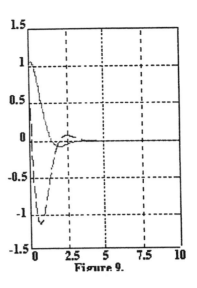

Figure 9.

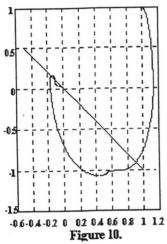

Figure 10.

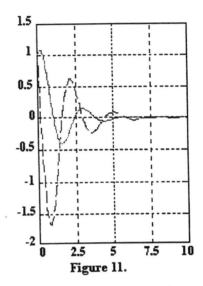

Figure 11.

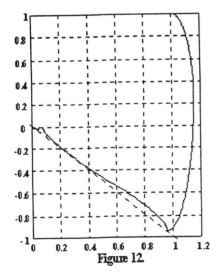

Figure 12.

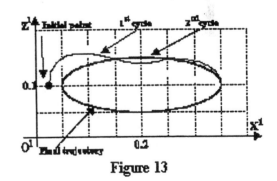

Figure 13

231

The inertial positions of the arms expressed in the inertial coordinate frame are presented in Table 2. The trajectory is defined by $x = x_0 + a \sin \omega t$; $z = z_0 + b \cos \omega t$ with $x_0 = 0.2$ m, $z_0 = 0.1$ m, $a = 0.3$ m, $b =$

Table 2.

TM^j	$q_1^j(0)$	$q_2^j(0)$	$q_3^j(0)$	$q_4^j(0)$	$q_5^j(0)$	$q_6^j(0)$	$q_7^j(0)$
TM^1	$\pi/6$	$\pi/8$	$7\pi/12$	$2\pi/8$	$\pi/15$	$15\pi/8$	0
TM^2	$5\pi/6$	$4\pi/5$	$4\pi/5$	$3\pi/4$	$3\pi/4$	$2\pi/3$	π

0.1 m, $\omega = 0.8$ rad/s. The trajectory lies inside the work envelopes of the both arms and does not go through any workspace singularities. The maximum force constraints are defined by

$$F_X^j \le F_{MAX} = 50 \text{ N} \ ; \ F_Z^j \le F_{MAX} = 50 \text{ N}$$

and the optimal indexes $\min\left(\sum_j F_X^{j2}\right), \min\left(\sum_j F_Z^{j2}\right)$ are used. The generalized trajectories $q^j(t), \dot{q}^j(t), \ddot{q}^j(t)$ are obtained by the constraints $q_7^1(t) = 0$; $q_7^2(t) = -\pi$ and by the condition $\min \|\dot{q}^j(t)\|$ The solution of this problem is given by solving the nonlinear differential equation (Schilling,1990),

$$\dot{q}^j(t) = \left[V^{jT}(q) \ V^j(q) \right]^{-1} V^{jT}(q) \dot{w}(t) \quad \text{where}$$

$w = (x,z)^T$ and $V^j(q)$ is the Jacobian matrix of the arms ($j = 1,2$). A decoupling control procedure (3.11)- (3.12) is used, where the following parameters are selected, $\alpha = 3; \xi_i^j = 2.5; \omega_i^j = 18, i = 1 \cdots 7; j = 1,2$

A fuzzy controller which applies the general rules of the DSMC procedure (Table 1) is used. Figure 8 shows the evolution of ξ_5^1 and the position error and the position error rate Δq_5^1, (dash line) are represented in Figure 9. Figure 10 presents the phase trajectory in $\left(\Delta q_5^1, \Delta \dot{q}_5^1\right)$ plane. The force error ΔF_x^1 and the force error rate $\Delta \dot{F}_x^1$ are presented in Figure 11 and Figure 12 exhibits the phase trajectory for the force F_x^1. We consider a plan-parallel motion of the object, without any rotations. Figure 13 presents the final trajectory. We can remark the error during the 1th cycle and the convergence to the proposed trajectory during the 2nd cycle.

6. CONCLUSIONS

The two-level hierarchical control procedure is constructed in this paper to solve the large-scale control problem of a chain robotic system formed by tentacle manipulators grasping a common object. The upper-level coordinator collects all the necessary information to solve the inter-chain coordination problem, the force distribution. Then, the problem is decoupled into j lower-level subsystems, for every arm. The local fuzzy controllers are assigned to solve the local control. A local decoupling controller is introduced and the conditions that assure the stability

of the motion are determined. A DSMC procedure is used and the fuzzy rules are established. The simulation problem for two closed-chain tentacle robotic system has also been studied.

REFERENCES

Ivanescu,M.,(1984) "Dynamic Control for a Tentacle Manipulator", *Proc. of Int. Conf.*, Charlotte, USA,1984.

Ivanescu,M.,(1986), "A New Manipulator Arm: A Tentacle Model", *Recent Trends in Robotics*, Nov. 1986, pp. 51-57.

Silverman,L.M.(1969) "Inversion of Multivariable Linear Systems", *IEEE Trans. Aut. Contr.*, Vol AC – 14,1969.

Cheng, Fan –Tien,(1995) "Control and Simulation for a Closed Chain Dual Redundant Manipulator System", *Journal of Robotic Systems*, pp. 119 – 133,1995.

Zheng, Y.F., J.Y.S., Luh, (1988) "Optimal Load Distribution for Two Industrial Robots Handling a Single Object", *Proc. IEEE Int. Conf. Rob. Autom.*, pp. 344 – 349,1988.

Mason, M. T.,(1981) "Compliance and Force Control", *IEEE Trans. Sys. Man Cyb.*, Nr. 6,1981, pp. 418 - 432.

Cheng, F. T., D. E., Orin,(1991) "Optimal Force Distribution in Multiple-Chain Robotic Systems", *IEEE Trans. on Sys. Man and Cyb."*, Jan.1991, vol. **21**, pp. 13 - 24.

Cheng, F. T., D. E ., Orin,(1991) "Efficient Formulation of the Force Distribution Equations for Simple Closed - Chain Robotic Mechanisms", *IEEE Trans on Sys. Man and Cyb.*, Jan.1991, vol. 21, pp. 25 -32.

Wang, Li-Chun T.,(1996) "Time-Optimal Control of Multiple Cooperating manipulators", *J. of Rob.Sys.*,'96, 229-241.

Khatib,D.E.(1996) "Coordination and Descentralisation of Multiple Cooperation of Multiple Mobile Manipulators, *Journal of Robotic Systems*, **13** (11) , 755 - 764.

Schilling, R. J.,(1991) "Fundamentals of Robotics",*Prent. Hall*, '90.

Ross,T.J.(1995) "Fuzzy Logic with Engineering Applications", *Mc.Grow Hill* , Inc.

Ivanescu,M., V., Stoian,(1995) "A Variable Structure Controller for a Tentacle Manipulator", *Proc. of the 1995 IEEE Int.Conf. on Robotics and Aut.*, Nagoya, Japan, May 21 - 27, 1995, vol. **3**, pp. 3155 - 3160.

Ivanescu,M., V., Stoian,(1996) "A Sequential Distributed Variable Structure Controller for a Tentacle Arm", *Proc. of the 1996 IEEE Intern. Conf. on Robotics and Aut.*, Minneapolis, April 1996, vol. **4**, pp. 3701 - 3706.

Ivanescu,M.,V.Stoian,(1996), *"A Micro-Tentacle Manipulator with ER-Fluids"*, Micro-Robot World Cup Soccer Tournament, MIROSOT, Proceedings, November 9-12,1996, Taejon, KOREA, vol. **1**, pp. 74 - 79.

FAULT-TOLERANT PROCESS CONTROL AND SOME FUTURE DIRECTIONS

JIANBO H. MENG*, CIYONG LUO, XUETONG ZHANG

THE INSTITUTE OF PROCESS AUTOMATION AND INSTRUMENT
CHONGQING UNIVERSITY
CHONGQING 400044, CHINA
FAX: +86-23-65106656, EMAIL: jmeng@cqu.edu.cn

Abstract: Fault-tolerant means the ability, when one or more than one key parts of system is faulty, the system could continuously, stably, and reliably operate and autonomously keep the specified function, or lower the function in an acceptable variable ranges, by taking corresponding method (Levis, 1987). The research of fault-tolerant technique, is originally come from the simulating to natural life's activities. Later, for the special requirement of defense and military industry, this technique was firstly applied in large-scale computer systems (Chen, 1981; Spitzer, 1988; Sadehi, 1995). Recent years, as the industrial developed and the systems' scale became larger and larger, manufacture (or production) safety and operation economics have been paid more and more attentions (Levis, 1987; Makansi, 1995). Thus lead to the higher requirement of reliability and effectiveness of process control systems, by which the safety and economics of manufacturing is ensured. This results in the research and development of fault-tolerant process control.

In this paper, we try to give some view and comments to the present state of fault-tolerant techniques used in process control, and propose some future directions for this research. *Copyright © 1998 IFAC*

Keywords: Fault-tolerant systems, process control, artificial intelligence, variable structures, safety, fieldbus, control equipment

1. THE STATE OF PROCESS CONTROL

Accompanying with the development and progress of science and technology, process control has stepped three stages. The first one is the Distributed and Local Control based on the local base instruments (about 1940's). The second one is the Centralized Control used mainly the analogue and some digital individual instruments. The third phase is the Centralized and Distributed Control (DCS), based on the microcomputers (present).

Recently, along with the fast progresses in artificial intelligence (AI) and computer technology, process control is forwarding to the new Field-bus Control (FCS) stage, which based on the network communication and could be characterized by the concept of "intelligent distribution" (Meng, 1997; Meng, 1998b). This development mode of process control from "distribution" to "centralized", and then come back to "distribution", is just more corresponding with the spiral development rule of human being's knowledge, which comes firstly from "simple" to "complicated", and then back to the higher "simple", and so on. Related to this new network mode of FCS, a great changes is being faced to process control area.

The so called field-bus instrument (FBI), is a new type of constitute standard for next generation instrument, which developed by using computer network technology as the reference. It emphasized that the instrument, as it is a node of process control network , should be an intelligent point having the autonomous management, communication, self-diagnosis, self-tuning functions, etc.. Each intelligent point in process control network is related by two-way digital communication. According to the control requirement of each process point, some sub-networks could be formed by required

intelligent instruments, in which each process equipment is taken as the center of each sub-network. The whole process control system network is then formed by these sub-networks. At the system level, the inter-coordinate is realized for each intelligent point and for each sub-network (Meng, 1997; Meng, 1998b; FF, 1997). This working and operating style is mostly like the human being's organization and management for manufacturing, in which the effectiveness for organization and the fast for response are both expected. We specially called this way as the "human beings' colony simulating" (Meng, *et al.*, 1998a)

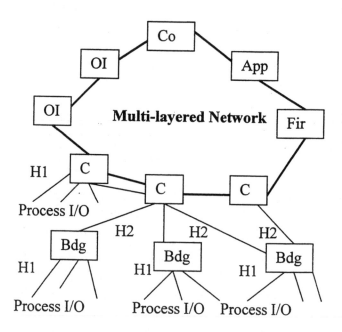

Co: Coordinator
App: Application
Fir: Firewall
OI: Operator Interface
Bdg.: Bridge
C: Controller

Fig.1 The New Field-bus Control Systems (Hodson, 1996)

2. PRINCIPLES OF FAULT-TOLERANT PROCESS CONTROL

The fault-tolerant process control means that, when one or more than one instrument is faulty in a process control system, such as the temperature, pressure, flowmeter, etc., and the controller or control valve is faulty, the process control system could autonomously switch to the other normal instrument, or by making use of appropriate algorithms, in order to ensure the ability of continuous and stable operation of process system, in the same performance quota or in a lower but acceptable variable ranges. According to the management methods for instrument failures, fault-tolerant process control could be divided into two types. One is the hardware fault-tolerant (hardware redundancy), and the other is the software fault-tolerant (software redundancy, or analytical redundancy) (Shinskey, 1981; Williams, 1984).

The premise of fault-tolerant process control is the fault detection and diagnosis the failure type. Then, appropriate processing for the failure could be made. For the former case, there were many researches have been done, esp., for computer systems (Chen, 1981; Wang, 1991). The present developing field-bus instrument is expected to have these two functions. But in loop (sub-network) and system network level, a specified inter-coordinate processing hardware and software should be needed to realize the fault-tolerant process control (Meng, 1997; Meng, 1998b).

The basic method for fault detection and diagnosis could be again divided into two major types, i.e., the parameter method and non-parameter method. The later one is also called fault-detection and diagnosis based on process models, and there are many researches for this type (Chen, 1981; Wang, 1991; Shu, 1992; Haegglund, 1996; Yuan, 1996; Zhou, 1995; Hu, 1994). The parameter method has been widely used in present DCS (Foxboro, 1992; Elsage-Bailey, 1994; Hodson, 1996).

Having the bases of intelligent instruments which could have the ability of self-detection, and diagnosis, etc., the control and/or management of the fault instruments would be the key problem for realizing fault-tolerant process control. The aim is that, deciding the processing scheme for faulty instrument, according to the faulty site, type, and degree, to isolating the fault parts, changing the system structure and re-configuration system (we called system re-configuration), in order to ensure the process system continuously and stably operate, and keeping the desired performance quota. It's obviously that, for this case the system redundancy is required.

From the structure aspects, the two types of redundancy of fault-tolerant process control system are all in parallel structure and consisted with more than one information processing sub-systems. Each information processing sub-system has the same aim, while the aim could be making a control signal (corresponding to the controllers or valves), or predicating a process parameter (corresponding to the measuring instruments), as shown in Fig. 2.

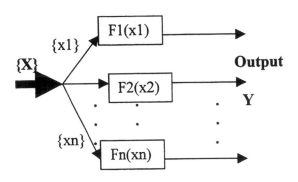

Fig.2 Parallel Structure of Fault-tolerant Process Control (Eryurek, 1995)

Here, {Xn} is the n-th sub-set of input parameters, Fn(Xn) is the algorithm of n-th processor, Y is the output.

From Fig.2, the hardware redundancy could be regarded as the simplest parallel structure, as follows,

$$\{X1\}=\{X2\}= \ldots\ldots =\{Xn\} \qquad (1)$$
$$F1(X1)=F2(X2)= \ldots\ldots Fn(Xn) \qquad (2)$$

In this case, each sub-set of input parameters and each process algorithms are the same respectively. It means that, the fault-tolerant ability is simply improved by physical isolating and switching for faulty instrument in this structure. This simple hardware redundancy method has been widely used in control, input/output, communication, power supply models, etc., of the present application DCS (Lu, 1996; Dong, 1992; Foxboro, 1992; Elsag-Bailey, 1994; Hodson, 1996).

Differing with the above hardware redundancy, the software redundancy is realized by making different sub-set process parameters, and plus different processing algorithms, i.e.,

$$\{X1\}\neq\{X2\}\neq \ldots\ldots \neq\{Xn\} \qquad (3)$$
$$F1(X1)\neq F2(X2)\neq \ldots\ldots \neq Fn(Xn) \qquad (4)$$

It's easy to understand that, in this structure, the n possible processing methods for making output Y could be provided. Thus the potential reliability of producing the output Y is larger than the hardware case. Besides, by making use of the different combination of process parameters, the fault-tolerant ability of whole system could be more improved. This is because that, in this way, one or more than one failures of input parameters couldn't affect the output production of correct result,

which is made by those processing algorithm who didn't make use of the faulty parameters.

Of course, as the some common parameters are probably used in each process algorithms. So, the completely fault-tolerant for every possible fault parameters couldn't be realized. Also, this structure is taking the requirement of adding more input redundancy as its premise.

For the output case, it also should be researched as the similar way. According to the development of Field-bus Instruments, the input and output would be combined with each other into a one for most cases. So, the input and output should be taken into consideration simultaneously.

Because of the improvement for fault-tolerant ability in this structure is achieved by making use of multiple processing scheme. So there is no need for physical isolating of fault hardware, and it's convenient to be used in process instrument and control systems to realize the variable-structure instrument and control (V-I&C). This direction should be researched more for the future field-bus control systems.

It's necessary to be noticed that, although the aim of fault-tolerant process control is making the effort to keeping the performance quota of re-configurated system is the same with the former system, while failure occurs. But, in order to avoid the larger economic lost caused by plant shutdown, etc., It's valuable that by taking a lower performance index, to exchange the continuous and stable operation of process system and making the conditions for doing some rush repairs and/or renew the system. This direction could be also paid more attention for the research along with the development of field-bus instruments.

The more details of network topology analysis of fault-tolerant process control system will be discussed in the following sections.

3. ANALYSIS OF SOME APPLICATION EXAMPLES

In this section, some application examples of fault-tolerant process instrument and control systems will be presented.

3.1 Fault-tolerant Mass Flowmeter (Meng, 1997; Meng, 1998b)

In case of requiring mass flowmeters, usually, we could use the in-direction (or called combination) scheme. It's shown as in Fig. 3.

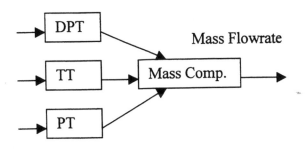

Fig. 3 Mass Flowmeter

For this type of mass flowmeter, there will be three typical failures,

1). Temperature transmitter (TT) failure
2). Pressure transmitter (PT) failure
3). Differential-pressure transmitter (DPT) failure

For the 1st and 2nd types, we could use a predicated value for each of them, by making use of the historical values of TT or PT. This means that, the mass computer (or processing unit) should and could have some "intelligence".

For the 3rd type, similar with the above two situations, we could pre-find a relationship function between the values of DPT, TT, and PT for certain mass flowrate. This means that, when fault is occurs in DPT, an algorithm results (or called estimated results) could be used to replace the real value of mass flowrate.

It should be noted that, in this case, software method is mainly needed. And also, the understanding of process which is measured and/or controlled is strongly needed. For all of our works, this later factor is always a very important key point, both for measuring and controlling.

3.2 Fault-tolerant level Control System of Boiler Drum

For most cases of boiler control, the level control system for its drum in the most important. Normally, there are three variables could be used for control system. Feed water flowrate (Fw), vapor flowrate (Fv), and the level of drum (Ld), as shown in Fig.4.

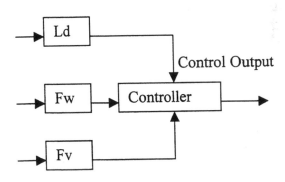

Fig. 4   Level Control System for Boiler Drum

While one of Fw, Fv, and Ld has problems, the control algorithms could be re-configured. For example, by simply used

Ld, while Fw, and/or Fv has problems. Also, we could find a relationship between Ld and the drum pressure (Pd). This could be an additional backup for Ld's fault. Even though, in these cases the performance of boiler will be low down in some scale, but its reliability could still be ensured.

For control vales of this system, there should be some backup by hardware redundant method.

E. Eryurek (1995) had studied a fault-tolerant level control system for boiler drum by using traditional instruments. For the new generation types of filed-bus instrument, we have done some researches (Meng, 1997; Meng, 1998b; Luo, 1998).

Sheng (1996) and Liu (1996) had used the similar scheme for paper production and chemical process control.

3.3 Fault-tolerant Process Control Systems based on Field-bus Instrument

With field-bus instruments, which should have real intelligence of self-diagnosis, self-repairing, communication and co-ordination capability, etc., the process control could be made as fault-tolerant while it is needed. This system could be divided into three layers,

● Instrument layers, this will be the measuring and transmission combination unit, control and actutering combination unit. Here, software method will be mostly used.

● Loop layers, in which the process equipment is taken to be as the core for this layer.

In this layer, except software method in each individual instrument, the network co-ordination method, topology analyzing method, etc., should be more used in this layer.

● System layers, this means that, in a plant level, there would be many loop layers by which the whole control system for plant automatic operation could be achieved. This will be a complex network, as shown in Fig. 1. In this case, the co-ordination for each loop layers is the most important while some loops have problems. It's similar with a human beings's formed colony organization and operation (Meng, 1998a).

Obviously, for each types of processes or objects to be controlled, the co-ordination ways would be more different. With the developing of instrument and loop levels' fault-tolerant function, there will be more and more possibility and ways to achieve the aim.

4. CONCLUSION

People have been trying to simulating, expanding, and developing themselves's ability and intelligence. Ship, trains, airplanes, computers, etc., could be considered as the typical important results. This effort, now with the computers and communication technologies well developed is fulled more and more possibilities. The simulation and developing human beings' intelligence, not only the individuals but also the colonies, could have more progresses. Fault-tolerant instrument and process control, could be considered as a part of this very important progresses. We have seen the more clearly sunlight of achievement, and there need more and more peoples to be involved.

REFERENCES

Banerjee, S. (1995). AI in chemical and process industries, *Chemical Eng. World*, **30(7)**, 105-108

Binder, E. (1996). Diagnosis and monitoring of electrical power plant components, *Elektrotech Int. Tech.*, **113(2)**, 133-138

Blanke, M. (1997). Fault-tolerant control systems - a holistic view, *Control Engineering Practice*, **5(5)**, 693-702

Borodani, P., et al (1995). Minimum risk evaluation methodology for fault tolerant automotive control system, *Proc. of the 1995 4th Int. Conf. On Applications of Advanced technologies Transportation Eng.*, Capri, Italy, 1995

Chai, T. (1995). Adaptive fault-tolerant control for randam and multivariable system, *Automatica Sinica*, **21(4)**, 476-479 (in Chinese)

Chang, T. et al (1995). Fault tolerant control of traffic networks, *Proc. Of the 1995 IEEE Conf. On Control Applications*, Albany, USA, 1995, 119-124

Chen, T. (1981). *Diagnosis and fault-tolerant for digital systems*, Defense Industry Press, Beijing (in Chinese)

Dong, C. (1992). PLC with triple-redundant structures, *Process Automation and Instrument*, **13(4)**, 244-247 (in Chinese)

Elsag-Bailey (1994). *INFI-90 DCS*,

Eryurek, E., et al (1995). Fault-tolerant control and diagnostics for large-scale systems, *IEEE Trans. On Control Systems*, **15(5)**, 34-42

FF (1997). *Foundation field-bus and its technology*,

Foxboro (1992). *I/A series open industrial systems*,

Grebe, E. (1996). Power station and network, a homogeneous system, *VGB Kraftwerktech*, **76(1)**, 23-26

Gruhn, P. (1996). The evaluation of safety instrumented system, *ISA*, **35(2)**, 127-136

Haegglund, T. (1996). Modern controllers supervise vales automatically, *InTech*, **43(1)**, 44-46

Harashima, S. (1995). Fault-tolerant subway passenger information control system, *Proc. of the 2nd Int. Symp. On Autonomous Decentralizal Systems*, Phoenix USA, 80-85

Hodson, W. R. (1996). How fieldbus will affect DCS architecture, *InTech*, November, 50-53

Hu, S. (1994). On fault-tolerant functional observer, *Proc. Of 1994 American Control Conf.*, Baltimale, USA, 1994, 260-264

Huang, X. (1996). Fault-tolerant control for discrete time systems, *Control Theory and Application*, **13(1)**, 1996, 36-39 (in Chinese)

Johnson, D. A. (1994). Automated fault insertion, *InTech*, **41(1)**, 42-43

Kopetz, H. (1995). Communication infrastructure for a fault-tolerant distributed real-time system, *Control Eng. Practice*, **3(8)**, 1139-1146

Levis, A. H. (ed.) (1987). Challenges to control: a collective view - report of the workshop held at the university of santa clara on September 18-19, 1986, *IEEE Trans. On Automatic Control*, **32(4)**, 274-285

Liu, W. (1996). Fault-tolerant control based on expert system and its application in synthesis process, *Chemical Industrial Automation and Instruments*, **23(5)**, 13-16 (in Chinese)

Lu, M. (1996). Fault-tolerant technology in process control systems, *Journal of Huabei Power University*, **23(1)**, 92-94 (in Chinese)

Luo, C. (1998). Fault-tolerant control based on fieldbus, *M. Sc. Thesis*, Chongqing University, Chongqing, (in Chinese)

Makansi, J. (1995). Special report: information technology for powerplant management, *Power*, **139(6)**, 41-72

Meng, J., et al (1998a). *Machinery intelligence*, Chongqing Press, Chongqing,

1998 (in Chinese)

Meng, J. (1997). Fault-tolerant instrumentation theory and its industrial realization, *Application Documents to National Outstanding Youth Scientist Foundation,* (in Chinese)

Meng, J. (1998b). Fault-tolerant process control theory and its application, *Application Documents to National Outstanding Youth Scientist Fundation,* (in Chinese)

Rochet, R., et al (1995). Efficient synthesis of fault-tolerant controllers, *Proc. of the European design and Test Conf. Paris,* France,

Sadehi, J., et al (1995). Integrated fault-tolerant control law, *Proc. of IEEE 1995 National Aerospace and electronics Conf.,* Dayton, USA, **Vol.2,**

Sheng, J. (1996). The design of robust linear controller for discrete systems, *Control and Decision,* **11(1),** 68-72 (in Chinese)

Sheng, Y. (1996). The approach to the fault-tolerant control of paper production, *Control Theory and Application,* **13(6),** 333-340 (in Chinese)

Shinskey, F. G. (1981). *Controlling multivariable processes,* ISA,

Shu, S. (1992). *Analysis and synthesis of reliability for control systems,* Science Press, Beijing, (in Chinese)

SIC (1997). Technical documents, Sichuan Instrument Corp., Chongqing, (in Chinese)

Spitzer, C. R. (1988). *Digital avionics,*

Wang, Z. (1991). *Reliability, redundancy, and fault-tolerant,* Air Industry Press, Beijing, (in Chinese)

Williams, T. J. (1984). *The use of digital computers in process control,* ISA, 1984

Zhou, D. (1995). New progresses of fault diagnosis technology of control system based on models, *Automatica Sinica,* **21(2),** 1995, 244-247 (in Chinese)

Zhu, L, Meng, J. (1998). *Instrumentation theory,* Mechanical Industry Press, Beijing, (in Chinese)

--- (1996). Feedback design in control reconfigurable systems, *Int. J. Robust Nonlinear Control,* **6(6),**

--- (1996). Non-intrusive valve diagnosis in power plant, *Power Eng.* **100(1),**

---, Field-bus instruments and field-bus control systems, *National Projects Documents,* (in Chinese)

NONLINEAR PROCESS MODELING USING A DYNAMICALLY RECURRENT NEURAL NETWORK

Shi-Rong Liu[1] and Jin-Shou Yu[2]

1. *Department of Automation and Computer Technology, Ningbo University, Ningbo 315211, P. R. China*
2. *Research Institute of Automation, East China University of Science and Technology Shanghai 200237, P. R. China*

Abstract: This paper describes the Elman network and the modified Elman network for nonlinear dynamic system modeling. A dynamic back-propagation algorithm is introduced to train the Elman network. The modified Elman network with the self-feedback links of the context units has been shown that it can model a nonlinear dynamic process using the standard back-propagation algorithm training. How to select effectively the self-feedback gains of the context units is discussed. The simulation results have shown that the modified Elman network has sufficiently approximate capability to nonlinear dynamic systems. *Copyright © 1998 IFAC*

Keywords: Neural networks; Elman network; modified Elman network; nonlinear system modeling.

1. INTRODUCTION

Both the theory and practice of nonlinear process modeling have advanced considerably in recent years. It is known that a wide class of discrete-time nonlinear systems can be represented by the nonlinear autoregressive moving average (NARMA) model with exogenous inputs. The mathematical function describing a real-world system can be very complex and its exact form is usually unknown so that in practice modeling of a real-world system must be based upon a chosen model set of known functions. Polynomial functions have the capability of approximating a system to within an arbitrary accuracy. This provides the foundation for modeling nonlinear system using the polynomial NARMA model and several identification procedures based on this model have been developed. Because there are some difficulties in perfectly identifying general

real-world nonlinear dynamic systems in the NARMA model, neural networks are an obvious alternative. Neural networks can therefore be viewed as just another class of functional representations. Multi-layered feedforward neural networks have been widely used in modeling and controlling complex nonlinear dynamic systems. Recently, the modeling of dynamic systems based upon dynamically recurrent neural networks has been exciting a great deal of research interests.

A dynamically recurrent neural network, also called a feedback neural network, is one in which self-loops and backward connections between nodes are allowed. One of consequences of these connections is those dynamic behaviors not possible with strictly feedforward networks, such as limit cycles and chaos, can be produced with dynamically recurrent neural networks. Dynamically recurrent neural networks with symmetric weight connections always converge to stable state in the well-known Hopfield network. The dynamically recurrent neural networks without the symmetry constraint have been shown that have more complex dynamic behaviors than symmetric dynamically recurrent neural networks. Another possible benefit of recurrent neural networks is that smaller networks may provide the functionality of much larger number of nodes. Partially recurrent neural networks are a class of simple recurrent neural networks with asymmetric connections. In partially recurrent neural networks, the weights on the feedback links are fixed, and so the standard back-propagation (BP) learning rule may be employed for network training. Such networks are also referred to as sequential networks, and the nodes receiving feedback signals are called context units. In these networks, at time step k the context units have signals coming from part of the network state at time $k-1$. Thus the state of the whole network at a particular time depends on an aggregate of previous states as well as on the current input.

The Elman network (Elman, 1990) is a type of partially recurrent network. There have been more research interests in this network and it has been applied to dynamic system identification (Pham and Liu, 1992) and financial prediction (Kamijo and Tanigawa, 1990). It has been shown that this network can learn to mimic closely a finite-state automaton (Elman, 1991; Servan-Schreiber et al., 1991). Pham and Liu found that the basic Elman network trained by the standard BP algorithm was able to model only first-order dynamic system and then proposed a modified Elman network for modeling high-order dynamic systems (Pham and Liu, 1992). Recently the dynamic back-propagation (DBP) algorithm has been used for training the basic Elman network. It has been shown that the modified Elman network is an approximation of the basic Elman network trained by DBP (Pham and Liu, 1996). This paper introduces DBP algorithm of the Elman network training and the modified Elman network. How to select the self-feedback link gains of the modified Elman network is a valuable issue, which obviously affect the network's dynamic behaviors. Through studying the modeling results of two nonlinear processes, some important points of view are discussed on nonlinear system modeling with the modified Elman network.

2. THE ELMAN NETWORK AND DYNAMIC BACK-PROPAGATION ALGORITHM

2.1 The Elman Network and Mathematical Model

Elman presented a simple recurrent neural network as shown in Fig. 1 (Elman, 1990), called Elman network. From Fig. 1 it can be seen that in addition to the input units, the hidden units and the output units, there are also the particular context units in a basic Elman network. As general multi-layered feedforward networks, the input layer units are only buffer units which pass the input signals without any change, and the output layer units are linear units which sum the signals fed to them in linear weighted form. The hidden layer units can have linear or nonlinear activation functions, and context layer units can be considered as one time step delay operators (e.g. Z^{-1} in Fig. 1) which are used only to memory the previous output state of the hidden layer units. The feedforward connections are modified in network training, but the recurrent connections are fixed, the Elman network is sometimes called a partially recurrent network.

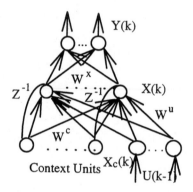

Fig. 1. The basic Elman network

At the specific time k, the network inputs include the input vector $U(k-1)$ and the previous output vector $X(k-1)$ of the hidden units, which is also the current output vector $X_c(k)$ of the context units. At this step, the network acts as a feedforward network and propagates these inputs forward to produce the network output vector $Y(k)$. The standard BP algorithm can then be employed for training the network. After the training step, the activated vector $X(k)$ of the hidden layer at time k is sent back to context units through the current links and saved there for next training step (time $k+1$). At the training beginning, the output initial values of the hidden units are generally set to one-half of their maximum range. In the applications of nonlinear system modeling, the activation functions of the hidden units often use hyperbolic tangent functions.

The mathematical relation of the Elman network can

be represented in state space model. Set the external input $U(k-1) \in R^r$, the network output $Y(k) \in R^m$, the hidden layer output $X(k) \in R^n$ and the context layer output $X_c(k) \in R^n$. The state space representations of the network can be writen as

$$X(k) = f\left(W^c X_c(k) + W^u U(k-1) + b^x\right)$$
$$X_c(k) = X(k-1)$$
$$Y(k) = W^x X(k) + b^y$$

$$(1)$$

where W^u is the link weight matrix from the input units to the hidden units, W^c is the link weight matrix from the context units to the hidden units and W^x is the link weight matrix from the hidden units to the output units respectively, b^x and b^y are the bias vector of the hidden units and the bias vector of the output units separately, $f(\cdot)$ is the nonlinear vector function of the hidden units. The biases of the network are not drawn in Fig. 1.

2.2 Dynamic Back-Propagation Algorithm

It has been discovered through simulation that a linear Elman network trained by the standard BP algorithm can only model first-order linear system (Pham and Liu, 1992). The cause is that the gradient only trace back one time step in the standard BP algorithm so that the learning stability of the link weights from the context units to the hidden units is not guaranteed. The learning rate in the training process is too small to make the network learning errors converge to an allowable approximate accuracy. The dynamic BP algorithm has been used to train the Elman network, otherwise the basic Elman network should be improved in modeling high-order dynamic systems (Pham and Liu, 1996).

Consider equation (1) in learning process, the state vector of the context units is

$$X_c(k) = X(k-1)$$
$$= f\left(W^c(k-1)X(k-2) + W^u(k-1)U(k-2) + b^x(k-1)\right)$$

$$(2)$$

From equation (2), it can be shown that $X_c(k)$ depends on the weights of the past different time instants, e.g. $W^c(k-1)$, $W^u(k-1)$, $W^c(k-2)$, ... When the BP algorithm is applied, the dependence of $X_c(k)$ on the weights should be taken into account. A BP algorithm that considers this dependence is the dynamic BP algorithm.

At time k, the squared error of the network output is defined as

$$E_k = \frac{1}{2}\left(Y_d(k) - Y(k)\right)^T \left(Y_d(k) - Y(k)\right), \quad (3)$$

where $Y_d(k)$ is the desired output of the network. Set the link weight matrices' elements $w_{ij}^x(\cdot)$, $w_{jl}^c(\cdot)$, $w_{jq}^u(\cdot)$ respectively. When the network being trained, the weights are modified at each time step k. For the weight matrix W^x from the hidden units to the output units, the error gradient with respect to $w_{ij}^x(k-1)$ is

$$\frac{\partial E_k}{\partial w_{ij}^x(k-1)} = -\left(y_{d,i}(k) - y_i(k)\right)x_j(k)$$
$$= -\delta_i^0 x_j(k)$$

$$i=1,2,...,m; \ j=1,2,...,n, \quad (4)$$

where

$$\delta_i^0 = y_{d,i}(k) - y_i(k). \quad (5)$$

For the weight matrix W^u,

$$\frac{\partial E_k}{\partial w_{jq}^u(k-1)} = -\sum_{i=1}^{m}\left(\delta_i^0 w_{ij}^x(k-1)\right)f_j(\cdot)u_q(k-1)$$

$$j=1,2,...,n; \ q=1,2,...,r. \quad (6)$$

For the weight matrix W^c,

$$\frac{\partial E_k}{\partial w_{jl}^c(k-1)} = -\sum_{i=1}^{m}\left(\delta_i^0 w_{ij}^x(k-1)\right)\frac{\partial x_j(k)}{\partial w_{jl}^c(k-1)}$$

$$j=1,2,...,n; \ l=1,2,...,n. \quad (7)$$

As discussed above, the context unit state $X_c(k)$ is dependent on $W^c(k-2)$, then

$$\frac{\partial x_j(k)}{\partial w_{jl}^c(k-1)} = f_j(\cdot)\left(x_l(k-1) + \sum_{i=1}^{n}w_{jl}^c(k-1)\frac{\partial x_i(k-1)}{\partial w_{jl}^c(k-1)}\right)$$

$$. \quad (8)$$

If the weight changes are assumed to be very small in each iteration, then (8) can be approximately represented as

$$\frac{\partial x_j(k)}{\partial w_{jl}^c(k-1)} = f_j(\cdot)\left(x_l(k-1) + \sum_{i=1}^n w_{jl}^c(k-1)\frac{\partial x_i(k-1)}{\partial w_{jl}^c(k-2)}\right)$$

(9)

Equation (9) expresses a recursive property of the dynamic BP method.

The general weight modification based on the gradient descent method is

$$\Delta w_{ij} = -\eta\frac{\partial E_k}{\partial w_{ij}}.$$ (10)

The dynamic BP algorithm for training an Elman network can be summarized as follows,

$$\Delta w_{ij}^x = \eta\delta_i^0 x_j(k)$$

$$\Delta w_{jq}^u(k) = \eta\left(\sum_{i=1}^m \delta_i^0 w_{ij}^x(k-1)\right)f_j(\cdot)u_q(k-1)$$

$$\Delta w_{jl}^c(k) = \eta\left(\sum_{i=1}^m \delta_i^0 w_{ij}^x(k-1)\right)\frac{\partial x_j(k)}{\partial w_{jl}^c(k-1)}$$

i=1,2,...,m; j=1,2,...,n,
q=1,2,...,r; l=1,2,...,n,   (11)

where the partial derivative of $x_j(k)$ with respect to $w_{jl}^c(k-1)$ can be calculated from equation (9).

3. THE MODIFIED ELMAN NETWORK

Pham and Liu put forward a modified Elman network as shown in Fig. 2 (Pham and Liu, 1992), which can be used to model high-order dynamic systems with the standard BP algorithm. The modified Elman network is those self-feedback links with one-step time delay are introduced to the context units of the basic Elman network and the self-feedback gain a is fixed.

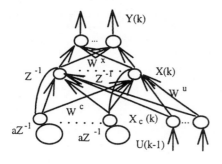

Fig. 2. The modified Elman network

The modified Elman network can described by the follow equation set,

$$X(k) = f\left(W^c X_c(k) + W^u U(k-1) + b^x\right)$$
$$X_c(k) = X(k-1) + aX_c(k-1)$$
$$Y(k) = W^x X(k) + b^y$$

(12)

The standard BP method is employed for training the modified Elman network and the weight modified formulae are

$$\Delta w_{ij}^x = \eta\delta_i^0 x_j(k)$$

$$\Delta w_{jq}^u(k) = \eta\left(\sum_{i=1}^m \delta_i^0 w_{ij}^x(k-1)\right)f_j(\cdot)u_q(k-1)$$

$$\Delta w_{jl}^c(k) = \eta\left(\sum_{i=1}^m \delta_i^0 w_{ij}^x(k-1)\right)\frac{\partial x_j(k)}{\partial w_{jl}^c(k-1)}$$

i=1,2,...,m; j=1,2,...,n
q=1,2,...,r; l=1,2,...,n,   (13)

where

$$\frac{\partial x_j(k)}{\partial w_{jl}^c(k-1)} = f_j'(\cdot)x_{c,l}(k).$$ (14)

If the lth variable $x_{c,l}(k)$ of $X_c(k)$ in the equation (12) is multiplied by $f_j'(k)$, then (14)

$$\frac{\partial x_j(k)}{\partial w_{jl}^c(k-1)} = f_j'(\cdot)x_l(k-1) + a\frac{\partial x_j(k-1)}{\partial w_{jl}^c(k-1)}.$$ (15)

If equation (15) is compared with equation (8), it can be observed that the structures of (15) and (8) are the same, and the difference is only in their coefficients. Therefor the modified Elman network trained by BP algorithm is similar in performance to the Elman network trained by DBP algorithm.

The internal dynamic behaviors of the context units with the self-feedback links can be described by a first-order linear discrete-time equation (see (12)). According to the impulse-response characteristic of first-order linear discrete-time system, the dynamic responses of the context units monotonically decrease when the self-feedback gain a is selected as $0<a<1$. In fact the dynamic characteristics of the context layer substantially affect the dynamic behaviors of whole network so that the dynamic characteristic of the modified Elman network may be very complex. The value of self-feedback gain a should be tested carefully in the training process of the network for a special real-world nonlinear process.

4. NONLINEAR PROCESS MODELING BASED ON THE MODIFIED ELMAN NETWORK

In this nonlinear process modeling study, only single-input single-output examples are considered. When a modified Elman network is applied to modeling a BIBO (bounded input bounded output) nonlinear process, the gain a must be determined according to the mononically decrement property of the dynamic behaviors of the context units in learning process.

4.1 Simulation Results

Example 1. The nonlinear second-order system to be modeled is expressed by

$$y(k+1) = \frac{y(k)(y(k-1)+2)(y(k)+2.5)}{8.5+y^2(k)+y^2(k-1)} + u(k)$$

(16)

where $y(k)$ is the output of the system at the kth time step and $u(k)$ is the plant input which is a uniformly bounded function of time. The plant is stable at $u(k) \in [-2,2]$. The structure of the network was selected as $3 \times 8 \times 1$, and the self-feedback gain a was 0.65 through several experiments. In the network training, the plant input $u(k)$ was an independent random noise of uniform distribution with $u(k) \in [-2,2]$, and 100 data pairs were used as training samples. The output error squared sum was used as training criterion. The BP algorithm with self-adaptive learning rate was adopted and the momentum constant α was 0.65. After 20000 iterations, the square root of quadratic error mean was 0.0154. After the modified Elman network is trained, its prediction power is tested for the plant input

$$u(k) = \begin{cases} 2\cos(2\pi k \times 0.01) & 0 \le k \le 200 \\ 1.2\sin(2\pi k \times 0.05) & 200 < k \le 500 \end{cases}$$

(17)

As shown in Fig. 3, the learned network can predict the nonlinear system output quite well.

Example 2. The nonlinear third-order system to be identified is described by

$$y(k+1) = \frac{y(k)y(k-1)y(k-2)u(k-1)(y(k-2)-1)+u(k)}{1+y^2(k-1)+y^2(k-2)},$$

(18)

where $u(k)$ is a uniform function of time and the system is stable at $u(k) \in [-1,1]$. The structure of the network was selected as $5 \times 8 \times 1$, and the self-feedback gain a was 0.65 through several experiments. In the network training, the plant input $u(k)$ was an independent random noise of uniform distribution with $u(k) \in [-1,1]$, and 100 data pairs were used as training samples. The output error squared sum was used as training criterion. The BP algorithm with self-adaptive learning rate was adopted and the momentum constant α was 0.65. After about 20000 iterations, the square root of quadratic error mean was 0.0149. After the modified Elman network is trained, its modeling power is validated by the plant input

$$u(k) = \begin{cases} \sin(2\pi k \times 0.01) & 0 \le k \le 500 \\ 0.8\sin(2\pi k \times 0.004) + 0.2\sin(2\pi k \times 0.04) \\ & 500 < k \le 1000 \end{cases}$$

(19)

The simulation result in Fig. 4 shows that the modeling power is satisfactory.

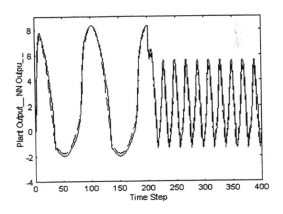

Fig. 3. Simulation result for a nonlinear second-order system identification using a modified Elman network.

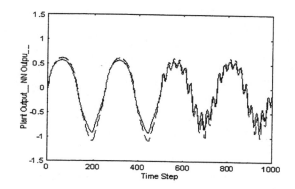

Fig. 4. Simulation result for a nonlinear third-order system modeling using a modified Elman network.

4.2 Discussion

Remark 1. From above simulation results, it is shown that a modified Elman network employed for modeling nonlinear systems is able to have the same approximate accuracy such as a three-layered feedforward network used to model nonlinear systems. The modified Elman network makes advantage of network structure simple and computing expense little. For example, to identify the same nonlinear plant in Example 2, a $5 \times 20 \times 10 \times 1$ three-layered BP network is adopted (Narendra and Parthasarathy, 1990) and after 10^5 iterations the simulation result is similar in performance to Fig. 4.

Remark 2. The self-feedback gain a of the modified Elman network is a important parameter. It is very sensitive to the stability of network training. For a BIBO nonlinear system, the concept of the impulse response of first-order discrete-time system may direct the selection of the gain. An appropriate self-feedback gain will be of benefit to smoothing and accelerating learning process.

Remark 3. The Elman network or the modified Elman network can be directly driven only by external input $u(k)$ theoretically. The dynamic characteristics of whole network may be activated completely by the internal behaviors of the network. The simulation practices have shown that if the network is directly driven by external input, it often lead to the learning iterations increasing greatly and sometimes the learning process unstable. Therefor, the effectively learning method should be carefully investigated too.

Remark 4. In order to activate a variety of system dynamic modes in nonlinear system identification, a

gaussian white noise sequence or a uniform distribution random noise sequence is used generally as the external exciting signal for nonlinear system to be identified. In fact it is unreasonable. Because nonlinear systems do not have the characteristic of frequency distribution like linear systems, a nonlinear system may be stable in some bands and unstable in other bands. If the stable operation bands of the system to be modeled can be estimated roughly, a sine signal with the addition of adding small magnitude random noise, which the frequency of the sine signal is equal to or over the upper limit frequency of the system, or a combination sine signal with multi-frequency may be selected as the external exciting signal. The effectiveness of this method has been validated in practice.

5. CONCLUSION

This paper has discussed in detail the Elman network, the modified Elman network and the modified Elman network's applications in nonlinear systems modeling. The simulations have shown that the modified Elman network will have a potential application prospect in nonlinear dynamic system modeling.

REFERENCES

Elman, J. L. (1990). Finding structure in time. *Cognitive Sicence*, **14**, 179-211.

Elman, J. L. (1991). Distributed representations, simple recurrent networks, and grammatical structure. *Mach. Learn*, **7**, 195-225.

Kamijo, K. and T. Tanigawa (1990). Stock price pattern recognition-a recurrent neural network approach. *International Joint Conference on Neural Networks*, **1**, 215-221.

Narendra, K. S. and K. Parthasarathy (1990). Identification and Control of Dynamical Systems Using Neural Networks. *IEEE Trans. Neural Networks*, **1(1)**, 4 -27.

Pham, D. T. and X. Liu (1992). Dynamic system modelling using partially recurrent neural networks. *J. of System Engineering*, **2**, 90-97.

Pham, D. T. and X. Liu (1996). Training of Elman networks and dynamic system modelling. *Int. J. of Syst. Sci.*, **27(2)**, 221-226.

Servan-Schreiber, D., A. Cleeremans and J. L. McClelland (1991). Graded state machines: The representation of temporal contingencies in simple recurrent networks. *Mach. Learn*, **7**, 161-193.

CONTROL ON THE BASIS OF NETWORK MODELS

V.Gladun, N.Vaschenko

*Institute of Cybernetics of National Academy of Sciences of Ukraine,
Prospect Akad. Glushkova, 40, 252022, Kiev-22, Ukraine
Email: glad@aduis.kiev.ua*

Abstract: If mapping of variants of a decision into their criterial estimations is not given directly, instead of traditional methods of linear, integer and nonlinear programming heuristic methods of decision making should be applied. As a result a problem of adequate knowledge representation arises. The paper describes methods and tools of decision making on the basis of a special environment model which is referred to as a balance network. *Copyright © 1998 IFAC*

Keywords: artificial intelligence, control systems, decision making, decision support systems, knowledge representation.

1.INTRODUCTION

Traditional consideration of a decision making process, as a choice among alternatives on the basis of known mapping between the alternatives and their criterial estimations, is somewhat idealized because this mapping is often multi-step and changeable and is not given directly. Dependencies of criterial estimations on the alternatives change under the influence of external and internal conditions of a system functioning. Let us define this quite a realistic situation of control more precisely.

The task is defined as $<P_{in}, X, P, Y, F, L>$, where

P_{in} is the set of input parameters of the system;

$X = \{x_i\}$, $i=1, 2,...,I$ is the set of control parameters (control actions) that can be used for changing system states;

$P = \{p_k\}$, $k=1, 2,...,K$ is the set of derivative parameters;

$Y = \{y_i\}$, $j=1, 2,...,J$ is the set of output parameters, $Y \subseteq P$;

$F = \{f_k\}$, $k=1, 2,...,K$ is the set of functions that reflect dependencies of derivative parameters on the other parameters;

$L = \{l_j\}$, $j=1, 2,...,J$ is the set of limitations of type $a_j \leq y_j \leq b_j$ that are defined for output parameters.

The goal consists in finding the point $x=(x_1,...,x_I)$ in the space of control parameters, where

$$(\forall j \in 1,2,...,J)\ a_j \leq y_j \leq b_j \qquad (1)$$

The state of the system for which condition (1) is satisfied is called a normative one. The problem of achieving the normative state will be referred to as the problem of parametric normalization.

The set of control parameters X is the set of alternatives among which the decision is chosen. The output parameters combined with their limitations play the part of criterial estimations.

If mapping $X \rightarrow Y$ would be given by some system of equations, the problem could be solved by methods of linear, integer or nonlinear programming. On condition that $X \rightarrow Y$ mapping is not given directly it is necessary to use heuristic methods (Gladun, 1987; Chaib-draa *et al.*, 1992) which are known to depend on the ways of knowledge representation.

The knowledge representation should be convenient for performing the following actions: formation of the subset of control parameters that are relevant to the output parameters under consideration; finding sequences of calculations for output parameters;

calculation of influence of decisions on the state of the system as a whole.

When fulfilling these actions, search processes prevail. Therefore, in complex systems to organize control processes in changeable situations, network models of application domains should be used.

2. BALANCE NETWORK

The domain model used for solution of parametric normalization problem should provide:
1) finding subsets of control actions influencing values of the output parameters;
2) finding direction of control actions change, bringing the output parameters nearer to their normative values;
3) formation of sequences of transformations connecting control actions with the output parameters.

Let us define a network structure meeting these requirements.

The balance network is an oriented graph $<V,U>$ in which
1) V is the set of vertices corresponding to parameters of all types;
2) connections between the vertices (set U) are regulated by the following rule: each vertex is connected by entering arcs with vertices which represent arguments of a function used for calculation of a corresponding parameter;
3) arcs are specially marked if the functions are growing or decreasing with respect to the parameters represented by vertices issuing the arcs.

Input, control and output parameters are represented correspondingly by input, control and output vertices of the network.

The balance network not containing cycles is called hierarchical (Fig.1). Hierarchical balance networks can be looked at as a certain development of structures suggested in (Tyugu, 1984).

Environments, in which decision making problems are solved, are usually represented by hierarchical balance networks.

Let us introduce some additional definitions connected with an application of the networks for modelling of applied domains.

An output parameter is called a normalized one if its value matches the limitation indicated for it. The output parameters that don't match their limitations are called unnormalized.

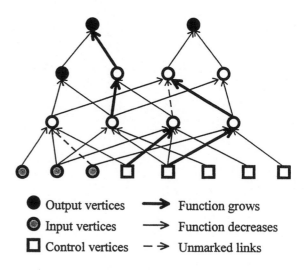

● Output vertices   → Function grows
◉ Input vertices   → Function decreases
□ Control vertices   ⇢ Unmarked links

Fig.1. Hierarchical balance network.

Normalized and unnormalized output parameters are represented correspondingly by normalized and unnormalized vertices of the network.

The situation in which all output parameters are normalized is called the normative situation. The balance network in which there are no unnormalized vertices is referred to as the balanced network.

The goal of a control consists in achievement of the normative situation.

3. THE SOLVING PROCEDURE

The problem of finding point $x=(x_1,...,x_l)$ meeting requirement (1) is solved by sequential selection of control parameters values. The strategy of solution is a cyclic process, where at each step "the best" control action is chosen and the model of the new situation, arising as a result of its application, is formed.

If there exists a possibility to use a generalized criterion for selection of "the best" control action providing minimization of the steps number in transformation of the initial situation into the normative one, the general scheme of control is reduced to the scheme of dynamic programming. In case the creation of such criterion is difficult, it could be replaced by a certain heuristic estimating function that should be approximated to the criterion. Naturally, in this case minimization of the steps number in transformation of the initial situation into a normative one is not guaranteed.

The probability of achievement of the normative situation and a necessary number of control actions depend on the quality of the heuristic function.

The following designations will be used when describing the process of control action selection:

α_1, α_2 designate the entering arcs of vertices representing functions of set F in cases when these functions are correspondingly growing or decreasing with respect to parameters represented by vertices issuing these arcs;

Y' is the set of unnormalized parameters;

R is the set of relevant control actions, i.e. the ones that influence the unnormalized output parameters;

$h_r^{y'}$ is the variable designating a necessary "direction" of a change of control action $r \in R$ for normalization of output parameter y';

H_r is the variable designating "direction" of a change of control action r taking into account all unnormalized output parameters.

In accordance with the above, the method of decision formation consists of the following operations.

<u>Operation 1.</u> The search of relevant control actions.

For every unnormalized output parameter $y' \in Y'$, the necessary change direction is defined. Assume, limitation $a_{y'} \leq y' \leq b_{y'}$ is given for y'. If $y' < a_{y'}$ in the moment of operation execution, it is necessary to increase the value of y', if $y' > b_{y'}$, to decrease.

For every unnormalized parameter $y' \in Y'$, entering ways of the corresponding vertex are traced to the vertices representing control actions. Before tracing $h_r^{y'} = 1$ for all $r \in R$, if the value of y' should be increased, and $h_r^{y'} = -1$ if the value of y' should be decreased. Later on the sign of variable $h_r^{y'}$ value is saved when an α_1 marked arc is traced and changes when tracing an α_2 marked arc. When tracing unmarked entering arc of a vertex representing a certain function, variable $h_r^{y'}$ takes on value 0.

As a result of operation 1 implementation, set R is formed and a change direction of relevant control actions is defined.

<u>Operation 2.</u> Calculation of values of variables H_r.

For each $r \in R$, variable H_r is calculated in such a way:

$$H_r = \sum_{y' \in Y'} h_r^{y'}.$$

When $H_r > 0$ it is expedient to increase the value of control parameter r, if $H_r < 0$, to decrease.

<u>Operation 3.</u> The selection of "the best" control action.

At present "the best" control action is the control action having $\max |H_r|$.

<u>Operation 4.</u> Simulation of control action application.

The value of selected control parameter r depending on the sign of H_r increases or decreases at some fixed portion that is indicated by experts a priori for every control parameter.

All parameters, corresponding to vertices connected by entering ways with the changed control parameter, are re-counted.

Application of a control action results in normalization of existing unnormalized vertices or sometimes in appearance of similar new vertices.

For a new state of the system if it is not a normative one, all the above operations repeat.

So, the general algorithm of solution formation looks as follows:
1. Is the current state of the network a normative one? If "no", go to p.2, if "yes", go to End.
2. Operation 1.
3. Operation 2.
4. Operation 3.
5. Operation 4. Go to p.1.
End.

In case of success the algorithm produces a sequence of control actions, i.e. a certain plan of reaching a normative situation.

If for every $r \in R$, function $h_r^{y'}$ values are both positive and negative it means that subspace of nondominating alternatives (the Pareto set) is achieved. In this case further improvement of selected alternatives is only possible on the basis of additional information on alternatives or criteria.

Preferences of a decision making person can be taken into account with the help of weight coefficients that are introduced for control actions or for output parameters and are used in selection of control actions.

4. APPLICATIONS

The above approach is implemented in the tool system MANAGER. The MANAGER signals parameter violations, recommends possible control actions for liquidation of these violations and helps in choosing decisions taking into account their multi-step influence on the system as a whole. There is a certain experience of the system application for administrative regions management, business activity support, decision making support in nonstandard situations when controlling complex technical objects, selection parameters providing the main output characteristics of turbo-generators, and for manager training. There is an intention to apply this approach for creation of technologies when designing the new materials.

Sometimes in multiloop control systems it is necessary to organize the distributed selection of control actions. The N control loops available, the set of parameters is divided in subsets corresponding to separate control loops.

$$Y' = \bigcup_{n=1}^{N} Y'_n, \; n=1, 2,...,N, \text{ where } Y'_n \text{ is the set of}$$

unnormalized parameters of control loop n. When $Y'_n = \varnothing$, loop n is considered as a normalized one. For sets Y'_n, Y'_m ($n{\neq}m$; $n,m=1,2,...,N$) relation $Y'_n \cap Y'_m \neq \varnothing$ is possible. Parameter $y'_{n,m} \in Y'_n \cap Y'_m$ is referred to as a common parameter of control loops n and m.

Decision making process is carried out for each control loop separately. In any control loop, the preferable control actions are the ones that do not change common parameters of normalized loops. The result of common parameters changing certain control loop becoming unnormalized, the decision making process is repeated for it. So, common parameters are of importance as connective links by which decision making within one loop influences the state of other loops.

In a multiloop control system, regulation of a separate control loop is often an isolated process and is carried out by different specialists having no contact with one another. Common parameters allow to organize cooperative interaction in multi-agent decision making.

5.CONCLUSION

The methods described in the paper are recommended to implement in strategic information systems to support the highest level of management at enterprises, plants and administrative districts.

The strategic information systems have to support solution of the following tasks:
1. Analysis of the opportunities and threats or constraints that exist in the external environment.
2. Analysis of the organization's internal strengths and weaknesses.
3. Establishment of the organization's mission and development of its goals.
4. Formulation of strategies (including corporate level, business unit level, and functional level) that will match the organization's strengths and weaknesses with the environment opportunities and threats.

The balance network is a suitable instrument for solution of such sort of problems.

REFERENCES

Chaib-draa,B., R.Mandiau and P.Millot (1992). Distributed Artificial Intelligence: An Annotated Bibliography. *SIGART Bulletin*, Vol. 3, N 2, pp.20-37.

Gladun,V.P. (1987). *Solution planning*. Naukova dumka, Kiev (in Russian).

Tyugu,E. (1984) *Conceptual programming*. Nauka, Moscow (in Russian).

INTELLIGENT AGENTS FOR DECISION SUPPORT

Patrick Race and Kara L. Nance

Department of Mathematical Sciences
University of Alaska Fairbanks

Abstract: Many environmental, geological and meteorological studies have been enthusiastically
undertaken in the Arctic Region producing vast quantities of data. Unfortunately, these are
frequently unassociated efforts and the methods of data retrieval and interpretation are, in most
cases, dissimilar. The SynCon Project is an ambitious effort to compile this data into a cohesive
geographical information system (GIS) and create a knowledge base system, that can be easily
used to aid in research and remediation planning for this sensitive environmental region.
Copyright © 1998, IFAC

Keywords: agents, artificial intelligence, computer applications, decision support systems, expert
systems, knowledge-based systems

1. INTRODUCTION

Due to an extreme sensitivity to global climactic
change, the Arctic region of Earth's ecosystem has
been recognized as an important location for
scientific studies pertaining to global environmental
conditions and contaminants. Studying and learning
more about the effects of environmental change in
the arctic will help show how other, less sensitive,
regions may eventually be affected. With the
recognition of the importance the Arctic plays as a
global indicator, many environmental, geological and
meteorological studies have been enthusiastically
undertaken producing vast quantities of data.
Unfortunately, these are frequently unassociated
efforts and the methods of data retrieval and
interpretation are, in most cases, dissimilar. One
result is that other members of the scientific
community can only apply the data to its original
hypothesis or goal, leaving it ineffectual for
alternative studies. However, by compiling this data
into a cohesive geographical information system
(GIS) and creating an expert system to reason with
the data, it can be easily used to aid in research or
remediation of sensitive environmental areas. The
goals of the SynCon project (George et. al, 1995;
George et al 1996; Nance et al, 1996; Nance, 1997a;
Nance, 1997b; Nance, 1997c; Parsons, 1997; Race
and Nance, 1998; Tedor and Nance, 1998), originally
funded by the office of Naval Research, include
undertaking the task of organizing many of the
heterogeneous data sets pertaining to the circumpolar
arctic region and the creation of a knowledge base

system which will be used to track and deal with
environmental contaminants in a sensible and
comprehensive manner. SynCon multidisciplinary
teams are assigned to investigate the many individual
components associated with this research effort

2. SYNCON PROJECT

The use of knowledge discovery system
methodologies is one approach to solving the data
organization dilemma present in many research
efforts including environmental research. The
SynCon Project focuses on synthesizing and
assessing arctic contamination data, although the
algorithms and tools being created will be readily
adaptable to other scientific fields. The SynCon
project has two major components: data organization
and knowledge discovery. The knowledge discovery
phase included the design of a spatial (georeferenced) knowledge base system to hold data
measurements and information on contamination
including radionuclides, persistent organics, heavy
metals, acidification, petrochemicals, human health
measurements, and other. Search, browse, and query
tools that allow complex questions and graphical
display of data are available. New tools are being
developed to aid in the visualization of data both to
enhance understanding and to improve remediation
planning by providing decision assistance to policy-
makers. The prototype system has been implemented
and is being populated. The lessons learned
especially those involving environmental metadata

development, provide insight for further organization of environmental data sets.

2.1 Metadata

In efforts to complete this project with the most powerful and beneficial knowledge base possible, an extensive set of metadata had to be created. Metadata is essential information about the original data sets used to organize and simplify retrieval and interpretation as well as information about how the data is stored. Among the most important information in the metadata are the time, date, and location that the data was collected. This information is used to temporally and spatially map the data with the aid of georeferencing software. Georeferencing software uses the latitude and longitude to associate the data to the correct location on a map; it also uses altitude in applicable cases such as air and water samples. Then it utilizes the time and date to pin down the exact moment the data was taken. This is important because older data may no longer appear relevant in an area but it can still be used to determine how long it took changes to occur and to analyze trends. Another advantage of the temporal data is that future predictions for an area's contaminant level can be extrapolated with a certain degree of accuracy. This prediction can then be used by decision-makers or scientists to determine how badly the area needs to be remediated. Of course as a decision making tool there are many more aspects of the project.

2.2 Decision Support Systems

The knowledge discovery and data mining techniques that are currently under development for the SynCon project facilitate the development of Decision Support System to aid responsible parties in remediation planning. Decision support tools help scientists and policy makers to make informed decisions. Primary advantages of using computers in the decision-making process include volume capabilities, processing speeds, and indefatigability. As stated by Rodgers, "computer technology in the hands of truly creative people is being used as it's meant to be: to amplify man's intelligence and provide a business life-style that is ore rewarding and productive." (Rodgers, 1986) The term *decision support system* has a variety of definitions that vary slightly depending on context. As aptly stated, the "distinguishing characteristics of DSS include interactive access to data and models that deal with a specific decision that cannot be solved by the computer alone, but that require human invention (Olson and Courtney, 1992)."

Using the simple method of predicting future levels of contaminant based solely upon temporal and spatial data lends itself to a "shopping list" style clean up method where the most contaminated areas are restored first and the least contaminated are dealt with last. This is not a valid approach in the real world, because there are so many other factors that require attention. This is where the heart of one of the components of the SynCon project lies, in trying to determine which are the most important elements of this very large equation and how heavily each should be weighed. For example, a "shopping list" method might indicate that one area has a much higher level of a contamination than another and this would by design be dealt with first, but what it doesn't take into account is the location and pathways of the contaminants. The actual contamination source could be thousands of miles away, thus leaving scientists treating symptoms rather than curing the illness. Without an intelligent system, the higher level contaminant in an uninhabited area would be dealt with first leaving a lower level contaminant free to mingle with a city's water supply. Along with human and wildlife populations affected by contaminants many other key factors play important roles. What type of contaminant is it? Where does it come from? Can it be remediated? Is it near a population center? How long does it take to restore the contaminated area? What species of plant and animal life could be lost? Is the contaminant in an accessible area? How much will clean up cost? Will the contaminant spread if it isn't dealt with immediately?

2.3 Pathways

One of the other real world problems that SynCon is attempting to deal with is that of pathways. Contaminants don't always just remain stable in one spot, they move around, and one of the more difficult tasks is to determine what is the probable carrier and what areas the contaminant will affect. One obvious pathway is a stream or river, which could quickly distribute a dangerous contaminant over a long distance. With the georeferenced data in the knowledge base of the SynCon project one can determine which rivers might serve as pathways for contaminants and predict where contaminants may be carried. However, there are other pathways that aren't so obvious and an important feature of our system is the incorporation of inverse pathway detection. This is the ability for the system to intelligently determine in which direction and at what speed a contaminant is being carried. This is then matched up with several possible carriers from which the user will be able to determine the pathway. Examples of possible carriers are migratory species of birds, herds of grazing land mammals, air currents, and water currents. Pathways provide valuable information regarding planning for remediation. In addition, data mining algorithms can be used to identify pathways or potential pathways that human researchers have not previously acknowledged or identified.

3. INTELLIGENT AGENTS

Team B of the SynCon project is responsible for developing the cooperative intelligent agents that will determine the relative need for remediation to assist decision-makers and policy-makers. These agents must be intuitive and flexible so that they can work with real world situations successfully. They must be able to use the knowledge derived from the spatial and temporal data to simulate realistically results or consequences of specific remediation plans. These agents are being developed using mathematic modeling techniques specific to this task. Two such methods are polygonal and point thresholding. These are being applied in the mapping procedure used to display contaminant data in a visual form. Much like topographical maps display height using different scale of color, our maps will represent contaminant information. Both thresholding techniques are used to generate the color code for a specific region of the map.

3.1 Point Thresholding

Point thresholding builds potentially overlapping concentric circles around measurement points creating a ripple effect. Points that are geographically near each other will have outlying circles that overlap each other resulting in a darker color or higher value assigned to the associated region. This is more accurate than polygonal thresholding for points that are spread out over large areas and can more accurately represent the falloff of a contaminant the farther one gets from the source. Or course problems arise when too many points are too close together because the overlapping circles will exaggerate the graphical representation of data. Putting a ceiling on levels around points can, to some degree, solve this problem. However, another problem that isn't so easy to deal with appears when one point represents a very high level of contaminant and the overlapping circles must cover a large area. This is inaccurate because in nature contaminants will rarely spread in a consistent radius from a point. In most cases pathways such as water, gravity, or air in a specific direction will move the contaminant. But by applying polygonal thresholding methods as well and using the tools available in the SynCon system we should be able to create an accurate graphical representation. Some of the problems associated with point thresholding can be remedied by choosing polygonal thresholding for particular sites.

3.2 Polygonal Thresholding

In order to understand the concept of polygonal thresholding, one must first understand triangulation. Triangulation is a method of dividing up a large area into a series of triangles. In this case, the large area is a region of a map and the data samples on this map are the points defining the vertices of each triangle. A greedy method of triangulation is used which is a fast, efficient way to apply the problem to our data set. Given a set of N points, the triangulation is obtained by finding the shortest possible path between any two points and creating an edge. An edge can only be created if it doesn't intersect any other edge. The result is a spider web of non-overlapping triangles ranging in shape and size based upon the location of the measurement points. It is then up to the polygonal thresholding agent to determine what color should be assigned to each region to best graphically depict the contaminant level. The color or shade of gray applied by the agent is derived from the data at each of the three vertices defining the triangular region. The polygonal thresholding agent controls this job of evaluating the data and determining what color value each triangle should be represented by.

In a simple sense, the polygonal thresholding agent could simply assign a value to the region representative of the average contaminant levels found at each of the triangle's vertices. Of course, the methods for determining a contaminant level are more complex and many problems are encountered. Some of the many variables that must be taken into account by the intelligent agent when deciding what value to assign a triangle are the proximity of the points, geographic phenomenon, known pathways, accuracy of measurement, relative population, and time needed to complete remediation. The proximity of the points is important because when very few points are distributed over a large region or a few outlying points exist, it doesn't necessarily mean that the entire area of the triangle can accurately be represented by one color. To solve this problem the triangle might then need to be divided into smaller sections.

Geographic phenomenon such as glaciers, mountains, and rivers must be interpreted because if a contaminated area is geographically isolated, it may not spread as easily. Likewise, if a contaminant is measured in a known pathway it must be dealt with. The accuracy of the measurement must be scrutinized so that less accurate or biased tests can't throw off the other more legitimate data gathered. The populations of the triangulated region as related to its size is evaluated to help determine the urgency of the remediation. The time needed to complete the clean up process is also an important factor because longer cleanup efforts may need to be mounted with more advance notice. Using this information, accompanied by a much more intensive core set of variables, a decision-making tree or truth table can be constructed upon which to base the intelligent agent.

3.3 Intelligent Reasoning

A decision tree allows the agent to follow a simple predetermined process while applying advanced reasoning techniques. This is a formula based way to approach the problem solving effort and although it would appear simple at first brush, it actually takes quite a while to set up an accurate tree without having a computer generate it. A decision tree starts at the root with an initial, and ideally the most important, question. Then depending on how each question is answered the agent will step down each branch to another question until ultimately reaching a final conclusion. Keeping in mind that each question often has many possible answers and that each answer leads to another question, the decision tree can become quite large quite quickly. In some cases, even using a computer to generate and transverse an accurate tree of considerable size is still beyond the limits of our technology. Chess is a game of limited moves and for each move one player makes it might provide his opponent with several dozen more. Needless to say this extreme growth rate makes the tree too cumbersome for computers to efficiently deal with in its entirety. However, a simpler game such as tic-tac-toe, can be solved easily and no well-programmed computer should ever lose to a human opponent.

One method used to simplify complex trees is to use a most desirable outcome method. Rather than using the entire tree at once when a question is encountered, the agent steps down a few levels and determines which answer will have the most favorable result assuming that this will be most likely to lead to the most favorable conclusion. Getting back to our problem of determining the contaminant level using the additional variables, it is important to try approaching the questions in order of importance. Starting with a question pertaining to the proximity of the vertices would be ideal because if it is determined that the points are too distant, the triangle can be broken down into more manageable size. From this point it is up to the programmer and their advisors to construct a realistic decision tree that will follow a logical series of questions and eventually evolve into what is called an expert system.

Expert systems are computer systems built specifically to supply a user with the knowledge and reasoning of an expert in a specific field. An expert system uses facts and rules in the form of a knowledge base to decide what assumptions or conclusions to make under a given circumstance. Computers are excellent tools for sifting through these rules and solving a problem it can comprehend, however, they often perform poorly with attempts to replicate the human approach. The human approach being an intuitive, intelligent guess process involving pattern recognition. Humans don't perform well in the number crunching, pixel pushing method computers employ and computers don't always perform well in the seat of the pants manner humans rely on. Neither works well in the shoes of the other but together they have both strengths. In the case of the SynCon project, a knowledge base takes existing knowledge from many sources and compiles it into one intelligent, reasoning system. Human experts in many scientific fields of study will then have the power to access and expand upon the system providing an amazing potential for growth. Scientific community aside, the biggest step this expert system takes is in the political world. Decision makers with no prior knowledge of a contaminated area can be visually presented with knowledge which will not only be informative but which will have already factored in several key political elements. Scientists will save time explaining complex charts and researching archived data which will allow for more time in the field.

4. FUTURE RESEARCH

Future additions to the SynCon project will further deal with the complex problems presented by pathways and other specific geographic features as well as addressing issues involved in more accurate extrapolation of temporal data. With a very useful product on its way, the question of distribution in a viable medium is also raised and answered. An Internet interface is being developed to provide scientists with an interactive forum for exchanging data while updating the knowledge base of the SynCon project. This is the ideal environment for it because it can exist as a dynamic, expanding system. However, some of the most affected populations, including the indigenous populations in many of the affected Arctic regions do not have timely and reliable Internet access. For these populations, a CD version is being produced. Unfortunately, with a CD version, most information will only flow in one direction but with the rapid advances in DVD technology it might actually be realistic to expect that a read/write disc will become a realistic option for these populations. With the combination of Internet distribution and a disc format this decision making tool may become one of the leading experts in the field of environmental conservation.

ACKNOWLEDGMENTS

The SynCon research team and subteams gratefully acknowledge the United States Office of Naval Research, Arctic Nuclear Waste Assessment Program and the Arctic Research Initiative Program for providing support for this project.

REFERENCES

Arctic Environmental Reference Database Workshop. Workshop Proceedings of the USGS/AEDD and UNEP/GRID Workshop, Arendal, Norway, Sept. 1-3, 1993.

Careau, H. and Eric Dewailly. The Northern Aquatic Food Chain Contamination Database: a research tool. The Science of the Total Environment 160/161 (1995) 530-543.

George, S.; S. Hills, K. Nance, C. Packett, O. Parsons and G. Weller. 1995. "Synthesizing and Assessing Arctic Contamination Data." American Association for the Advancement of Science Conference. 1995.

George, S.; K. Nance; and O. Parsons. 1996. "Development of a Prototype Georeferenced Knowledge Base for Monitoring and Assessing Contaminants in Arctic Alaska." Proceedings of the American Society for Photogrammetry and Remote Sensing, Alaska Region Annual Surveying and Mapping Conference.

Nance, K.;O. Parsons; and S. Hills. 1996. "SynCon - A System for Synthesizing and Assessing Arctic Contamination Data." Proceedings of the International Conference on Circumpolar Health.

Nance, K. (1997a) "Applying AI Techniques in Developing Plans for Formerly Used Defense Sites (FUDS)" Journal of Mathematical Modeling and Scientific Computing, Vol. 8. (In Press)

Nance, K. (1997b) "Synthesis of Heterogeneous Data Sets" Proceedings of the 9th Annual Software Technology Conference.

Nance, K. (1997c) "Decision Support for Data Mining" Proceedings of the SCS 1997 Simulation Multiconference. 1997.

Olson, D.L. and J.F. Courtney, Jr. 1996. Decision Support Models and Expert Systems. Macmillan Publishing Company, New York, New York.

Parsons, O. "SynCon - A System for Synthesizing and Assessing Arctic Contamination Data." Department of Mathematical Sciences Technical Report 97-01. 1997.

Race, P., and K. Nance. "Synthesizing and Assessing Arctic Contamination Data." Proceedings of the American Society of Business and Behavioral Sciences: Information Systems. February 20-26, 1997.

Tedor, J. and K. Nance. "Developing Decision Support Tools for Environmental Data Organization." Proceedings of the American Society of Business and Behavioral Sciences: Decision Sciences. February 20-26, 1998.

PHYSIOLOGICALLY-BASED PATTERN ANALYSIS

Claudia Kropas Hughes

*Air Force Research
Laboratory, Materials and
Manufacturing Directorate
AFRL/MLMR
2977 P Street
Wright Patterson AFB, OH
45433-7765*

Steven K. Rogers, Ph.D.

*Battelle Memorial Institute
505 King Ave.
Columbus, OH
43201-2693*

Matthew Kabrisky, Ph.D.

*Air Force Institute of
Technology, Department of
Electrical Engineering
2950 P Street
Wright-Patterson AFB, OH
45433-7765*

Mark E. Oxley, Ph.D.

*Air Force Institute of
Technology, Department of
Mathematics
2950 P Street
Wright-Patterson AFB, OH
45433-7765*

Abstract: The digital computer is a wonderful tool for numeric calculations at incredible speeds, but humans are still vastly superior in pattern recognition and classification of complex visual image patterns. An understanding of how our visual system operates has the potential to inspire us to better, more efficient means of automatic, computer aided pattern analysis of images. Scale, shift and rotations, are easily compensated for within the Human Visual System, and the scene is understood even when presentation is made for the first time. A study of the models for the Human Visual System provides insight into pattern analysis techniques for automatic pattern recognition and classification.
Copyright © 1998 IFAC

Keywords: Human perception, Image analysis, Visual pattern recognition, Models,

1.0 Introduction

Image analysis is a very complex process. The idea of manipulating data values, that represent 'objects' within an image, is difficult and typically requires specific a priori information to make valuable judgments. Mathematically, one means to automate this process, is to map the images to a space where the relevant complexities are maintained, and the redundant information is eliminated. Finding that space, where the features of interest are fully discriminated, is the challenge.

Knowing that the human is the best system for pattern recognition and classification of complex visual images, is the inspiration for automated/computerized pattern analysis. Humans can readily recognize, classify, down select and correlate features from different images, in spite of translations, rotations and scalings of the original data. This paper explores the concepts of the Human Visual System as a means to find a working feature space, and the means to perform various computational image/pattern analysis.

1.1 Mathematical Modeling of the Eye

The Human Visual System is the best system for recognizing shapes and interpreting shapes and forms, pattern recognition and classification. With this as the

inspiration, much work has been done into the understanding of how the Human Visual System works in conjunction with image processing.

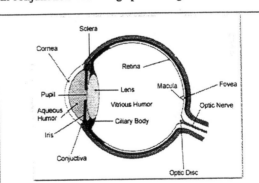

Figure 1. Anatomy of the Eye, top, (NOSTRA) and Visual System Pathway, bottom, (Bednar, 1977).

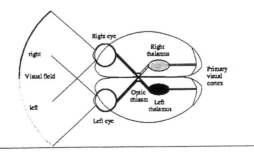

Figure 1 shows the anatomy of the eye and the visual system path through the brain back to the cortical area.

The optics of the eye, which includes the pupil, the lens, and the ocular media, will attenuate any high spatial frequencies in the input signal. At the back of the eye, the optical signal is sensed by an array of neurons in the retina. The retina has coarse and fine resolution sampling. The fovea is the high resolution/fine focus area on the retina. In the foveal region, there is a one-to-one correspondence between received signal and the transmission of the signal to the thalamus and beyond to the primary visual cortex. The off-foveal region is a coarse sort, and the signals are multiplexed through a limited number of nerves back to the thalamus and primary visual cortex (compression on the order of 100:1 in relation to the number of photoreceptors versus the number of optic nerve fibers). The split at the thalamus and into the primary visual cortex is a one-way information flow from the eyes to the different regions of the primary visual cortex and distributed to the cortex.

The work that has been done in this century to understand the functionality of the visual system was spearheaded by the work of W.J.S. Krieg (1953), H.G. Dusser de Barenne (1942) and W.S. McCulloch (1951). Krieg developed a dissection and reconstruction technique which enabled the diagramming of the major pathways in the visual system. Around that same time, H.G. Dusser de Barenne and W.S. McCulloch developed the technique for neuronography which determines the direction of information flow in the pathways. This work validated the work by Krieg, and the two techniques together were instrumental in understanding the one-way information flow, and the complex interconnections within the cortical regions as well as those between thalamus and cortex. (Kabrisky, 1966)

From the work by Krieg, Dusser de Barenne and McCulloch, the information flow from the eyes to the thalamus to the primary visual cortex out to the areas of the cortex (the great mass of brain) has been diagrammed, as shown in Figure 2. (Kabrisky, 1966)

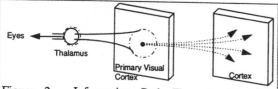

Figure 2. Information Path Through the brain (Kabrisky, 1966)

The information paths are in a two dimensional space, and a single dimensional cross-view is shown in Figure 3.

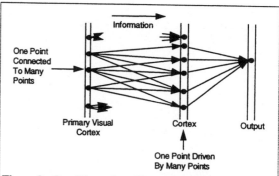

Figure 3. One Dimensional Representation Information Paths Through the Brain (Kabrisky, 1966)

The important point to notice is that from the primary visual cortex, one input is connected to many points in the cortex, and many points in the primary visual cortex are connected to one point in the cortex. In both directions, the areas of interconnection are well localized. A combination of these many points results in a singular output. This understanding of the method of interconnection of the cortical regions, made a strong case for a model representation of the Human Visual System. The model for the Human Visual System that was proposed is that the connectivity between the primary visual cortex to a cortical area of the brain and finally to another region of the cortex, bringing it back to a single point, enables the system to perform a two dimensional cross-correlation (output) (Kabrisky, 1966).

2.0 Perceptual Space Concepts

The properties that have been demonstrated experimentally (Krieg, 1953, Dusser de Barenne, 1942, McCulloch, 1951 and McLachlan, 1962) are the starting point for a mathematical model of the Human Visual System. One of models, postulated by M. Kabrisky, states the properties that the functional transformations must include. One such transformation that will effectively mimic the processes of the Human Visual System is the Fourier transform. In other words, according to Kabrisky, the Human Visual System works in Fourier space. The use of Fourier analysis as an interpretation of the Human Visual System has been accepted and expounded. (Kabrisky, 1966; Maher, 1970)

In Fourier analysis, the information in one part of a signal is spread throughout the entire system. A local characteristic of the signal becomes a global characteristic of the transformed signal.

Examination of Fourier analysis has led to further investigations with other transform medium, specifically wavelet transforms. Whereas, Fourier analysis transforms an image from a function of one variable (e.g. time), to a function of one variable (e.g. frequency), the wavelet transform produces a function of two variables (e.g. time and frequency). Wavelets are perceived as a means to mimic the Human Visual System because of the multiresolution concept: going from coarse resolution for the overall shapes, to higher resolutions for increasingly finer detail.

2.1 Fourier Analysis

In the 60s, optics was viewed as a means to non-computationally provide a fast, efficient technique to correlate images for pattern recognition. The Fourier transform is readily adaptable to an optical homolog. The optical homolog is a configuration of optical ray tracings, on corresponding collecting screens, and the distribution of light on the collecting screen describes a phase and amplitude distribution. This optical technique performs the well-known cross-correlation function. The process will output an image where the bright spots highlight where the pattern from one image is present in the second image. Since the mathematics of diffraction is the same mathematics that is used in Fourier analysis, the optical technique correlates with the cross-correlation that is achieved in Fourier analysis. This is the same as the calculation of the Fourier transform of a specific pattern, and cross-correlated to another image; maximum values will appear where the patterns are the same. If displayed as an image, the output image will have bright spots in the locations where the patterns match. So you can immediately see where there is a match between two image features. That is, take a feature of an image and convolve it (in Fourier space with the original image) and then a bright spot appears where that feature exists in the image (McLachlan, 1962). ·

Based on the Kabrisky model, and again translating to an optical homolog, it has been shown that this Fourier model, using a two dimensional Fourier transform as an approximation for the optical distribution, effectively correlates to a subjective human evaluation/comparison of forms. The experiment performed involved a series of test subjects quantifying the distance between the 'similarity' of images of animal crackers. The machine, using low-pass filtered two dimensional Fourier transforms, calculated the distances between the animal cracker image patterns. The extent of correlation between the test subject mean values, and the Fourier outputs was calculated to be between 0.721 and 0.970, indicating a high correlation between the human outputs and the computer simulations. This work demonstrated a statistically significant correlation of human pattern perception and machine pattern classification using Fourier transforms (Maher, 1970).

Since the cross-correlation function produces a maximum where the functions are the same, this forms a definition of 'similarity' between the images. With this concept, the Fourier transform of two separate images, cross-correlated, is an indication of how similar or dissimilar. Since similar images will have the same (or approximate) transform values, then the distance between the Fourier transform of different images is an indication of similarity. In other words, the Fourier coefficients of images will cluster together when they are of similar form.

Specifically, given two images, taking the Fourier transform of each, and then measuring the Euclidean distance between the normalized Fourier coefficient vectors will give a measure of how 'similar' or 'dissimilar' the images are to each other. It is important to normalize each of the vectors to lie on the unit sphere. This will allow a comparison on an "even playing field", and also, when normalized to the unit sphere, ensures that the maximum distance between the images will be 2.0. Therefore, thresholds can be applied that determine 'similar' versus 'dissimilar'. So, a simple distance metric can be used to determine the similarity of two images. The Euclidean distance between images x and y is:

$$Dist(X,Y) = \sqrt{\sum_{i,j}(X_{i,j} - Y_{i,j})^2}$$

where $X = \mathcal{F}\{x\}$ and $Y = \mathcal{F}\{y\}$ and the i, j values incorporate only the dominant low frequency Fourier coefficients of the original images.

Clustering in the Fourier Domain. One area where there has been extensive work on clustering or classifying to different classes, is the area of handwritten character recognition. Figure 4 shows representative images of handwritten digits. The data is 32x32 pixel images of the digits, 196 exemplars of each digit, taken from several different writers. A subset of 20 images of each of these first four digits were used to demonstrate some of the Fourier concepts.

Using the concept of close in Fourier space correlates to close in the Human Visual System space, a simple metric to check for 'similarity' between the images, is Euclidean distance.

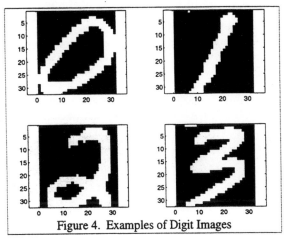

Figure 4. Examples of Digit Images

Calculating the distances between the various digits provides the following table.

Digits	0	1	2	3
0	0	1.1948	0.3573	0.4716
1		0	1.15	1.1127
2			0	0.3603
3				0

The distance between the normalized Fourier coefficients can only have a maximum value of 2.0.

The distance values closest to 0.0 would indicate identical images; distance values close to 2.0, would indicate completely dissimilar. The images above demonstrate this concept. The identical images have a distance of 0.0, and the 0 - 2 combination of digits, and the 2 - 3 combination of digits, which are somewhat similar in shape, have a distance much closer to 0.0 than to 2.0, and the very different shapes, such as 0 - 1 have a distance value over 1.0. As shown in the experiment by Maher, these distance measures correlate to the way the human would view and quantitatively measure similarity in the images.

This concept can also be used to classify various images as similar or dissimilar. A clustering algorithm is a very effective classifier for Fourier coefficients, since close in Fourier means close in Human Visual System. A k-means clustering algorithm, using a Euclidean distance metric is sufficient to identify which images are similar to another. A k-means clustering algorithm processes the data as follows. The first codeword is estimated by the mean of all the data. Each data vector is compared to the first codeword, distance-wise. Then the cluster center is split in two. Each of the data vectors are then distance measured to each of the new cluster centers, and the data vectors are partitioned to the cluster center to which they are closest. The new cluster centers are calculated by taking the center (average) of the data vectors in the partition set associated with that cluster center.

So, to demonstrate the classifying/clustering of images using a k-means algorithm, images of each digit, from the same database as those shown in Figure 4, were used as input to the clustering algorithm.

The input data samples were 20 different images of each of the four digits: 0, 1, 2, and 3. Each 32x32 pixel image was transformed to Fourier space. Each Fourier image was truncated to the 3x3 kernel of coefficients centered around the DC value. Because of the symmetry of the Fourier Transform, each input data vector needed to contain only the 9 distinct real

and imaginary coefficients. So, the input to the clustering algorithm was 80 data vectors, 9 coefficients in length. When four cluster centers are stipulated, Figure 5 shows the reconstruction from the cluster center vectors.

The clustering algorithm very effectively discriminates between the different forms of the different digits, while using a severely truncated Fourier representation. This is as expected, since Fourier coefficients are supposed to cluster to similar forms.

When eight cluster centers are stipulated, Figure 6 shows the reconstruction from the cluster center vectors. Notice that the codewords reconstruct to multiple codewords of the same digit. In other words, the codewords are duplications of each other, and the clustering is broken into more cluster centers than needed for accurate clustering. This makes sense in the way that the clustering algorithm is performed, namely that the clustering algorithm forces a splitting of the codewords, when the number of codewords is overspecified.

In summary, the energy present in the Fourier coefficients 'clusters' to similar forms and patterns, close in Fourier energy space, means close in the Human Visual System. This is true both for using all Fourier coefficients and for using a truncated kernel of Fourier coefficients, with distance as the metric to determine similarity.

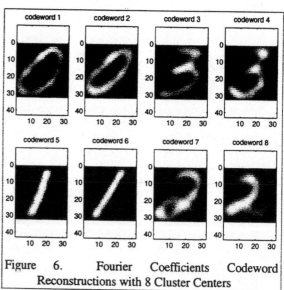

Figure 6. Fourier Coefficients Codeword Reconstructions with 8 Cluster Centers

Resizing of Images. Another advantage of using the Fourier space, is the ability to reconstruct the images to the same spatial planes, with the same pixel sizes between dissimilar sized images. This would assist on the problem of the dissimilarity in spatial resolution, pixel spacing and sampling between multiple images from multiple sources. For example, suppose that one sensor image has 512x512 pixels covering 270mm square, and a second sensor image of the same object, has 256x256 pixels. To allow easier computation between these images, they should be the same matrix size. To make an image a different matrix size, take the

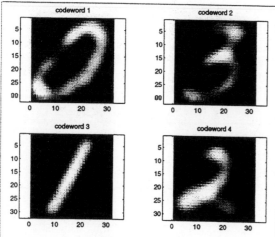

Fig. 5. Fourier Coefficients Codeword Reconstructions with 4 Cluster Centers

Fourier transform of the image, and reconstruct the inverse Fourier transform to the matrix size of interest. For reducing larger images, there is some loss associated with truncating the highest frequency terms, but this loss is insignificant in the Human Visual System sense, as the reconstructed image will look the same. Figure 7 shows images of a CT head scan. The original image, on the left, is a 512x512 pixels matrix size, and the image on the right is the reduced image to 256x256 pixels matrix size.

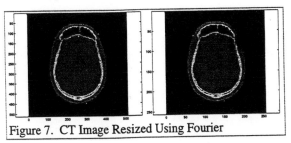

Figure 7. CT Image Resized Using Fourier

This method of changing matrix sizes is only changing the matrix size of the image, and that means changing the pixel size of the representation of the image. This method does not effect the underlying resolution of the data, which is based on the sensor equipment that generated the data.

2.2 Wavelet Analysis

Evidence has been presented that the Human Visual System performs a sort of wavelet transformation to represent the visual environment. Wavelets are a mathematically-based intermediate compromise of the human eye's capability of performing Fourier analysis of the visual scene, as well as responding/ picking out specific edges. (Field, 1990) Whereas Fourier Transforms transform a signal of one variable, such as time or space, into a function of one variable, such as frequency, Wavelet transforms produces a function with 2 variables, such as time and frequency. At heart, Wavelets are performing the same signal analysis as Pyramid Algorithms do for Image Processing, Subband Coding performs for Signal Processing, and the Quadrature Mirror Filters do for Digital Speech Processing. The underlying commonality is the application of a succession of filters. (Hubbard, 1996)

By the application of a succession of filters, multiresolution theory, wavelets were linked to the filters used in signal processing. Multiresolution refers to breaking a signal into a coarse resolution image that provides the lowest frequency components of the signal, and then defining higher and higher resolution to provide the finer details of the signal. Each 'higher' resolution is referred to as an octave, which means from one resolution to the next resolution, it is twice as fine and will double the frequency of the wavelets, so that they encode frequencies twice as high. Wavelets are of constant shape with changing dilation, which results in resolution, scale and frequency changes, all at once. (Hubbard, 1996) As a result of the application of a succession of filters, wavelets are harder to interpret than Fourier; there is no direct physical correlation of the data. (Field, 1990)

Gabor Transforms. Gabor expansion, like wavelet analysis, is considered a special frame. Gabor expansions are used in image processing because they reveal the time-frequency distribution of the image signal (Pei and Yeh, 1997). A Gabor transform is basically a windowed Fourier transform when the envelope for the window is a Gaussian. For this analysis, the size of the 'window' is fixed for each analysis, but the frequencies inside the window varies. For this filter the variables are frequency and the position of the window. The smaller the window, the more time information that is obtained, but some low frequency information is lost. The larger the window, the more frequency information, but the less precise the information about time. (Hubbard, 1996) The Gabor filter is basically a series of band pass filters that cover the entire image and the size of the window determines the predominant precision in the frequency or time arena.

The following Figure 8 shows an example of the series of band pass filters that make up a Gabor filter. Each window of the filter incorporates only a limited band of frequency and spatial information. Figure 8 also shows the overall Gabor filters.

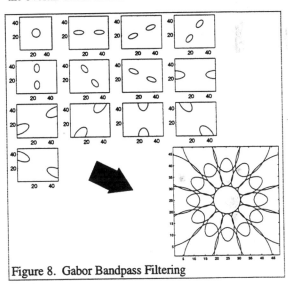

Figure 8. Gabor Bandpass Filtering

Visual Difference Predictor (VDP). An application of the Gabor filter in image processing, inspired by the Human Visual System concepts, is determining a quantitative value for image fidelity. There are two means of measuring images: (1) absolute fidelity or quality of one image, and (2) a relative measure of fidelity between two images, i.e. measuring how close a distorted image is to the original.

Extensive work has been done in the area of measuring perceptual image fidelity, using a Human Visual System model approach. (Daly, 1993 and Martin, 1996 and Heeger and Teo, 1995, and Westin, et al 1995) The earliest work was based on generating a single number that represented the comparative image fidelity between two overall images. The later work is based on an approach to produce a map that specifies the perceptual differences between the two images. This provides the ability to determine which

261

regions/sections of the image are similar and dissimilar. There are several recent approaches to quantifying the perceptual image fidelity between two images. One of these approaches is the Visual Difference Predictor (VDP) (Daley, 1993 and Martin, 1996).

A common basic processing structure for the fidelity measures between two images is depicted in Figure 9 (Martin, 1996). As can be seen in the figure, the images are processed by the psychophysical Human Visual System model; the retinal stage that processes the image by an amplitude nonlinearity and then a contrast sensitivity filter (linear filter). Following the retinal stage is the cortical stage, which consists of a space/spatial frequency transform across multiple frequency bands, these are called the cortex bands. These cortex bands are calculated using Gabor filters as the means to achieve the bandpass filtering. This is followed by differencing, masking, and detection operations, within each cortex band, which is the comparative stage between the two images. The output is then a map or Fidelity Measure between the two images being compared.

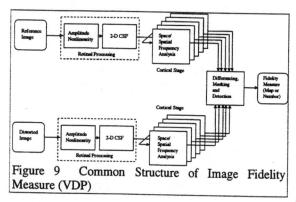

Figure 9   Common Structure of Image Fidelity Measure (VDP)

The VDP algorithm produces a map where the visual differences can be perceived between two images. The result of the VDP algorithm, *P(c)* is a probability of detecting a signal of contrast *c*. Where this contrast, c, is detectable, is the psychophysical threshold where a human would subjectively say those pixels are not the same.

This Fidelity Measure is a quantitative metric to evaluate the 'similarity' between two images, an original and a distorted version, on a pixel-by-pixel basis. This map highlights where the associated pixels between the images are similar by a human perceptual threshold, i.e. where a human would evaluate the pixels as 'close enough to be the same'.

3.0 Conclusions

Using the understanding of perception and Human Visual System, provides us a basis for using Fourier analysis and Wavelet theory, as a starting point for feature extraction. Processing of the pixel data directly, allows an association/correlation of techniques and keeps the entire process very understandable, as well as computationally   intensive,   and   sometimes

computationally intractable. Logic dictates, that if only specific features are of interest or concern, then image data should be reduced to just those necessary features. Data reduction , however, is not a trivial problem. The means for feature extraction utilizing a priori information and established understanding of certain techniques can alleviate some of the problems, both with the pattern analysis and the computational tractability.

4.0 References

Bednar, James A. "Tilt Aftereffects in a Self-Organizing Model of the Primary Visual Cortex", Master's Thesis, University of Texas at Austin, January 1977.

Dusser de Barenne, H. G., et al. "Physiological Neuronography of the Cortico-Striatal Connections", Association for Research in Nervous and Mental Disease, 21: 246-266, 1942.

Field, D.J. "Scale-invariance and Self-similar 'Wavelet' Transforms: and Analysis of Natural Scenes and Mammalian Visual Systems", Wavelets, Fractals, and Fourier Transforms, eds. M. Farge, J.C.R. Hunt, and J. C. Vassilicos, Oxford: Clarendon Press, 1990. Pp. viii, 151-193.

Hubbard, Barbara Burke. The World According to Wavelets. Wellseley, MA: A K Peters, Ltd., 1996. Pp. 137-152, 187-202.

Kabrisky, Matthew. A Proposed Model for Visual Information Processing in the Human Brain. University of Illinois Press, Urbana and London, 1966.

Krieg, W. J. S. Functional Neuroanatomy. New York: The Blakiston, Co., Inc., 1953.

Maher, Frank A. "A Correlation of Human and Machine Pattern Discrimination", NAECON '70 Record, pp. 260-264.

Martin, Curtis E., Captain USAF. 1996. Perceptual Fidelity for Digital Color Imagery. Ph.D. Dissertation, Dept. of Electrical Engineering, Air Force Institute of Technology.

McCulloch, W.S. "Why the Mind is in the Head". The Hixon Symposium. John Wiley and Sons, Inc., 1951.

McLachlan, Dan. "The Role of Optics in Applying Correlation Functions to Pattern Recognition", Journal of the Optical Society of America, April, 1962, Volume 52, Number 4, pp. 454-459.

NOSTRA - Naval Ophthalmic Support and Training Activity: http://nos40.med.navy.mil/main.htm

Pei, S.C. and M. H. Yeh. "An Introduction to Discrete Finite Frames", IEEE Signal Processing Magazine, Nov 1997, pp. 84-96.

Westen, SJP, et al. "Perceptual Image Quality Based on a Multiple Channel HVS Model", Proceedings of the 1995 International Conference on Acoustics, Speech, and Signal Processing, pp. 2351-2354.

ENHANCING FLIGHT SAFETY: RECOVERY FROM WINDSHEAR DURING TAKE-OFF

M. Aznar Fernández-Montesinos[1], G. Schram[2], H.B. Verbruggen[2], R.A.Vingerhoeds[1*]

[1] *Delft University of Technology, Faculty of Technical Mathematics and Informatics*
Zuidplantsoen 4, 2628 BZ Delft, The Netherlands. Fax: (31-15) 2787141
[2] *Delft University of Technology, Faculty of Electrical Engineering*
Mekelweg 4, 2628 CD Delft, The Netherlands

Abstract: Flight safety during normal as well as during hazardous weather conditions is a key issue in modern flight control systems. From all weather phenomena, low-altitude wind shear remains a particular concern to flight safety. Aircraft might not be able to cope with the energy loss due to a windshear encounter. A fuzzy flight controller that employs total energy concepts has been extended to improve safety during windshear. Through incorporating a special windshear recovery procedure, the fuzzy controller provides aggressive thrust management and flight envelope protection. The application of this generic windshear recovery concept in a take-off scenario is demonstrated by simulation examples. Copyright © *1998 IFAC*

Keywords: Windshear Recovery, Aircraft Control, Fuzzy Control.

1. INTRODUCTION

Adverse weather conditions such as windshear are a major concern to flight safety. Windshear has been the cause or contributing factor for several major accidents and many incidents. The generic term windshear refers to a rapid change in the wind speed and/or direction (Fujita 1980).

The low-altitude *microburst*, the worst and most dangerous form of windshear, is produced by a strong sudden downdraft of cool air, which strikes the ground, producing winds that spread out in all directions. An aircraft flying through a microburst during take-off will go through three stages, as depicted in figure 1. First, the aircraft will encounter a performance increasing headwind as it enters the forward outflow of the microburst. The pilot typically responds by reducing the engine power to maintain the proper flight glidepath and airspeed. As the aircraft approaches the core of the microburst, it will experience a strong downdraft which will force the aircraft to descend. This downdraft is followed by a sudden strong tailwind. This results in a sudden loss of either altitude and/or airspeed.

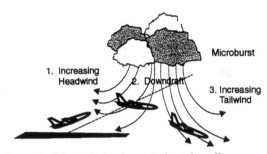

Figure 1: Effect of microburst during take-off

Numerous methods, techniques and technology in terms of how to avoid, escape and recover from these phenomena have been proposed by the airlines, aerospace industry and research institutes (FAA 1987, Bowles 1990, Miele et al. 1987, Oseguera and Bowles 1988, Zhao and Bryson 1990, Psiaki and Stengel 1991, Psiaki and Park 1992). As windshear may exceed the performance capability of the aircraft, windshear avoidance has been emphasized. Recent advances in forward-looking detection systems, which identify the hazard in advance, make timely recognition and avoidance possible. Despite the advances in the detection area and due to the short-

---

[*] R.A. Vingerhoeds is currently with Siemens Automotive SA, Toulouse, France.

term transient nature of the phenomenon, it is essential to help pilots and aircraft respond properly in the case of an encounter. Guidance and control should be provided to cope with the energy loss associated with windshear.

In this paper, a generic windshear recovery concept is proposed using fuzzy logic (Aznar et al. 1997, Schram et al. 1998). The assumption is made that the microburst is detected but too late to avoid an encounter. A fuzzy logic controller based on energy principles has been extended to cope with the problem. The energy-based controller suits well the windshear problem as windshear is basically an energy management problem. The recovery module performs a dual role: flight path guidance and flight envelope protection. The protection mechanism divides the available energy into airspeed and altitude, depending on the margins with respect to minimum airspeed and altitude. The advantage of using fuzzy logic is that the fuzzy controller provides a framework in which special windshear recovery procedures can be easily incorporated to enhance performance and hence safety.

2. WINDSHEAR PROBLEM: ENERGY MANAGEMENT

Since the windshear affects in particular the longitudinal motion of an aircraft, the total energy balance in the vertical plane is considered (Miele et al. 1987). The airplane total energy is defined as the sum of air-mass relative kinetic energy and the inertial potential energy. The air-mass kinetic energy is used since airspeed describes the aircraft capability to climb or maintain altitude. Then, the aircraft specific energy (energy per unit weight), or *potential altitude* h_p, is defined as:

$$h_p = \frac{E}{mg} = \frac{V_A^2}{2g} + h \qquad (1)$$

where V_A is the airspeed, mg is the aircraft weight and h is the altitude. The potential climb rate is obtained by differentiating the above expression:

$$\dot{h}_p = \frac{\dot{E}}{mg} = \frac{V_A}{g}\dot{V}_A + \dot{h} \qquad (2)$$

The rate of change of airspeed and climb rate can be formally derived from the force equations of motion of the aircraft. By substituting the rate of change of airspeed and climb rate, the potential altitude can be written as:

$$\dot{h}_p = V_A \frac{T\cos\alpha - D}{mg} - \frac{\dot{w}_x}{g}\cos\gamma_A - \frac{\dot{w}_h}{g}\sin\gamma_A + \frac{w_h}{V_A}\sqrt{\ } \qquad (3)$$

where T is the thrust in the direction of the x-axis of the body frame, D represent the aircraft drag, $\dot{w}_x$ is the horizontal time derivative of wind velocity, $\dot{w}_h$ is the vertical time derivative of wind velocity, α is the angle of attack (the angle between the x-axis of the body frame and the air-mass direction of V_A), and γ_A is the air-relative flight path angle. The first term is the airplane's excess thrust-to-weight ratio. The subsequent three wind terms describe the windshear impact on the aircraft energy state. By combining the three terms and for small flight path angles, a single quantity called the F-factor (Bowles, 1990) can be defined as:

$$F = \frac{\dot{w}_x}{g} - \frac{w_h}{V_A} \qquad (4)$$

The F-factor indicates the influence of air-mass movement on the rate of change of specific energy. The parameter can be physically interpreted as the loss in available excess thrust-to-weight ratio (excess thrust) required to maintain steady flight conditions due to wind variations. Hence, it is a measure of the loss in rate of climb capabilities of the aircraft because of the windshear.

Considering equations (2) and (3), to maintain a constant climb rate despite wind variations (as required during landing and take-off missions), *tight control* of air-relative energy is generally recommended. Direct thrust compensation of the energy, up to full thrust is a straightforward way to cope with energy changes. However, a positive thrust change is generally limited by the maximum thrust-to-weight ratio

Since windshear is an energy management problem, the available energy and its *distribution* between altitude and airspeed play an important role. Different approaches have been proposed. The generic FAA recommended guideline of applying 15 degree pitch and full thrust can result in a small altitude loss, but at the expense of considerable speed loss which might be below the normal minimum. This might also lead to angle-of-attack excursions beyond the ones for stall warning. In other approaches, altitude is also clearly preferred above airspeed (Psiaki and Stengel 1991). On the other hand, only control of airspeed at the expense of altitude is also not desirable. This means descending to a low altitude with high airspeed like proposed in (Zhao and Bryson 1990). This is, however, not feasible (obstacles) and unacceptable for pilots. Therefore, it is clear that a *compromise* must be found like recommended in (Bailey and Krishnakumar 1987). However, the energy partition in their controller is constant, i.e., an a-priori a choice is made beforehand on how to distribute the available energy. Fuzzy logic can be used to provide a smooth and proper energy distribution according to each flight mission.

Apart from considering the energy distribution, the performance limits of the aircraft are crucial. Possible margins with respect to a stall situation and a minimum altitude (deviation) should be preserved.

3. A GENERIC FLIGHT RECOVERY CONCEPT

The basic philosophy of this recovery concept is that a compromise between airspeed and altitude is pursued continuously during the encounter with respect to the associated safe operating altitude and stall speed margins, *regardless* of the specific flight mission, the guidance strategy, and the microburst intensity (Aznar et al. 1997). This protection mechanism can be easily integrated in a fuzzy controller based on total energy principles (Schram and Verbruggen 1997). This energy-based controller incorporates control engineering and pilot (total energy) concepts through the use of fuzzy logic. Following the total energy concepts (Lambregts 1983), thrust (T) (see equation 3) is used to increase the aircraft total energy (E) while pitch angle changes controls the exchange between kinetic and potential energy. Another advantage of this controller is that through the use of linguistic descriptions, the controller laws are easy to interpret. And, due to its rule-based structure, special procedures to cope with exception handling can be easily incorporated. The fuzzy controller has been extended with a special recovery procedure for the windshear problem.

The main idea of this windshear recovery concept is to monitor and guard the flight envelope independent of any flight mission. This is achieved by the introduction of a priority scheduling function, which indicates to what extent altitude or airspeed have priority. Depending upon the altitude and airspeed limits, energy distribution takes place. While this priority is used to divide the energy, thrust is used to increase the energy of the aircraft. In this way, proper energy management and distribution are conveyed. Furthermore, the use of fuzzy logic provides a smooth energy distribution.

In Figure 2, the extended controller for windshear recovery is shown. The aircraft block represents the aircraft dynamics including an inner loop pitch angle controller. The fuzzy controller block represents the longitudinal fuzzy controller that has been developed

in (Schram and Verbruggen 1997). Two inputs to the controller are velocity error V_e (the difference between a reference airspeed $V_{A,ref}$ and the actual aircraft speed V_A) and altitude error h_e (the difference between a reference altitude h_{ref} and the actual altitude h). The airspeed is given by V_A while inertial speed or groundspeed is given by V. The control actions are the throttle settings δ_{TH} (= δ_{TH1} = δ_{TH2}) which result in changes in thrust and the pitch angle command θ_C. The modifications of the controller are explained in the next two sections.

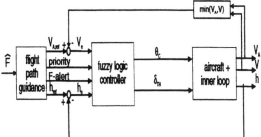

Figure 2: Block-schematic representation of the extended longitudinal outer loop controller.

3.1 Tight control

The tight control of energy is initiated by the new input F-alert. This variable indicates the encounter of a microburst windshear based on an estimate of the F-factor (from a reactive detection system). This variable is used to select either the normal or a new defined aggressive rule base for the thrust defined for the case of a microburst encounter. The new rule base guarantees a tight tracking of the air-relative energy. Assuming that a windshear detection system is present, the F-alert is activated if the estimated F-factor exceeds a certain threshold. The FAA suggests a threshold value of 0.10. When the core of the microburst is passed and the F-factor decreases below the threshold, the alert is slowly deactivated.

For the F-alert, two membership functions are defined such that a smooth interpolation is achieved for values between 0 and 1. The membership functions are chosen such that the sum of the membership values equals one and a linear interpolation is achieved. In Figure 3, the two membership functions are depicted:

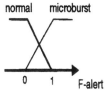

Figure 3: Membership functions for F-alert controller input

Then, as a function of the windshear alert, the throttle rule base is selected by means of fuzzy rules. If F-alert = 0, the normal rule base is selected. On the other hand, if F-alert = 1, the aggressive rule base is chosen. The aggressive rule base has the same antecedent conditions as the normal case but the consequents are more aggressive. In fact, the gains of the control laws are implicitly increased. As an example, given the following rule from the normal controller (Schram, 1998):

IF normal situation AND airspeed error is positive medium (PM), i.e., too slow AND altitude error is zero (ZE),
THEN throttle command is positive small (PS).

By using similar antecedent conditions but choosing more aggressive consequents, more aggressive throttle commands are given in case of a microburst encounter. Considering the same example, the following rule is obtained if the alert is activated:

IF microburst encounter AND airspeed error is positive medium (PM) AND altitude error is zero (ZE),
THEN throttle command is positive big (PB).

By employing the triangular shaped membership functions and by modelling the AND connectives by a product operator, a gradual and linear interpolation between the rules is achieved. In this way, a smooth gain-scheduled controller is implicitly obtained.

Another modification of the controller is the inclusion of the minimum airspeed/groundspeed strategy (Psiaki and Park 1992). The minimum of airspeed and groundspeed (see figure 2) is controlled in order to prevent the initial thrust reduction due to the energy increase associated with the headwind. As a result, some extra energy is preserved and can be used at a later/critical stage of windshear.

3.2 Flight envelope protection

In a windshear situation, the flight envelope has to be monitored and guarded independent of any guidance. Critical situations when the aircraft is near stall speed and near a minimum altitude must be avoided by all means. Therefore, a priority scheduling function is integrated to distribute the energy between altitude and speed depending upon these safety limits.

In order to find such a priority function, the specific energy of the aircraft is considered, see equation (1). With respect to altitude and airspeed changes Δh and ΔV_A, the specific energy changes by:

$$\left(\Delta h_p\right)_h = \Delta h$$

$$\left(\Delta h_p\right)_{V_A} = \frac{1}{g}\left(V_A \cdot \Delta V_A + \frac{1}{2}\left(\Delta V_A\right)^2\right) \quad (5)$$

Considering a minimum airspeed V_A and a minimum altitude h_{min}, the priority function $f(h, V_A)$ is defined as:

$$f(h, V_A) = \frac{4}{\pi}\arctan\left(\frac{\left(\Delta h_p\right)_h}{\left(\Delta h_p\right)_{V_A}}\right) - 1 \quad (6)$$

with $f(h, V_A) = 0$ in case $\left(\Delta h_p\right)_{V_A} = 0$. The new variable $f(h, V_a)$ indicates the need for altitude ($f(h, V_a) = -1$) or airspeed priority ($f(h, V_a) = 1$). Clearly, the energy distribution is proportional to the ratio $\left(\Delta h_p\right)_h / \left(\Delta h_p\right)_{V_A}$ for the values of $f(h, V_A)$ which are in between -1 and 1.

For the priority variable, three membership functions are defined, see Figure 4. Each membership function belongs to a rule base for pitch angle commands. The membership functions are defined such that a linear interpolation is achieved between one of the special rule bases and the normal rule base.

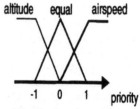

Figure 4: Membership functions for priority controller input.

Two special rule bases are defined for altitude and airspeed priority. The rule bases have the same antecedent conditions as in normal case (no windshear), but the consequents are different to convey proper energy distribution according to the priority activated at the encounter. The consequents are pitch angle changes. In case of airspeed priority, airspeed is controlled by pitch whereas with altitude priority, altitude is controlled by pitch.

4. SIMULATION EXAMPLE: RECOVERY DURING TAKE-OFF

In take-off, the pilot has no choice but to fly through the windshear once the aircraft is airborne. The application of this recovery concept for landing and abort landing scenario's have been addressed in (Aznar et al. 1997).

4.1 Windshear scenario

For the microburst simulation, the model of Oseguera and Bowles is used (Oseguera and Bowles 1988). In the simulations, the microburst parameters of this model are chosen in order to result in a maximum F-

factor of 0.26. In Figure 5, the horizontal and vertical wind velocities are shown as well as the F-factor along the intended *inertial flight path*.

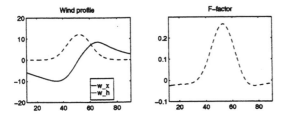

Figure 5: Horizontal and vertical wind velocities (left), F-factor (right).

The wind profile consists of a horizontal wind, a headwind (defined as negative) which rapidly changes into a tailwind (defined as positive). The vertical wind, the downdraught (defined as positive), is also depicted.

4.2 Determination of priority and F-factor

The minimum airspeed is defined as the stall speed (airspeed under this speed result in large drop in lift). The minimum altitude is initially determined as a slope deviation. When the aircraft reaches an altitude of 305 m, this altitude is chosen as the minimum altitude. In this way, obstacles near the airport can be avoided. In Figure 6, the take-off scenario is depicted. The slope of the flight path is 6 degrees. The minimum altitude is shown as a dashed line.

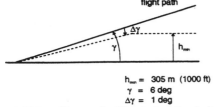

h_{min} = 305 m (1000 ft)
γ = 6 deg
$\Delta\gamma$ = 1 deg

Figure 6: Take-off scenario. The dashed line indicates the minimum altitude.

The windshear F-factor is directly computed from the simulated wind velocities and delayed by six seconds.

4.3 Aircraft response

For the simulations, the Research Civil Aircraft Model (RCAM) is used (Schram and Verbruggen 1997). In the simulations, only the longitudinal motion in the vertical plane is considered. An aircraft model with mass of 150,000 kg is trimmed on a flight path angle of 6 degrees with a groundspeed of 80 m/s. The initial altitude is 100 m. The stall speed for this particular configuration is 57.9 m/s. The excess thrust-to-weight ratio in this example is 0.04 since 90% of maximum thrust is already used. This is

much smaller than in the landing configurations but the aircraft is climbing with a large slope and airspeed. If the F-factor exceeds the threshold of 0.10, the tight energy control and the flight envelope protection mechanism are activated. If the F-factor is below the threshold value, a smooth transition to the nominal controllers is performed by smoothly scaling the priority with F-alert. In Figure 7, the simulation results are shown.

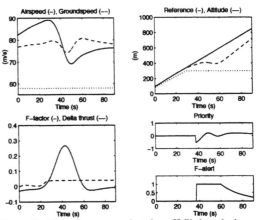

Figure 7: Aircraft response in take-off flight mission. The dotted lines indicate the stall speed and the minimum altitude in the respective figures.

In the first 20 s, air-relative energy is build up since groundspeed is controlled instead of airspeed. From 20 s to 60 s, the aircraft enters the microburst center. As the F-factor indicates the equivalent specific excess thrust needed to maintain steady flight and the maximum F-factor is 0.26 while the excess thrust-to-weight ratio is 0.04, the aircraft is forced to lose energy. During the encounter, energy is exchanged from the airspeed reserve to altitude as indicated by the priority values. An initial negative priority value of -0.5270 indicates altitude priority. Then, the priority switches smoothly from altitude to airspeed and goes smoothly to an equal distribution at the end of the encounter. Although energy loss is inevitable, both margins are preserved and flight envelope protection is achieved. Since the excess thrust-to-weight ratio is very small, application of full thrust remains after the microburst encounter to compensate the energy loss.

267

4.4 Varying microburst strength

In the following, the results of different microburst strengths are presented and discussed for the take-off scenario. The microburst strength is varied and the corresponding peaks of the F-factors vary between 0.25-0.35, respectively. In Figure 8, the airspeed and altitude responses are shown together with their reference and minimum allowed values.

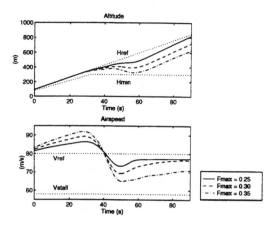

Figure 8: Aircraft responses for varying microburst in take-off scenario.

For all microburst strengths, both the airspeed and altitude margins are respected because of the flight envelope protection mechanism. For the increasing microburst strength, more energy is again automatically built up by tracking the groundspeed and the aggressive throttle. This energy is then used to compensate for the energy loss. However, because of the small excess thrust-to-weight ratio, large altitude drops cannot be avoided. Nevertheless, the aircraft stays within the flight envelope that has been defined in this take-off scenario.

5. CONCLUDING REMARKS

Low-altitude windshear remains a concern to aircraft safety. The problem is a loss of air-relative energy (potential and kinetic) due to the microburst windshear that has to be compensated for. A novel recovery concept shown in this paper conveys proper energy management. A fuzzy controller based on total energy principles is extended to provide flight envelope protection and aggressive thrust management during a windshear encounter. The modified fuzzy energy controller and the flight envelope protection are integrated as one fuzzy logic control system for both normal and windshear conditions. The recovery concept has been demonstrated for a take-off mission.

REFERENCES

Aznar Fernández-Montesinos, M., G. Schram, R.A. Vingerhoeds, H.B. Verbruggen and J.A. Mulder (1997). Windshear Recovery using fuzzy logic guidance and control. *AIAA Guidance, Navigation and Control Conference.* AIAA 97-3629. Also to appear in: *Journal of Guidance, Control and Dynamics* (1998).

Bailey, J.E. and K. Krishnakumar (1987). Total energy control concepts applied to flight in windshear, AIAA 87-2344.

Bowles, R.L. (1990). Windshear detection and avoidance: airborne systems survey, *29th Conference on Decision and Control,* Honolulu. Hawaii, U.S.A., pp. 708 - 736.

FAA (1987). Windshear training aid. U.S. Department of Transportation, Federal Aviation Administration.

Fujita, T.T. (1980). Downbursts and microbursts- An aviation hazard. *19th Conference on Radar Meteorology,* American Meteorology Society Honolulu, Hawaii, U.S.A., pp. 94 - 101.

Lambregts, A.A. (1983). Vertical flight path and speed control autopilot design using total energy principles, *Technical report,* AIAA 83-2239.

Miele, A., T. Wang and W.W. Melvin (1987). Optimization and acceleration guidance of flight trajectories in a windshear. *Journal of Guidance, Control and Dynamics,* vol. **10**, no. 4, pp. 368 - 377.

Oseguera, R.M. and R.L. Bowles (1988). A simple, analytic 3-dimensional downburst model based on boundary layer stagnation flow. NASA Technical Memorandum 100632.

Psiaki, M.L. and K. Park (1992). Thrust laws for microburst windshear penetration. *Journal of Guidance, Control and Dynamics,* vol. **15**. no. 4, pp. 968 - 975.

Psiaki, M.J. and R.F. Stengel (1991). Optimal aircraft performance during a microburst encounter *Journal of Guidance, Control and Dynamics,* vol. **14**, no. 2, pp. 440 - 446.

Schram, G., and H.B. Verbruggen (1997). RCAM: A Fuzzy Control Approach. In: *Robust Flight Control, A Design Challenge,* J.F. Magni, S. Bennani, J. Terlouw (eds.), Springer Verlag, Lectures notes in Control and Information Sciences 224, pp. 397 - 416.

Schram, G., M. Aznar Fernández-Montesinos, and H.B. Verbruggen (1998). Enhancing flight control using fuzzy logic, in. H.B. Verbruggen and H.Z. Zimmermann (eds.), *Fuzzy algorithms for control,* Kluwer, to be published.

Zhao, Y. and A.E. Bryson (1990). Optimal paths through downbursts. *Journal of Guidance, Control and Dynamics,* Vol. **13**, No. 5. Sept.- Oct. 1990., pp. 813-818.

MIMICKING A FUZZY FLIGHT CONTROLLER USING B-SPLINES

M. Aznar Fernández-Montesinos, R.A.Vingerhoeds[*] and H. Koppelaar

*Delft University of Technology, Faculty of Technical Mathematics and Informatics
Zuidplantsoen 4, 2628 BZ Delft, The Netherlands. Fax: (31-15) 2787141*

Abstract: A fuzzy logic flight controller is mimicked by using a B-spline neural network model where the inputs are represented by B-spline functions. Choosing the order or distribution of the basis functions are important steps in the design of a B-spline neural model. A set of weights, which is optimised during the learning process, controls the mapping performed by the multi-dimensional basis functions. The weights can be seen as fuzzy singletons representing the consequent of the rules, the local control actions. The fuzzy controller can be seen as a B-spline interpolator whose working can be also interpreted by linguistic rules. Copyright © *1998 IFAC*

Keywords: B-spline modelling, Fuzzy control, Neurofuzzy systems.

1. INTRODUCTION

Trying to learn a control function or generating a model of a process or identify fault conditions forms a complex problem. In many cases, implementation and maintenance of the knowledge used to solve these problems becomes complicated if only one technique is used. Instead, a combination of techniques, a hybrid approach, has to be used. The principle behind hybrid systems is that they enable a combination of the strengths of each technique if tacitly applied. In this paper, particular attention is given to the synergy of fuzzy logic and neural networks, known as neurofuzzy approaches.

B-splines have been proposed as basis for neurofuzzy systems in the so-called B-spline networks (Brown and Harris, 1994). The representation of the network inputs using B-splines produces a continuous piecewise polynomial output that is ideal for function approximation. The B-splines basis functions can be regarded as membership functions with some optimal properties, whose representation can be interpreted linguistically.

2. B-SPLINE NETWORK

A B-spline neural network is constructed from a linear combination of multi-dimensional basis functions, which are piecewise polynomial of order k. This type of model belongs to a class of general function approximators, called the basis function expansion (Friedman, 1991).

2.1 B-spline basis functions

Splines have been used for data-fitting problems and for representing curves and surfaces in computer aided geometric design (Bartels et al. 1987). B-splines, also referred to as basis spline curves, are piecewise spline polynomial functions (de Boor 1972, Schumaker 1981). In order to define a B-spline function, the order (or degree) must be defined as well as the knot set defining the intervals over which the basis functions are defined. The knot set is particularly of importance as it defines the range of values over which the basis functions are defined. A uniform B-spline is a B-spline whose internal knots are equally spaced. The shape of a univariate B-

---

[*] R.A. Vingerhoeds is currently with Siemens Automotive SA, Toulouse, France.

spline is defined by its order and the multiplicity of the knots. In Figure 1, B-splines of different orders (equivalent to degree+1) are shown. Order 1 corresponds to piecewise constant approximation (degree 0), order 2 to piecewise linear (degree 1) and order 3 to piecewise quadratic approximation (degree 2). B-splines provides a way of choosing and representing a set of membership functions. In fuzzy applications, 2nd order B-splines (triangular membership functions) are often used.

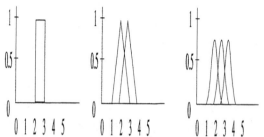

Figure 1: Examples of univariate splines of different order

It is the shape (order) and distribution of the basis function (partition of the input space) which determines the smoothness and flexibility of the resulting model.

B-spline networks produce approximate functions that are differentiable and vary smoothly. Given a set of simple for B-splines of order k, the spline is continuous up to and including its $(k-2)^{th}$ derivative. Hence, the order of a B-spline is chosen depending upon a required continuity. B-splines have the property that they form a partition of unity. As B-spline functions have a compact support, only a fixed number of adjacent B-splines functions contribute to the network. This reduces the computational complexity.

Multidimensional splines are constructed by tensor multiplication of n univariate splines. In this sense, the properties of the univariate B-spline functions are carried over to the multivariate case.

Spline functions can be described as linear combinations of weighed B-splines basis functions. Given a set of knots

$$x^{min} = \lambda_0 \leq \lambda_1 \leq \cdots \leq \lambda_{r_i} < \lambda_{r_i+1} = x^{max}$$

a spline function can be represented as:

$$S(x) = \sum_{j=0}^{n} c_j N_k^j(x) \qquad (1)$$

where $N_k^j(x)$ is the jth B-spline basis function of order k and the c_j are the B-spline coefficients. These coefficients are in fact the so-called control points of

the curve. The basis function $N_k^j(x)$ represents the weights or strength of each control point.

2.2 B-spline learning

The modelling capabilities of the B-spline network are determined by the non-linear mapping performed by these multi-dimensional basis functions. The output of such a network, the spline function, is the linear combination of basis functions:

$$y(x) = \sum_{i=1}^{k} a_i(x) w_i \qquad (2)$$

where a is the vector of multi-dimensional basis functions output, $(a_0(x), \cdots, a_k(x))$ when excited by the present input $x = (x_1, \cdots, x_n)$ and w is the vector of weights associated with a. The representation of a spline function as a weighed combination of normalized basis splines is a general B-spline interpolation through a specified set of points (de Boor 1978).

There is a direct correspondence between the control points of a curve and weights of a network. When training the network and adjusting the weights, the position of the control points is being optimized. By moving individual control points, localized changes are being performed (local control property).

Given that the set of knots and the order of the functions are fixed, the w coefficients are the only determining parameters of the spline function. Training is done to adapt these weights so that the spline function y approximates the model using well-established training algorithms. The mean square output error can be applied to measure model performance. This criterion can be expanded to weight adaptation algorithms such as the least mean square (LMS) and normalized least mean-square algorithm (NLMS). The LMS is given by:

$$w(t) = w(t-1) + \delta \left(y(t) - a^T w(t-1) \right) a(t) \qquad (3)$$

where δ (>0) is the learning rate. Instead of using a constant leaning rate, the stochastic LMS training rule (Brown and Harris 1994) assigns a learning rate to each basis function. The rate can be adapted as confidence in a particular weight increases. These changes to the learning rate retain the initial convergence capability while filtering out measurement and modelling noise.

2.3 Neurofuzzy interpretation

Multivariate B-splines basis functions may be seen as antecedents of fuzzy production rules while the weights can be seen fuzzy singletons of the consequent model. Training the weights is equivalent to determining the position of the fuzzy singletons of the rules. The network can be translated to rules of the form:

rule (i): IF x is A_1^i THEN y is w_i

where A^i is a linguistic description of the ith-multivariate fuzzy set or B-spline basis function and w_i is its corresponding weight. As each multivariate function is constructed by tensor multiplication of univariate basis functions and the tensor product is equivalent to taking the logical intersection (AND) in fuzzy systems, the system can be described with rules of the form:

rule (i): IF (x_1 is A_1^i) AND (x_2 is A_2^i) ... AND (x_n is A_n^i)
THEN y is w_i

To enhance interpretability, this type of model can be translated into an extended linguistic Mamdani fuzzy model with rules the form:

rule (i): IF (x_1 is A_1^i) AND (x_2 is A_2^i) ... AND (x_n is A_n^i)
THEN y is B_1 with ρ_{1i}, and, .., and y is B_m with ρ_{mi}

where B_j is the linguistic description of the jth fuzzy output sets and ρ_{ij} is the degree of confidence of the jth output set to the consequents of the i^h particular rule r_i. The consequent of each rule consists of multiple output sets (Brown and Harris 1994) and, for each rule a parameter ρ_{ij} is assigned to all output sets describing the output. If the fuzzy output membership functions are chosen as symmetrical B-splines of order k, the following holds:

$$w_i = \sum_q \rho_{ij} y_c^j \qquad (4)$$

where y is the centre of the jth output set and

$$\rho_{ij} = \mu_{B_j}(w_i) \qquad (5)$$

The degree of confidence is a number in the interval [0,1] representing the different grade of membership of the output to the various output sets. The weights associated with the multivariate membership functions are used to evaluate the ρ_{ij} by the intersection of the weights with the output B-spline membership functions. By training these weights, the confidences are therefore modified. The consequent parameters are normalized and there are only a fixed number of B-splines, which are non-zero.

3. MIMICKING A FUZZY CONTROLLER

The optimal properties of a B-spline network (e.g. , continuity, smoothness, convex hull, convergence, locality, transparency, efficient evaluation, etc.) can be used to yield satisfactory results in a control application. The neurofuzzy approach is used to mimic a fuzzy flight controller, in this case to identify the fuzzy rule-base used by the controller. A fuzzy multi-input multi-output (MIMO) controller (Schram and Verbruggen 1997) that has been based on total energy principles (Lambregts 1983) is mimicked. This fuzzy controller has been designed and extensively tested with respect to other control law designs in a recently formulated civil aircraft benchmark problem (Lambrechts et al. 1997). One may ask why mimicking the working of an existing can be useful:

♦ The neurofuzzy approach can be used to translate (validate) the working of the controller linguistically. In the case of a fuzzy controller, the same rules may be obtained starting from numerical values.
♦ If the controller requires extensive computation time, a speed-up may be obtained using the neural networks advantage.
♦ The transparency of the controller is enhanced.
♦ The controller's smoothness may be improved.
♦ Learning may be performed on-line without interfering with the process.

The fuzzy logic controller consists of a fuzzy outerloop controller and a classical pitch inner-loop controller (Schram and Verbruggen 1997). The fuzzy outer-loop controller is based on the total energy control concept (Lambregts 1983). In the total energy concept, thrust is used to increase the energy of the aircraft while pitch regulates the exchange between potential (altitude) and kinetic (airspeed) energy. For example, if aircraft is flying too low and too slow, the pilot would increase the thrust to increase the energy of the aircraft. On the other hand, if the aircraft is flying too low, but velocity is too high, the pilot would increase the pitch angle (using the elevator). Based on these principles, a rule-base was set and tuned. In Figure 2, the controller is shown. The aircraft block represents the aircraft dynamics including an inner loop pitch angle controller. Two inputs to the controller are velocity error V_e (the difference between a reference airspeed $V_{A,ref}$ and the actual aircraft speed V_A) and altitude error h_e (the difference between a reference altitude h_{ref} and the actual altitude h). The derivatives of airspeed and altitude are also inputs to the controller. The control actions are the changes in throttle settings $\Delta\delta_{TH}$ ($= \Delta\delta_{TH1} = \Delta\delta_{TH2}$) which result in changes in thrust and the pitch angle command θ_C For more details, see (Schram and Verbruggen 1997).

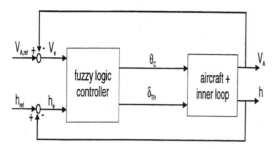

Figure 2: Fuzzy flight controller

The approach taken is to approximate the fuzzy outer-loop controller output using input/output flight data generated from the controller. In these simulations, the difference between an actual response and desired response are also chosen as controller inputs. For example if the pilot likes to descend to 70 meters, he would be alert to the change in height and to the difference between the actual and desired height. Depending upon this value, a big or small control action would follow. The output of the network is the desired change in a variable, for example change in throttle or change in pitch.

A B-spline network was trained to approximated the outerloop longitudinal controller by using the a-priori knowledge (structure). First the MIMO controller is composed in two multi-input single output (MISO) controllers. The four inputs of the outerloop controller were provided regarding the outerloop controller. The inputs are:

- altitude error *h_error*: the difference between the altitude of the aircraft and the reference altitude
- climb rate error *hdot_error*: the difference between the climb rate of the aircraft and the reference one
- airspeed error *v_error*: the difference between the altitude of the aircraft and the reference altitude
- horizontal acceleration *nx*

This inputs are pre-processed and combine to represent the aircraft state. The rule-base conveys the appropriate control action. The output classes were given as changes to the means value:

- change in throttle (with respect to a mean value): Δth
- change in pitch(with respect to a mean value): $\Delta\theta_C$

3.1 Input data

Training data sets were generated using the fuzzy flight controller. Data representing the landing of an aircraft by using different height and speed steps was generated using the RCAM-model and controller. The aircraft started from a height of 300 meters and airspeed was set to 75 m/s. The aircraft was trimmed for a glideslope (3 degrees glideslope). To reflect the aircraft dynamics, a step in height and speed was generated in each simulation. The steps in height and speed varied between -30 and 30, which resulted in different errors in airspeed and height. These errors lead to control actions to return the airplane to its reference values. Each simulation concerned 20 steps (a set of 6081 patterns from which a B-spline network was constructed). Each pattern was randomly presented to the network.

Of importance in the design of a B-spline controller model is choosing the shape (order) and the number of basis functions. This will determine the smoothness and accuracy of the controller. A-priori knowledge can be used to set the knots which define the B-spline functions. Increasing the number of knots increases the accuracy but also the number of rules.

3.2 Using a priori-knowledge

The input domain was divided into seven fuzzy sets defined by second order B-splines (triangular membership functions). Training can be used to establish the position of the knots of the B-splines (the antecedents of the rules). However, if enough B-splines are used for the input space, the local tuning of the knots has a small effect on the approximation capabilities of the network. Hence, the knots remain fixed and did not change during simulation. A uniform knot distribution was used. Basis splines were define to fulfill the partition of unity criterion in the whole input domain (min., max.) of the input set.

3.3 Network training

After representing the input space with B-splines and representing the output as fuzzy singletons, the weights are trained to obtain an optimal mapping. The stochastic approximation training algorithm LMS is used. The learning rate was experimentally set to $\delta = 1/ (1+t_i/100)$ where t_i is the number of times that the weight i has been updated.

Two mappings were studied: the mapping for change in throttle and the mapping for change in pitch. The results for both mappings are good. The change in throttle has a root square mean (RMS)-error of

0.0000419. To show the approximation capabilities of the B-splines, the data fitting for the control action, change in throttle setting change after presenting 100 times the network with 500 randomly chosen samples is depicted in figure 3:

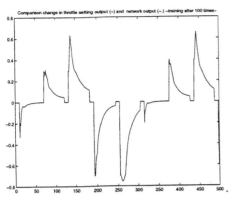

Figure 3: Change in throttle and network output after 100 steps (first 500 samples)

The mapping can be translated into rules; the behaviour of the controller can be described by a set of linguistic rules. First, each B-spline is given a linguistic label: The labels are: negative big (NB), negative medium (NM), negative small (NS), zero (ZE), positive small (PS), positive medium (PM), and positive big (PB). In order to allow for local refinement of the control action, additional B-splines of order 2 are defined for the output. The labels are: Negative extreme (NE), negative very big (NVB), negative big (NB), negative medium (NM), negative small (NS), zero (ZE), positive very small (PVS), positive small (PS), positive medium (PM), positive big (PB), positive very big (PVB) and positive extreme (PE). Note that the consequents of the rules represented by B-splines are normalized. On a knot span and a degree, there are at most 2 non-zero basis functions. The rule-base obtained from the network is given in table 1 whereas table 2 presents the rule base of the fuzzy controller (Schram and Verbruggen 1997).

For example *RULE 2* of the obtained rule-base would be:

IF aircraft is fast (airspeed error is NM) AND aircraft is too high (altitude error is NB)
THEN reduce throttle very much (change in throttle is NVB) (0.9989)
OR
IF aircraft is fast (airspeed error is NM) AND aircraft is too high (altitude error is NB)
THEN reduce throttle much (change in throttle is NB) (0.0011)

The rule base can be easily validated in this case as the same rule based as defined by (Schram and Verbruggen 1997) should be obtained. There is a rapid convergence towards the same labels of the defined rule-base.

In the same way, the mapping for change in pitch can be obtained. The mapping for change in pitch angle has a RMS-error of 0,0001088. The mapping can be translated into rules such as *RULE 7*:

IF aircraft is very slow (airspeed error is PB) AND aircraft is too high (altitude error is NB)
THEN nose much down (change in pitch is NE) (1.000)

In both cases, a fast convergence is achieved and the same rule-bases can be obtained.

4. CONCLUDING REMARKS

A B-spline network can be used for function approximation as shown for the case of mimicking a fuzzy flight controller. The working of the network can be interpreted using linguistic rules. The weights which control the mapping of these functions can be seen as local control actions which represent the consequent part of the rules. These rules can be transformed into extended Mamdani rules. Experimental work presented in this paper show the design of the used B-spline model and the validity of the approach. Future work will now concentrate on extending this approach to mimic control behaviour of human pilots.

REFERENCES

Bartels, R.H., J.C. Beatty, and B.A.Barsky (1987). An introduction to splines for use in computer graphics and geometric modelling. Morgan Kauffman.

Brown, M and C.J. Harris (1994). Neurofuzzy adaptive modelling and control. Prentice Hall International.

de Boor, C. (1972). On calculating with B-splines, Journal of Approximation Theory, vol. **6**, pp. 50-62.

de Boor, C. (1978). A Practical Guide to Splines. Springer-Verlag, New York.

Friedman, J.H. (1991). Multivariate adaptive regression splines. The Annals of Statistics **19**(1), pp. 1-141.

Lambrechts P., S. Bennani , G. Looye and D. Moormann (1997).The RCAM design challenge problem description. In: Robust Flight Control, A Design Challenge, J.F. Magni, S. Bennani, J. Terlouw (eds.), Springer Verlag, Lectures notes in Control and Information Sciences 224, pp. 149-179.

Lambregts, A.A. (1983). Vertical flight path and speed control autopilot design using total energy principles, *Technical report*, AIAA 83-2239.

Schram, G., Verbruggen H.B. (1997). RCAM: A Fuzzy Control Approach. In: Robust Flight Control, A Design Challenge. J.F. Magni, S. Bennani, J. Terlouw (eds.), Springer Verlag, Lectures notes in Control and Information Sciences 224, pp. 397-416.

Schumaker, L. (1981). Spline Functions Basic Theory. John Wiley and Sons, New York.

Table 1: Rule-base from the network where the control action is for throttle change. V_e indicates error in airspeed while h_e indicates error in altitude. The control action, change in throttle for each combination of antecedents is given in the table.

V_e ($\Rightarrow$) / h_e ($\Downarrow$)	NB	NM	NS	ZE	PS	PM	PB
NB	NE 1.00	NVB 1.00 NB 0.00	NVB 0.00 NB 1.00	NM 1.00 NS 0.00	NM 0.00 NS 1.00	NS 0.00 NVS 1.00	ZE 1.00 PVS 0.00
NM	NVB 1.00 NB 0.00	NVB 0.00 NB 1.00	NM 1.00 NS 0.00	NM 0.00 NS 1.00	NS 0.00 NVS 1.00	ZE 1.00 PVS 0.00	ZE 0.00 PVS 1.00
NS	NVB 0.00 NB 1.00	NM 1.00 NS 0.00	NB 0.00 NM 1.00	NS 1.00 NVS 0.00	ZE 1.00 PVS 0.00	ZE 0.00 PVS 1.00	PS 1.00 PM 0.00
ZE	NM 1.00 NS 0.00	NM 0.00 NS 1.00	NM 0.00 NS 1.00	NVS 0.00 ZE 1.00	PVS 0.00 PS 1.00	PS 1.00 PM 0.00	PS 0.00 PM 1.00
PS	NM 0.00 NS 1.00	NVS 1.00 ZE 0.00	NVS 0.00 ZE 1.00	PS 1.00 PM 0.00	PS 0.00 PM 1.00	PM 1.00 PB 0.00	PM 0.00 PB 1.00
PM	NS 0.00 NVS 1.00	ZE 0.99 PVS 0.01	ZE 0.01 PVS 0.99	PS 0.99 PM 0.01	PS 0.00 PM 1.00	PB 1.00 PVB 0.00	PB 0.00 PVB 1.00
PB	ZE 1.00 PVS 0.00	ZE 0.00 PVS 1.00	PS 1.00 PM 0.00	PS 0.00 PM 1.00	PB 1.00 PVB 0.00	PB 0.00 PVB 1.00	PE 1.00

Table 2: Rule-base from the fuzzy flight controller for throttle change: V_e indicates error in airspeed while h_e indicates error in altitude. The control action, change in throttle for each combination of antecedents is given in the table.

V_e ($\Rightarrow$) / h_e ($\Downarrow$)	NB	NM	NS	ZE	PS	PM	PB
NB	NE 1.00	NVB 1.00	NB 1.00	NM 1.00	NS 1.00	NVS 1.00	ZE 1.00
NM	NVB 1.00	NB 1.00	NM 1.00	NS 1.00	NVS 1.00	ZE 1.00	PVS 1.00
NS	NB 1.00	NM 1.00	NM 1.00	NS 1.00	ZE 1.00	PVS 1.00	PS 1.00
ZE	NM 1.00	NS 1.00	NS 1.00	ZE 1.00	PS 1.00	PS 1.00	PM 1.00
PS	NS 1.00	NVS 1.00	ZE 1.00	PS 1.00	PM 1.00	PM 1.00	PB 1.00
PM	NVS 1.00	ZE 1.00	PVS 1.00	PS 1.00	PM 1.00	PB 1.00	PVB 1.00
PB	ZE 1.00	PVS 1.00	PS 1.00	PM 1.00	PB 1.00	PVB 1.00	PE 1.00

Printed and bound by CPI Group (UK) Ltd, Croydon, CR0 4YY

09/10/2024

01042636-0001

AUTOMATIC CLASSIFICATION OF GEAR-UNITS BY NEURAL NETWORK TECHNOLOGIES

Ingomar Wascher
Volkmar H. Haase

University of Technology, Graz, Austria
IICM-Software Technology
Muenzgrabenstrasse 11
A-8010, Graz

Abstract: Noise emitted during test runs of mitre-gear units was analyzed using a neural network based tool. Gears can be classified dependent on how noise levels vary with changes of revolution speed. Faults can be detected and necessary adjustments can be identified. The kernel of the work is the appropriate selection, reduction and preprocessing of input data, parameterization of the clustering (unsupervised learning) algorithms, and the building of the classification model (supervised learning). - *Copyright © 1998 IFAC*

Keywords: Automobile industry, Data processing, Data reduction, Mechanical systems, Model-based control, Neural networks, Noise analysis, Statistics

1. THE PRACTICAL PROBLEM

This paper contains the main results of a diploma thesis done for Zurk GmbH, see (Zurk, 1998), at IICM - Software Technology at University of Technology in Graz, Austria.

The purpose of the thesis is to support the product testing of mitre-gear units for large trucks to shorten test cycles and to avoid disassembly of gear-units after testing.

Mitre-gear units are immediately after the production tested in test-beds where the number of revolutions is varied during a 948 sec. test run simulating real road situations.

Test runs deliver a set of parameters (noise, temperature, oil pressure, etc.) each second. Noise is most promising to classify types of gears. It is measured to get information about possible faults or necessary adjustments.

The noise level can therefore be expressed with the following "formula":

Noise level = f(number of revolutions/sec., possible defects, necessary adjustments)

Figures 1 and 2 show the noise level and the corresponding number of revolutions of a typical test run.

Based on measurements of approximately 150 gear units neural networks were used to cluster the measurement records to find groups of gear units behaving in a similar way.

If types of gear units can be identified in this manner any new test run of a gear unit will deliver a result e.g. gear unit is of type 1 . Now engineers know from experience that this means e.g. tighten ... (without disassembling the whole gear unit, as it has been necessary before).

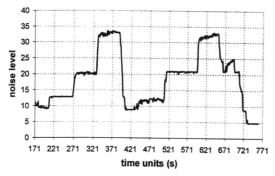

Fig 1. Noise-level of a typical shortened test run

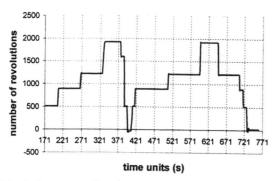

Fig 2. Corresponding number of revolutions

2. THE DATA ANALYSIS TOOL

The tool used for this work is Business Advisor from AIWARE, now acquired by Computer Associates. Business Advisor runs under Win95/NT and is a complete modeling and decision support system.

It is based on neural networks and can be used either for unsupervised learning, i.e. learning by similarities, or for supervised learning. Furthermore it offers simple to use statistical functionality.

3. PREPROCESSING OF DATA

From several thousand sets of test data (different types of gears), each containing some 900 records with 8 parameters a set of approximately 200 test runs (one type of gear) was chosen.

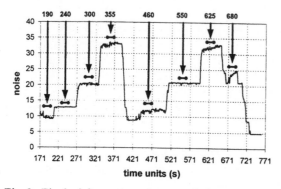

Fig 3. Final eight mean values used for the further process

In the next step the incomplete sets of data have been eliminated which led to 151 remaining records. Afterwards these records including 948 time units were shortened to 589 time units each by cutting off the early (time < 171) and late phases (time > 760). They were omitted because of external disturbances.

Also powerful computers (300 MHz Pentium) and neural network tools are unable to process some 500 parameters in reasonable time. The number of test points used from the test run was therefore reduced to 8 covering typical time-spans (where the revolution speed is constant for some 50 seconds) within the run. These main levels, exactly speaking the mean values of the levels, have been taken for further analysis. Figure 3 shows these 8 points.

4. DATA ANALYZING AND CLUSTERING

The method used for data analyzing was unsupervised learning, see (Pao, 1989). Business Adviser supports this method in the form of clustering by similarities and offers the possibility to vary the following parameters: Maximum Cluster Levels, Desired Maximum Clusters Per Level, Cluster Radius Change Rate (%), Desired Maximum of "$\underline{1}$" Population Clusters (%).

In the experiment these parameters have been varied to find out which clustering results were more or less stable over various parameter settings.

In this step similarities between the different clustering results obtained were identified. This was done by comparing the several results and computing the intersecting clusters using a self made c++ program.

As known unsupervised learning is very sensible to outliers. Therefore it was necessary to eliminate them which was easily done by removing the very small clusters.

The remaining smaller set of data was clustered once again with different values of the parameters leading to an iterative process.

The iteration has been carried out until the resulting clusters were compact and stable and no more outliers could be identified, i.e. no data item did change between two clusters.

The unsupervised learning algorithm finally resulted in 104 data items split up into 8 different clusters as can be seen in figure 4. The size of the clusters varies from 7 items to 27 items each as shown in table1. There are 3 clusters with a size smaller than 10. Omitting them as done in Fig 5 is increasing the readability.

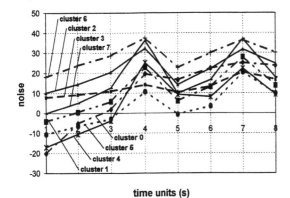

Fig 4. Mean values of the all eight resulting clusters

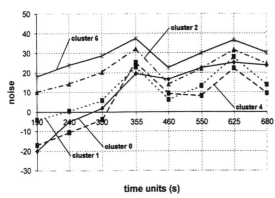

Fig 5. Mean values of the five main clusters for better readability

Table 1 The resulting 8 clusters and their size (number of members)

Cluster	0	1	2	3	4	5	6	7
Size	18	13	12	8	27	7	12	7

The result shows that there are 3 clusters (1, 2 and 6) which show the same time dependent behavior during the test run. The difference between them is that noise levels in cluster 6 are higher than in cluster 2 which are higher than in cluster 1.

But if cluster 0 is regarded one can see that it starts with low values and approaches high values at the end of the run. Obviously the behavior of these gear units changes with time - adjustments seem necessary.

5. BUILDING A MODEL

The aim of the next experiment was to build a model for mitre-gear units with supervised learning to be able to predict the type (i.e. the cluster number) for any new gear unit, that has gone through the testing process in the test beds. This would inform the mechanical engineers about possible defects or help them adjusting the gear parameters.

Table 2 Typical result of the classification

Data Item	Class	Predicted Class	Error
record 19	1	0.999429	ok
record 42	2	1.86426	ok
record 64	7	6.88321	ok
record 66	0	-0.0443008	ok
record 67	0	-0.044644	ok
record 103	3	2.75161	ok
record 110	4	3.82175	ok
record 122	1	0.119704	not ok
record 126	5	4.59796	(ok)
record 139	6	5.9722	ok

Using the preceding results a model has been build in order to be able to classify new data items, see (Pao, 1989). Again Business Advisor was used for this purpose with the following parameters: Two hidden layers with 5 and 3 nodes, Gaussian RBF nodes as functional link option, system error of 0.0001, independent outputs, node outputs use sigmoidal slope of 0.1 and the learning algorithm was accelerated backpropagation.

The training set R^2-value ranges from 98 to 99.8 depending on the selected set of testing data and the test set R^2-value ranges from 80 to 98.

A typical result is shown in table 2. There the real cluster number (Class) of the gear units (Data Item), the classification by the learned model (Predicted Class) and the error done by the classification can be seen. With the exception of record 122 the error value is satisfying.

CONCLUSION

These experiments demonstrate that analysis of noise data during test runs of mitre-gear units is a powerful instrument to identify faults and necessary adjustments. Classification of gear units informs the mechanical engineers of repair measures without being obliged to disassemble each gear unit after the test.

REFERENCES

Pao, Y.H. (1989). *Adaptive pattern recognition and neural networks*. Addison-Wesley, Reading.
Zurk, A.P. (1998). *Private Communication*. Graz.

AUTHOR INDEX